Biology and Knowledge Revisited

From Neurogenesis to Psychogenesis

The Jean Piaget Symposium Series
Available from LEA

Overton, W. F. (Ed.): The Relationship Between Social and Cognitive Development.

Liben, L. S. (Ed): Piaget and the Foundations of Knowledge.

Scholnick, E. K. (Ed): New Trends in Conceptual Representations: Challenges to Piaget's Theory?

Niemark, E. D., DeLisi, R. & Newman, J. L. (Eds.): Moderators of Competence.

Bearison, D. J. & Zimiles, H. (Eds.): Thought and Emotion: Developmental Perspectives.

Liben, L. S. (Ed.): Development and Learning: Conflict or Congruence?

Forman, G. & Pufall, P. B. (Eds.): Constructivism in the Computer Age.

Overton, W. F. (Ed.): Reasoning, Necessity, and Logic: Developmental Perspectives.

Keating, D. P. & Rosen, H. (Eds.): Constructivist Perspectives on Developmental Psychopathology and Atypical Development.

Carey, S. & Gelman, R. (Eds.): The Epigenesis of Mind: Essays on Biology and Cognition.

Beilin, H. & Pufall, P. (Eds.): Piaget's Theory: Prospects and Possibilities.

Wozniak, R. H. & Fisher, K. W. (Eds.): Development in Context: Acting and Thinking in Specific Environments.

Overton, W. F. & Palermo, D. S. (Eds.): The Nature and Ontogenesis of Meaning.

Noam, G. G. & Fischer, K. W. (Eds.): Development and Vulnerability in Close Relationships.

Reed, E. S., Turiel, E. & Brown, T. (Eds.): Values and Knowledge.

Amsel, E. & Renninger, K. A. (Eds.): Change and Development: Issues of Theory, Method, and Application.

Langer, J. & Killen, M. (Eds.): Piaget, Evolution, and Development.

Scholnick, E., Nelson, K., Gelman, S. A. & Miller, P. H. (Eds.): Conceptual Development: Piaget's Legacy.

Nucci, L. P., Saxe, G. B. & Turiel, E. (Eds.): Culture, Thought, and Development.

Amsel, E. & Byren, J. P. (Eds.): Language, Literacy, and Cognitive Development: The Development and Consequences of Symbolic Communication.

Brown, T. & Smith, L. (Eds.): Reductionism and the Development of Knowledge.

Lightfoot, C., LaLonde, C. & Chandler, M. (Eds.): Changing Conceptions of Psychological Life.

Parker, S. T., Langer, J. & Milbrath, C. (Eds.): Biology and Knowledge Revisited: From Neurogenesis to Psychogenesis.

Biology and Knowledge Revisited

From Neurogenesis to Psychogenesis

Edited by

Sue Taylor Parker
Professor Emeritus, Sonoma State University

Jonas Langer
University of California, Berkeley

Constance Milbrath
University of California, San Francisco

Routledge
Taylor & Francis Group

LONDON AND NEW YORK

First published 2005 by Lawrence Erlbaum Associates, Inc.

2 Park Square, Milton Park, Abingdon, Oxfordshire OX14 4RN
52 Vanderbilt Avenue, New York, NY 10017

Routledge is an imprint of the Taylor & Francis Group, an informa business

First issued in paperback 2019

Copyright © 2005 Taylor & Francis

Cover design by Kathryn Houghtaling Lacey

Cover photograph taken by Constance Milbrath at a special exhibit on Jean Piaget in honor of his centennial during the Growing Mind Centennial of Jean Piaget's birth, Geneva, Switzerland, Sept. 14–18, 1996.

Library of Congress Cataloging-in-Publication Data

Jean Piaget Society. Meeting (31st : 2001 : Berkeley, Calif.)
    Biology and knowledge revisited : from neurogenesis to psychogenesis / edited by Sue Taylor Parker, Jonas Langer, Constance Milbrath.
        p.   cm.
    Includes bibliographical references and indexes.
    ISBN 0-8058-4627-1 (alk. paper)
    1. Developmental neurobiology—Congresses. 2. Cognitive neuroscience—Congresses. 3. Piaget, Jean, 1896–Biologie et connaissance—Congresses. I. Parker, Sue Taylor. II. Langer, Jonas. III. Milbrath, Constance, 1943– IV. Title.

QP363.5.J436   2005
153—dc22   2004053193
                                                                          CIP

ISBN 13: 978-0-8058-4627-0 (hbk)
ISBN 13: 978-1-138-01279-0 (pbk)

We dedicate this book to Elizabeth Bates,
who was taken from us too soon,
but whose great influence on the issues raised
here can only continue to grow.

Contents

Contributors   ix

Preface   xi

1   Piaget's Legacy in Cognitive Constructivism, Niche
Construction, and Phenotype Development
and Evolution   1
Sue Taylor Parker

2   Piaget's Phenocopy Model Revisited: A Brief History
of Ideas About the Origins of Adaptive Genetic Variations   33
Sue Taylor Parker

3   Beyond Piaget's Phenocopy: The Baby in the
Lamarckian Bath   87
Terrence W. Deacon

4   Human Brain Evolution: Developmental Perspectives   123
Kathleen Rita Gibson

5   Cerebellar Anatomy and Function: From the Corporeal
to the Cognitive   145
Carol Elizabeth MacLeod

6 From Mirror Neurons to the Shared Manifold Hypothesis:
A Neurophysiological Account of Intersubjectivity   179
Vittorio Gallese

7 Plasticity, Localization, and Language Development   205
Elizabeth Bates

8 From Ontogenesis to Phylogenesis: What Can Child
Language Tell Us About Language Evolution?   255
Dan I. Slobin

9 The Emergence of Nicaraguan Sign Language:
Questions of Development, Acquisition, and Evolution   287
Richard J. Senghas, Ann Senghas, and Jennie E. Pyers

10 Can Developmental Disorders Be Used to Bolster
Claims From Evolutionary Psychology?
A Neuroconstructivist Approach   307
Annette Karmiloff-Smith and Michael Thomas

Author Index   323

Subject Index   337

Contributors

Elizabeth Bates   Formerly of the Department of Psychology, University of California at San Diego, LaJolla, California

Terrance W. Deacon   Department of Anthropology, University of California at Berkeley, Berkeley, California

Vittorio Gallese   Istitutuo Fisiologia Umana, Universita di Parma, Parma, Italy

Kathleen Rita Gibson   Department of Neurobiology and Anatomy, Medical School, Department of Orthodontics, Dental Branch and Graduate School of Biomedical Sciences, University of Texas Health Science Center, Houston, Texas

Annette Karmiloff-Smith   Neurocognitive Development Unit, Institute of Child Health, London, England

Jonas Langer   Psychology Department, University of California at Berkeley, Berkeley, California

Carol Elizabeth MacLeod   Anthropology Department, Langara College, Vancouver, BC, Canada

Constance Milbrath   Psychiatry Department, University of California at San Francisco, San Francisco, California

Sue Taylor Parker   Professor Emeritus, Department of Anthropology, Sonoma State University, Rohnert Park, California

Jennie E. Pyers   Department of Psychology, University of California at Berkeley, Berkeley, California

Dan I. Slobin   Department of Psychology, University of California at Berkeley, Berkeley, California

Richard J. Senghas   Department of Anthropology, Sonoma State University, Rohnert Park, California

Ann Senghas   Department of Psychology, Barnard College, New York, New York

Michael Thomas   Neurocognitive Development Unit, Institute of Child Health, London, England

Preface

Comparative cross-species and cross-cultural approaches to studying the evolutionary and developmental relations between biology and knowledge have a long and rich history. Key, on the one hand, is inquiry into how conceptual, perceptual and linguistic behavior grows out of yet extends beyond its roots in the evolution of brain development. This has received the most research attention. Key, on the other hand, is inquiry into how behavior influences and enters into regulating the development and evolution of the brain.

These are the two reciprocal inquiries that Piaget (1971) explored in his seminal, if at times controversial examination entitled *Biology & Knowledge: An Essay on the Relations between Organic Regulations & Cognitive Processes*. Accordingly, the evolutionary theme of the 31st annual Meeting of the Jean Piaget Society in 2001 was inspired by our desire to revisit ideas Piaget developed in *Biology and Knowledge*. The organizers sought to stimulate reconsideration of these ideas in light of recent comparative research in evolutionary developmental biology, neurobiology, and cognitive development. In particular we wanted to focus on epigenetic models of cognitive and language development in relation to the evolution of brain development.

The 2001 Meeting returned to the broad evolutionary theme of the 25th annual Meeting on Piaget, evolution, and development held in 1995 (Langer & Killen, 1998). It differed from the earlier meeting, however, in

being organized around ideas about the relationship between development and evolution raised in Piaget's books. It also differed in focusing
on specifically constructivist approaches to neurogenesis and language
acquisition, and their evolution.

ISSUES ADDRESSED IN CONTRIBUTIONS
TO THIS VOLUME

As a uniquely human characteristic, language has long fascinated psychologists, anthropologists, and biologists. It is both species-specific and
developmentally plastic, and completely dependent on participation in a
linguistic environment. It is both a product of, and a contributor to cultural niche construction (Laland, Odling-Smee et al., 1999; Odling-Smee,
1988). As such, it has driven the increasing pace of human biological and
cultural evolution. Recognizing this and the power of constructivist models, all the contributors to this volume approach the evolution and development of human linguistic and cognitive abilities, and their neural substrates, from epigenetic constructivist perspectives. They all emphasize
environment-contingent plasticity of behavioral and brain development.
They all employ comparative methodologies in their analyses, whether
comparative linguistic studies (Slobin, chap. 8, and Senghas, Senghas, and
Pyers, chap. 9), comparative studies of developmental disorders (Bates,
chap. 7, and Karmiloff-Smith & Thomas, chap. 10), or comparative species studies (Gallese, chap. 6, Gibson, chap. 4, and MacLeod, chap. 5).

Two chapters focus on language development and language change,
addressing implications these phenomena may have for understanding
the evolution of language capacity. Dan Slobin (chap. 8) addresses the
following hotly debated questions: whether linguistic ontogeny recapitulates its phylogeny, whether language change recapitulates its ontogeny,
and whether children create grammatical forms. He concludes that there
is no universal form of early child language that clearly reflects a biologically specified proto-language. Second, he concludes that innovations in
historical change in existing languages come from older speakers rather
than from preschool aged children as Bickerton (1981, 1990) and others
have suggested. Finally, he concludes that because languages are sociocultural products, studies of individuals alone cannot illuminate the phylogeny of linguistic abilities as evolutionary psychologists have suggested.
On a more positive note, however, he says that "Children's homesign systems suggests a human capacity to create something like a proto-

language. . . . However, for such a language to develop further, a community of users is needed."

The question of what role language users of different ages play in historical change in language is also addressed by Richard and Anne Senghas and Jennie Pyers' (chap. 9). These investigators describe the emergence of the new Nicaraguan Sign Language (NSL) among deaf children given increasing opportunities to interact socially with other deaf children during the past 25 years. They describe three historical periods: (1) the Pre-Emergence period characterized by social isolation of deaf people and use of isolated homesigns; (2) the Initial Contact period in 1977, which established vocational programs for deaf adolescents; and (3) the Sustained Contact period in the mid-1980s, in which a Deaf Association formed and began to assume control of learning and established a dictionary project and brought signing deaf models to schools.

They divided the signers into two historical cohorts: The first cohort entered the community between 1978 and 1983; the second cohort, between 1984 and 1990. They also divided signers to three age grades according to their age at first year of exposure to language: (1) late exposed, more than 10 years of age at exposure; (2) middle-exposed, between 6 and 10 years of age at exposure; and (3) early exposed, less than 6 years of age at exposure. In order to tease apart historical and age variables, they compared the grammar of individuals exposed to signing at different ages in different periods.

When they examined the use of spatial co-reference, they discovered that in the 1980s the first cohort began employing these modulations more frequently, and that children in the late 1980s began to impose a new constraint by which signs produced in the same location had a common referent, making the signing more specific. Further analysis revealed that whereas middle and early exposed signers in the second cohort used the common referent modulations, late-exposed signers from both cohorts were unlikely to produce spatial co-reference. From this they conclude that spatial co-reference is not as easy to learn for older signers. Consistent with Slobin's conclusions, they emphasize that ". . . language genesis requires at least two cohorts of the community in sequence, the first providing the circumstances that the second can exploit."

In a complementary approach to arguments about language acquisition and modularity, Karmiloff-Smith and Thomas (chap. 10) employ a "neuro-constructivist" approach to assessing claims that children with William's Syndrome provide evidence for innate mental modules. This claim is based on the supposition that these children display linguistic and social skills far beyond those typical of their retarded mental age. The authors emphasize the importance of going beyond descriptions of adult disabilities to trace the development of infants and children with this disorder across both apparently normal and abnormal behavioral systems. Their analysis of Williams Syndrome (WS) shows that, contrary to claims of evolutionary psychologists, language comprehension and production of children with WS as compared to normal children was either very delayed or that behavioral scores "in the normal range" were sustained by atypical cognitive processes. In fact, their vocabulary, syntax, and usage were all deficient. Moreover, these deficits and those in their social cognition including face recognition are consistent with their degree of mental retardation, rather than strikingly superior to their general mental abilities, as suggested by evolutionary psychologists. Taken together, their studies suggest that children with these syndromes follow atypical trajectories and display atypical brain development.

In another powerful approach to understanding language acquisition, Liz Bates (chap. 7) , to whom this volume is dedicated, describes the implications of prospective studies of language development in children from different language communities who have suffered early focal brain injuries. Rather than supporting a model of innate representations based on a universal architecture, these studies reveal that localization is plastic and modifiable by experience, therefore it is variable. Competitive pressures and relatively simple biases in computation style underlie specific localizations of function that develop through experience in both normal and abnormal development. As she says, ". . . most of the brain participates in linguistic activity, in varying degrees, depending on the nature of the task and the individual's expertise in that task." This suggests that "language facilitating mechanisms" are widely distributed in the brain, and predated language evolution. Both Bates (chap. 7) and Karmiloff-Smith and Thomas (chap. 10) argue that any language modularity that can be seen in adult brains is a distributed, contingent product of development rather than an innate organ.

In combination, these chapters reveal the plasticity and contingency of language development. Among other things, they reveal that the course of language acquisition depends on the dynamic interplay of internal factors (e.g., auditory and/or visual perception, brain injury or lack of injury, and/or typical or atypical genotypes) and external factors (e.g., the existence and nature of the linguistic community, as well as the organization of the local language). They also reveal how aspects of language acquisition and localization of these aspects depend upon the age at which children experience these various factors. On the other hand, they reveal the resilient and robustly multimodal nature of the symbolic capacity, which has led some investigators to misinterpret it as an innately programmed neural module.

In a chapter on evolution, Kathleen Gibson (chap. 4) takes a comparative developmental approach to the brain. She contrasts the increasingly complex, hierarchical mental constructional capacities of humans with the lesser capacities of great apes, who achieve cognitive and linguistic abilities comparable to those of 2-to 3-year-old humans. These greater capacities are reflected in both the increased size not only of the human neocortex, but of the basal ganglia and cerebellum, as well as the more prolonged period of brain development. Gibson also emphasizes that models of human evolution must recognize continuities between the minds of humans and those of great apes, with whom we share a recent common ancestor. Like other contributors to this volume, she rejects the model of "genetically determined, functionally dedicated neural modules" for language, noting that brain development is epigenetic and highly contingent on experience. She also notes, however, that similar environments and self-generated behaviors combine to channel development into predictable, species-specific patterns.

In a more specific approach to brain evolution, Carol MacLeod (chap. 5) uses the comparative species perspective to focus on the evolution of the cerebellum. She explains that "the cerebellum functions as a partner with the neocortex, processing information, but never sending direct commands to the body except through intermediaries." She notes that whereas the phylogenetically older, medial and anterior part of the cerebellum is involved in the execution of movement, the newer, lateral part is involved in planning movements, and hence in cognition. In her comparative study of ape and human brains, she discovered that greater and lesser apes as compared with monkeys show a significant increase in the size of the lateral cerebellum. She argues that this "grade level change" (shared by several sequential ape lineages) provided a springboard for superior cognitive and linguistic adaptations of hominoids.

In a daring new approach, Vittorio Gallese (chap. 6) addresses the human capacity for intersubjectivity, which underpins both language and such elements of social cognition as imitation and empathy. He describes his discovery of the "mirror neurons" in the ventral premotor cortex of macaque monkeys. These neurons respond specifically and uniquely to the sight or sound of grasping actions (by the hand or mouth) performed on object by the self or another individual: "Such a neural mechanism enables the monkey to represent the end-state of the interaction independently from the different modes of presentation. . . ." On the basis of these mirror neurons, Gallese proposes the "shared manifold hypothesis" that ". . . a similar mechanism could underpin our capacity to share feelings and emotions with others."

Taken together, these various contributions suggest the power of comparative epigenetic constructivist models to illuminate the development and evolutionary history of human linguistic and cognitive capacities. As such, they follow in the larger tradition of Piaget's constructivist paradigm and begin to address some of his questions without resorting to Lamarckian mechanisms.

Chapter 1 examines Piaget's constructivist paradigm in relation to some historic and recent models for the role behavior and development play in evolution. These include the Baldwin effect or organic selection (Baldwin, 1896), niche construction (Laland, Odling-Smee et al., 1999; Odling-Smee, Laland et al., 1996), and phenotype centered evolution (West-Eberhard, 2003). All these models reflect a growing trend in evolutionary biology to emphasize the phenotype rather than the genotype.

In contrast, chapter 2 of this volume focuses specifically on the phenocopy model, the mechanism Piaget (1978) proposed to explain the role of behavior and development in the origin of heritable adaptive variations. Piaget's ideas are examined in the context of his own intellectual history, and the history of ideas about this important subject going back to Lamarck (1984), through Darwin and the modern synthesis (Mayr & Provine, 1989), and beyond to the growth of developmental molecular biology. Although Piaget's phenocopy model does not stand up in light of modern developmental molecular biology, it addresses important questions about the origins of variations that have only recently begun to be investigated.

Likewise, Terrance Deacon's chapter 3 analyzes the epistemology of Piaget's attempt to devise a unified theory of development and evolution in a cybernetic model of auto regulation. He argues that even though Piaget was unable to achieve this synthesis, he recognized "critically incomplete" aspects of neoDarwinism, which are only now being addressed. He notes that Piaget's model, like those of Spencer (1872) and Lamarck (1984), is based on active adaptive agency in both developmental and evolutionary domains. He says that Piaget saw Baldwin's (1896) organic selection and Waddington's phenocopy (1975) model as showing indirect ways that development could affect the germ plasm. However, Deacon argues that, contrary to Piaget's phenocopy model, Waddington's (1975) genetic assimilation, Weismann's (1892) intraselection, and Baldwin's organic selection all involve developmental influences on evolution that work through Darwinian processes of selection rather than Lamarckian processes. Moreover, Deacon argues that development itself relies on Darwinian selection-like processes. He summarizes studies demonstrating how activity-dependent competition between signal pathways can create emergent properties in developing neurological systems. He concludes that, "Piaget's appeal to both Baldwin and Waddington can now be seen as insightful anticipations of a necessary complexification of evolutionary theory, though neither a repudiation of Darwinian mechanisms nor a return to Lamarckian paradigms. To explain the apparent autoregulatory power of biological evolution, does, as he suspected, require incorporating the role of epigenetic processes as mediators between genotype and phenotype. Yet there turns out to be a far more prominent role for Darwinian over cybernetic mechanisms of regulation in both development and evolution than Piaget could ever have imagined."

Finally, we note with sadness the death of one of the plenary speakers at the 2001 Meeting, Elizabeth Bates, who died in 2003 while this book was in preparation. This volume is dedicated to her many insightful contributions to our understanding of language acquisition and the behavior contingent plasticity of brain development.

REFERENCES

Baldwin, J. M. (1896). A new factor in evolution. *American Naturalist, 30,* 441–451.

Laland, K. N., Odling-Smee, F. J. et al. (1999). Evolutionary consequences of niche construction and their implications for ecology. *Proceedings of the National Academy of Science, USA, 96*(18), 10242–7.

Lamarck, J. -B. (1984). *Philosophical zoology.* Chicago, IL: University of Chicago Press.

Langer, J., & Killen, M. (Eds.). (1998). *Piaget, evolution, and development.* Mahwah, NJ: Lawrence Erlbaum Associates.

Mayr, E. & Provine, W. (Eds.). (1989). *The evolutionary synthesis.* Cambridge, MA: Harvard University Press.

Odling-Smee, J. (1988). Niche-constructing phenotypes. In H. C. Plotkin (Ed.), *The role of behavior in evolution* (pp. 73–132). Cambridge, MA: MIT Press.

Odling-Smee, J., & Laland, K. et al. (1996). Niche construction. *American Naturalist, 147(*4), 641–648.

Piaget, J. (1978). *Behavior and evolution.* New York: Pantheon Books.

Spencer, H. (1872). *Principles of psychology*. London: Williams and Norgate.

Weismann, A. (1892). *Das Keimplasma: Eine Theorie der Verebung*. Jenna: Gustav Fischer.

West-Eberhard, M. J. (2003). *Developmental plasticity and evolution*. New York: Oxford University Press.

1

Piaget's Legacy in Cognitive Constructivism, Niche Construction, and Phenotype Development and Evolution

Sue Taylor Parker

Professor Emeritus, Sonoma State University

JPS 2001 was organized around implications of Piaget's epigenetic concept of *cognitive constructivism*—his view that the developing child constructs sequentially more powerful schemes as he adapts (through assimilation and accommodation) to feedback he receives from his own actions on himself, on others, and on objects (Inhelder & Piaget, 1964; Piaget, 1952, 1954, 1962; Piaget & Inhelder, 1967), and as he reflects on this feedback (Piaget, 1985). The seminal idea is that the developing child constructs much of his own environment, which transforms him, and which he, in turn, transforms, and so on. (Likewise, of course, parents, siblings, and others construct much of the child's environment and transform it in response to his changing nature.)

The term *epigenetics*, derived from the Aristotelian word *epigenesis*, was introduced in 1947 by Conrad Waddington (1975) to describe the "branch of biology, which studies the causal interactions between genes and their products which bring the phenotype into being" (p. 218). He describes the elementary processes of epigenetics as having two aspects: "changes in cellular composition (cellular differentiation, or histogenesis) and changes in geometrical form" (p. 219). Although Waddington focused on the embryological period of phenotype development, Piaget and others have extended the usage to later developmental stages because the interactions between genes and their products in transformations of the phenotype continue throughout the life cycle.

Piaget argues that *construction* is an epigenetic phenomenon that continues processes of embryogenesis into postnatal life. It is this conception of epigenesis that we have adopted in the subtitle of our book. This view of epigenesis is not unique to Piaget. It was also emphasized by the neuropsychologist Colwyn Trevarthan (1973), who noted that psychogenesis is a continuation of embryogenesis outside the womb. Whereas Piaget's research on cognitive epigenesis began with the reflexes of neonates, some of his contemporaries studied continuities between behavioral development inside and outside the womb. The research pediatrician Arnold Gesell (1945), for example, noted that fetal infants continue their developmental course outside the womb much as they would have inside the womb had they not been born prematurely.[1]

In a related study, the embryologist, Viktor Hamburger (1964), demonstrated that normal brain development in the fetus depends on feedback to the sensory nerves from spontaneous motoric output of the developing fetus, which proceeds sensory nerve development. Taken together, these approaches suggest that brain development is driven by the assimilation of stimuli to developing schemes, whether these stimuli are produced solely by direct proprioceptive feedback from the child's own actions during prenatal life or by feedback from social and physical objects acted on by the child's vision, vocalizations, prehension and/or other motoric patterns during postnatal life. They demonstrate the genesis of the phenomenon of experience- and activity-contingent brain development, which continues to operate throughout development (Deacon, 1997; Edelman, 1987; Gibson & Peterson, 1991). Behavioral embryology is part of Piaget's larger interest in the relationship between development and evolution.

Piaget (1971) points to this neurological substrate when he notes that, ". . . the fundamental truth on which we base our argument is that the nervous system alone constitutes a specialized organ of functional regulation as well as being the instrument of cognitive function" (p. 30). On this topic he concludes

---

[1]Gesell (1945) described seven related principles of "morphogenesis of behavior" including the following four: (1) "individuating fore-reference" of motor system anticipating subsequent adaptations, for example, prehension; (2) "developmental direction" from head to tail, and from proximal to distal segments; (3) "spiral reincorporation" in which behavioral complexes repeat themselves at higher levels of organization; and (4) "reciprocal interweaving," in which counterbalanced sensorimotor functions fluctuate in dominance (i.e., flexor vs. extensors, right vs. left, ipsilateral vs. contralateral).

One process that strikes us as being an intermediate point between the hereditary organization and the laws governing thought is cortical functioning, which has the
double quality of being hereditary functioning . . . but having almost no hereditary
programming by way of cognitive structure. In fact the functioning of the brain is
hereditary, since the progress made in cerebral and cortical development among primates and hominids, including man, rather precisely determines the progress of intelligence. . . . (Piaget, 1971, pp. 327–328)

This chapter focuses primarily on Piaget's ideas about behavioral development and evolution. The first section presents a summary of Piaget's
ideas; the second section focuses briefly on a history of ideas about the
role of behavior in evolution. It begins in the 17th century with a discussion of Jean Baptist Lamarck and Pierre Jean Canabis in France, and
Erasmus Darwin and his contemporaries in England, continues with a
discussion of the naturalist-clerics who influenced Charles Darwin and
Herbert Spencer in the 19th century, and ends with a discussion of influential figures at the turn of the last century—Lloyd Morgan, James Mark
Baldwin, and William James.

Then, following a review of critiques of the neglect of the role of behavior in the modern synthesis of evolution, I argue that these and related
strands culminate in the concepts of *phenotype-centered evolution* and
niche construction. The final section of this chapter describes the contributions to this volume in light of the notions of epigenetic construction, experience-contingent brain development, phenotype-centered evolution,
and niche construction.

PIAGET'S CONSTRUCTIVIST MODEL
AND THE EVOLUTION OF BEHAVIOR

In the first part of his introductory chapter of *B&K*, Piaget challenges the
modern synthesis of evolution to explain the creative role of behavior in
evolution. This challenge pervades the following constructivist themes in
B&K: The first is the continuity between embryogenesis and psychogenesis; the second is the role of behavioral feedback and autoregulation
in epigenesis and subsequent development; the third is the acquisition of
knowledge through assimilation of and accommodation to environmental information by pre-existing structures; the fourth is construction of
powerful logical-mathematical models of reality, which facilitate environ-

mental transformations; the fifth is the neurological basis of cognitive construction and its historical/evolutionary continuity in brain evolution.

In his three books on evolution, *Biology and Knowledge* (*B&K*; Piaget, 1971), *Behavior and Evolution* (*B&E*; Piaget, 1978), and *Adaptation and Intelligence* (*A&I*; Piaget, 1980), Piaget questions the adequacy of neo-Darwinian theory to explain the origins and evolution of logical-mathematical cognition. He proposes a nexus of related problems expressed at different levels of generality including (a) how humans evolved the capacity for constructing logical-mathematical structures; (b) how these structures can so accurately describe the world; (c) how organic and cognitive regulations correspond; and the most general question, (d) how behavior and morphology could have co-evolved. The core of Piaget's argument is that none of these phenomena can be explained by orthodox neo-Darwinian evolutionary models.

In the Introduction to *B&K*, Piaget (1971) says: "The aim of this work is to discuss the problem of intelligence and of knowledge, in general (in particular, logical-mathematical knowledge) in the light of contemporary biology" (p. xi). Later in *B&K*, Piaget describes the following three forms of knowledge: instinct, perception, and a third category including conditioned behavior, habits, memory, and levels of intelligence. The latter category includes empirical and logical-mathematical intelligence. He then reiterates his well-known view that "knowledge is not a copy of the environment but a system of real interactions" (Piaget, 1971, p. 27), which are ". . . originally set off by spontaneous activity of the organ as much as by the external stimuli" (Piaget, 1971, p. 28). As elsewhere, he argues that all knowledge is based on assimilation of environmental information to previous structures through action schemes: Physical knowledge arises from actions on objects, logical-mathematical knowledge arises from reflection on outcomes of coordinations of actions, which, because they are necessarily on objects, inform physical knowledge.

Throughout *B&K*, Piaget (1971) emphasizes continuities and discontinuities between higher and lower forms of cognition as well as the problem of explaining their origins. Early on, he argues that there are no innate ideas:

> On the contrary, heredity and maturation open up new possibilities in the human child, possibilities quite unknown to lower types of animals but which still have to be actualized by collaboration with the environment. These possibilities, for all they are opened up in stages, are nonetheless essentially functional . . . in that they represent a progressive power of coordination. (p. 21)

Toward the end of *B&K*, Piaget (1971) stated

> ... although every kind of knowledge, including instinct, includes information
> about the environment, it nevertheless presupposes, in learning as in everything else,
> some "structuration" imposed upon it as a previous and necessary condition by in-
> ternal functioning allied to the subject's organization. This structurization, however,
> takes two forms, though remarkably isomorphic one to the other: first a hereditary
> form innately programmed down to the last detail the structures ... whose inner
> logic we have shown to be closely connected with the forms and schemata of
> sensorimotor intelligence; ... the second form, not programmed in detail by hered-
> ity, intervenes with a kind of assimilatory mechanism, wherever learning, however,
> elementary, is going on. (p. 264)

In the section titled "Bursting of Instinct," he elaborates further on the purported connection, saying

> ... intelligence does inherit something from instinct although it rejects its method of
> programmed regulation in favor of constructive autoregulation. The part of instinct
> that is retained allows the intelligence to embark on two different but complemen-
> tary courses: interiorization in the direction of its sources, and exteriorization in the
> direction of learned or even experimental adjustments. (Piaget, 1971, p. 366)

He argues that the externalization or phenotypic accommodation to the environment produces physical knowledge, whereas the internalization or formal structuration through reflective abstraction produces l-m knowledge.[2] Logical-mathematical knowledge interacts with and informs physical knowledge, but cannot be reduced to it.

Piaget expresses the crux of his major theme in *B&K*, when he asserts that it is

> ... unthinkable that the human brain's capacity for constructing logical-math-
> ematical structures that are so admirably adapted to physical reality should be ex-
> plained away by mere selection, as the mutationists do, for factors of utility and sur-
> vival would have led only to intellectual instruments of a crudely approximate kind.
> ... (Piaget, 1971, p. 274)

---

[2]Reflective abstraction is the "... process of reconstructions of new combinations, which allows for any operational structure at any previous stage or level to be integrated into a richer structure at a higher level" (Piaget, 1971, p. 320).

Subsequently, in *B&E*, Piaget raised the question of the role of behavior in evolution, noting that discussion of this issue had been dominated by two alternatives, which he characterizes as either *chance or auto regulation*:

> Either organs come into being independently of behavior, both being the result of mutations, so that we have two more or less autonomous sets of chance occurrences which it falls to selection alone to reconcile . . . or else organs and behavior are of necessity correlated from the very moment of their inception, in which case behavior must play the principle role in this process . . . *either chance or selection can explain everything* or *else behavior is the motor of evolution.* (Piaget, 1978, p. 147, italics added)

Although this may be a false opposition, both the theory of phenotype-centered evolution and the concept of *niche construction* to be described support the idea that development, especially behavioral development, is the motor of evolution!

HISTORICAL VIEWS OF THE ROLE
OF BEHAVIOR IN EVOLUTION

Pre-Darwinian and Darwinian Ideas About the Role of Behavior in Evolution

Many of the ideas Piaget expresses about instinct and intelligence and the evolution of behavior have a long history going back at least to the Enlightenment. As is well known (see discussion in chap. 2), Lamarck (1984) believed that habitual behavioral adaptations to changing environments preceded and engendered heritable changes in body form and behavior. Less often appreciated is the fact that Lamarck's contemporaries and subsequent early evolutionists (or transformists) also argued that behavior, both instinctive and intelligent, played an active role in evolution.

As Robert Richards (1987) described it, "Early evolutionists, such as Lamarck and Cabanis, through Darwin, Wallace, Spencer, and later Darwinians—all proposed, though a variety of ways, that behavior and mind drove the evolutionary process" (p. 6). Richards (1987) continues by arguing that these thinkers regarded instinct as "the paradigm of evolved behavior, the model for weaving other aspects of mind into the evolutionary scheme" (p. 7). He notes that Darwin and Spencer considered moral behavior to be a kind of instinct brought under the guidance of reason: "Instinct thus formed the evolutionary hinge linking the minds of lower animals with that of man" (Richards, 1987, p. 8).

Surprisingly perhaps, in the 18th and 19th centuries, ideas of the mental continuity between animals and humans were common to both evolutionists and creationists who espoused the doctrine of *sensationalism*. This doctrine held that human knowledge depended entirely on sensations, the same resources animals depended on for their knowledge of the world. In contrast, the competing Aristotelian and Cartesian doctrines both held that the human soul differed critically from the animal soul (Richards, 1987).

Richards (1987) reported that during the 17th century in France, various scholars of the sensationalist school, including Julian Offrey de la Mettrie, Charles-Georges Le Roy, Jean-Pierre Cabanis, and Jean Baptist Lamarck, wrote detailed monographs on animal behavior and argued over definitions of instinct. These arguments involved questions of the degree of automaticity or flexibility of instinctive behavior, its degree of variability within species, and the distinction between instincts present at birth and those appearing in later stages of development. Continuity between instinctive and intelligent behavior was also suggested by several naturalists of the time. The idea that habitual behavior arose in response to new environmental challenges and was passed on to offspring was widely believed in France. In England, Charles Darwin's grandfather, Erasmus Darwin, another sensationalist thinker, discussed similar ideas.

Analyzing Darwin's notebooks and other published records, Richards (1987) showed that Darwin's ideas about instinct were influenced by these thinkers, and even more strongly influenced by such British Natural Theologians as John Fleming, Algernon Wells, and especially William Kirby, William Spence, and Henry Lord Bougham. Specifically, Roberts argues that Bougham's emphasis on variability in instincts within a species and their idea that instincts are manifested without experience or teaching led Darwin to believe that instincts arose through natural selection rather than through inheritance of acquired habits. Darwin also reacted to Reverend Wells' idea of the perfect articulation of instinct and structure with the notion that instincts vary before related structures do.

Likewise, Richards (1987) argued that Kirby and Kirby and Spence's entomological studies of slave-making ants challenged Darwin's theory of natural selection by making him aware of the existence of neutered castes, which could not pass their helping instincts to offspring. He shows that Darwin's solution to this problem was inspired by his reading of William Youatt's work on cattle breeding, which showed artificial selection of families rather than individuals.

> This principle of selection, namely not of the individual which cannot breed, but of the family which produced such individual, has I believe been followed by nature in regard to the neuters amongst social insects; the selected characters being attached exclusively not only to one sex, which is a circumstance in the commonest occurrences, but to a peculiar & sterile state of one sex. (Richards, 1987, pp. 149–150, quoting from Darwin's manuscript, *Natural Selection*)

Darwin thereby solved the problem that he foresaw could falsify his theory of natural selection, anticipating what was later described as kin selection (Hamilton, 1964).

Charles Darwin (1859) also wrote entire chapters on the topics of both instinct and development in *The Origin of Species*. Moreover, I believe that he prefigured ideas of niche construction is his discussion of the feedback relationships involved in the evolution of bipedal locomotion, tool use, hunting, and increased mental powers in early man in *The Descent of Man* (1871). These elements of his work, however, were relatively neglected by the neo-Darwinist architects of the modern synthesis of evolutionary theory.

The Baldwin Effect and Organic Selection

The pioneering developmental psychologist, James Mark Baldwin, emphasized the role of behavior in evolution in his concept of organic selection, also known as the *Baldwin effect*.[3] Modern biologists and historians of science interested in his work have interpreted his evolutionary ideas quite differently.

Richards (1987), in his biography of Baldwin, noted the complexity of the history of the idea of organic selection. He traces Baldwin's changing use of the term. Early on, Baldwin conceived of accommodation to new environments through organic selection as the means by which children

---

[3]In this discussion of organic selection in *B&K*, Piaget cites Hovasse's characterization of organic selection as "the possibility of replacing an *accomodat* by a mutation" (Piaget, 1971, p. 298). Piaget notes the ambiguity of the idea of organic selection but relates it to Waddington's genetic assimilation (as have several other commentators). In his discussion of organic selection, Waddington, however, says that as it is generally understood, organic selection differs from genetic assimilation. Indeed, he says, "the theory seems to be an impossible one" (Waddington, 1975, p. 89).

traversed the various stages of intellectual development from imitation to intelligence. He believed that persistent imitation acted in a manner akin to the positive action of natural selection, retaining successful elements and eliminating copied elements that did not work. He emphasized the autonomous role of social heredity (as contrasted with biological heredity) in transmission of social behaviors, criticizing William James for ignoring this factor and for seeing no constraints on natural selection of mental variations. His ideas included the notion of social confirmability of ideas and the notion that ideas become part of the environment of thought, which influence selection of other thoughts.

In the spring of 1896, Baldwin proposed a new factor in evolution to explain the mystery of complex coadaptive variations without resorting to the inheritance of acquired characteristics:

> The solution of organic selection supposed that an animal's conscious intelligence might, in response to an environmental need, initially produce a coadaptive behavior system which would stave off extinction. But then natural selection could begin to save up those congenital variations that chanced to appear; the selective value of the system's elements would be maintained, while physical evolution replaced learned traits with instinctive ones. (cited in Richards, 1987, p. 483)

Although organic selection is known as the Baldwin effect, James McKeen Cattell, who reviewed Baldwin's paper, noted that three scholars had presented the idea simultaneously in 1896: J. M. Baldwin, Conway Lloyd Morgan, and Henry Fairfield Osborn. Without charging him with intentional deceit, Richards (1987) showed that Baldwin defended his priority with ambiguous quotations of his earlier use of the term, *organic selection*, even though his meaning had changed over time and even though he was influenced by Morgan's talk on the principle of organic selection at the New York Academy of Sciences in 1896, which preceded Baldwin's own talk on the subject.

Griffiths (2003) also traces the history of Baldwin's ideas on the subject. He argues that Baldwin's use of the term, organic selection, originally referred to the ability to acquire behaviors through reinforcement, which arose through natural selection, and later changed to refer to "mechanisms by which acquired characteristics could become hereditary without violating Weismann's doctrine of the continuity of the germ plasm" (Griffiths, 2003, p. 196).

In any case, most commentators agree that the pre-genetics formulation of the concept of organic selection renders it difficult to interpret.

Some interpret the newly acquired behaviors as relying on existing hidden genetic variation; others interpret it to rely on subsequent mutation. Richards (1987) noted that, following Romanes, later commentators tellingly criticized the idea of organic selection on the grounds that individual accommodations sufficient to preserve organisms would obviate the need for subsequent selection.

Deacon (2003) made a similar critique of the concept of organic selection. In his discussion of language evolution, he describes a process of change whereby demands of language itself generated selection pressure for symbol learning and phonological control along with increased capacity for automatization and imitation. In contrasting this process with Baldwinian organic selection, Deacon introduces the distinction between the phenomenon of *masking*, or shielding characters from selection, which occurs in organic selection, and *unmasking*, or unshielding characters from selection. Deacon notes that several factors including dominant alleles, developmental canalization, reduced competition, as in domestication, can mask selection. Conversely, novel selection pressure in new or changing ecological, social, or developmental conditions can expose previously masked traits to selection; this can happen. Novel experimental environments involving heat shock or ether treatment, for example, can cause breakdown of developmental canalization in fruit fly embryos (Waddington, 1975). (See chaps. 2 and 3, this volume, for discussion of genetic assimilation.)

In contrast, based on passages from *Evolution and Development* (Baldwin, 1902), rather than Baldwin's 1896 paper, West-Eberhard (2003) argued that Baldwin believed that organic selection could produce increased phenotypic plasticity, and thereby either increase or decrease the pace of evolution. She argues that the Baldwin effect has been consistently misrepresented by Waddington and others as depending solely on new mutations occurring subsequent to phenotypic plasticity.

Neo-Darwinism and Its Purported Neglect of Behavior and Evolution

Although Darwin's theory of natural selection provided a mechanism for adaptive evolutionary changes in behavior, his attempts to understand the origins of variation failed (Darwin, 1868). The sources of new hereditary characteristics did not begin to be understood until Mendel's work on the particulate nature of heredity was rediscovered in 1901 and was joined by DeVries' work on mutations.

In the 1930s and 1940s, after decades of wrangling, the two contending schools of thought, mutationists and selectionists, agreed on an integrated model of evolution according to which (a) mutation is the ultimate source of genetic variation producing new chromosomes, genes, and alleles; (b) recombination during meiotic generation of haploid sex cells and reformation of diploid cells through fertilization is the process that produces recombined chromosomal variants; (c) selection (differential reproduction and survival); and (d) genetic drift or sampling error in fertilization and survival are the means by which the proportions of competing variants change through generations. Selection leads to the proliferation of adaptations, whereas genetic drift does not. Speciation is thought to occur primarily through the evolution of reproductive isolating mechanisms (generally during geographic isolation of a small population from its larger mass), which ultimately prevents fertile mating between individuals of different species (Mayr, 1998).

Some historians of biology argue that because the modern synthesis was forged primarily by geneticists, taxonomists, paleontologists, and zoologists, it neglected certain other areas of biology. For at least 25 years, biologists and historians of biology have been criticizing the modern synthesis (Mayr & Provine, 1998) for neglecting three major areas of concern: development, behavior, and ecology. The next part of this chapter focuses on further critiques of the neglect of the role of behavior and ecology in evolution.

It is important to note that these accounts neglect to mention that ethologists studied the evolution of behavior before, during, and after the modern synthesis. (Darwin himself was a pioneer in ethology [Darwin, 1965].) Therefore, it is interesting to note that neither the terms *ethology* and *instinct*, nor the names of the three Nobel prize winners in ethology occur in the index of Mayr and Provine's (1998) history of the modern synthesis.[4]

---

[4]In part, the omission of ethology reflects the separate research programs: Ethology developed primarily in prewar Germany, whereas genetics, taxonomy, and paleontology developed primarily in the United States and England. In addition, political factors negatively influenced Lorenz's reputation in the United States. His ideas about instinct and learning were first introduced to English speakers by Daniel Lehrman in a hostile review in the *Quarterly Review of Biology*. Moreover, as Richards (1987) says, ". . . public association of Nazism with human evolutionary psychology froze any enthusiasm for the discipline immediately after the war, and continues to chill its development within contemporary biology of behavior as well as within the social sciences" (p. 536).

Ethology is a European branch of animal behavior. In Lorenz's (1981) words, "Ethology . . . is based on the fact that *there are mechanisms of behavior which evolve in phylogeny exactly as organs do, so that the concept of homology can be applied to them as well as morphological structures*" (Lorenz, 1981, p. 101). Consistent with this definition, ethologists have done comparative studies of species-specific behaviors, both as a means for taxonomic classification and for reconstructing the evolutionary history of particular behavior patterns. In contrast to some earlier formulations, this approach sees behavior as the product of evolution rather than as the pacemaker of evolution (Eibl-Eibesfeldt, 1970; Lorenz, 1965; Tinbergen, 1963).

Recent Critiques of the Modern Synthesis

More recently, critics of the modern synthesis have charged that its architects neglect behavior's causal role in evolution, relegating it to the subsidiary status of a product of evolution. Plotkin (1988a) said that many of the criticisms of neo-Darwinism come from developmentalists, ecologists, and cognitive scientists, who complain that neo-Darwinism is too narrow to explain how evolution works. This echoes Piaget's argument that behavior is the motor of evolution.

Plotkin (1988a) attributes evolutionary biologists' alleged neglect of behavior to lack of an adequate conceptual framework. In particular, he attributes it to the tendency to relate behavior to genetics, rather than to evolution. Plotkin (1988b) noted, however, that Piaget, J. M. Baldwin, and several biologists including Conrad Waddington, Ernst Mayr, Konrad Lorenz, and Allan Wilson considered behavior to be the pacemaker or driving force of evolution. Plotkin claims that these figures had a paucity of analysis and/or empirical support for their claims.

In his critique of the modern synthesis, Odling-Smee (1988) argued that neo-Darwinian theory (NDT) makes the following counterproductive assumptions: (a) that all heritable traits are genetic, (b) that only natural selection can nonrandomly influence random mutations, and (c) that biotic and abiotic forces in the environment are the sole sources of selection. Consequently, NDT lacks a mechanism for including environmental changes as integral parts of the evolutionary process.[5] He says that this

_____

[5]This leaves out Darwin's mechanisms of sexual selection (male competition and female choice) by which organisms act as selective agents favoring particular characteristics among members of same or the opposite sex.

omission results in two separate disciplines: one that handles environmental change (ecology), and the other that handles changes in organisms (evolutionary biology) (Odling-Smee, 1988, p. 75). Moreover, he says, the consequence of this linear modeling in both fields is that the modern synthesis fails to take into account the possibility that ". . . the outputs of active organisms are capable of modifying their own subsequent inputs in evolutionarily significant ways" (Odling-Smee, 1988, p. 75).

Like Plotkin (1988a), Oding-Smee (1988) credits Waddington with the realization that evolution involves at least four major subsystems: a genetic system of hereditary transmission, natural selection, an epigenetic system, and an "exploitive system" by which animals select and modify their habitats and hence influence selection pressures acting on them (Waddington, 1975).

In his review of Roe and Simpson's (1959) edited volume, *Behavior and Evolution*, Waddington (1975) described this feedback relationship as follows:

> Behaviour . . . is at the same time a producer of evolutionary change as well as a resultant of it, since it is the animal's behaviour which to a considerable extent determines the nature of the environment to which it will submit itself and the character of the selective forces with which it will consent to wrestle. (p. 170)

As Plotkin (1988a) noted, many theorists (ethologists, anthropologists, and psychologists) have emphasized the feedback between behavior and evolution, but they have lacked a compelling, generally accepted paradigm that explicitly expresses their intuitions. Recently two models of this relationship have been proposed. The first is the concept of *niche construction* (Laland, Odling-Smee, & Feldman, 2000, 2001; Laland, Odling-Smee, & Feldman, 1999; Odling-Smee, Laland, & Feldman, 1996, 2003) based in large part on Lewontin's concept of *construction* and Waddington's idea of *behavioral feedback*. The second is Mary Jane West-Eberhard's (2003) "phenotype-centered theory of evolution."[6]

---

[6]Richard Dawkins' concept of the *extended phenotype* also focuses on the phenotype, but only as a tool of the genes. His idea is ". . . the replicator should be thought of as having *extended* phenotypic effects, consisting of all its effects on the world at large, not just its effects on the individual body in which it happens to be sitting" (Dawkins, 1982, p. 4). After discussing the caddis fly larva's use of stones on its back, the spider's use of its web, the beaver's construction of dams, Dawkins argues that these phenomena are "conceptually negligible" steps from morphology to behavior. He notes that from his viewpoint, ". . . an animal artefact, like any other phenotypic product whose variation is influenced by a gene, can

Phenotype-Centered Evolution
and Niche Construction

A focus on the phenotype rather than the genotype is the centerpiece of both these models. Both West-Eberhard (2003) and Laland et al. (2001, 2002) cite Richard Lewontin's (1974, 2001a, 2001b) emphasis on the phenotype. This is clearly expressed, for example, in Lewontin's (2001) recent argument that *construction* is a better metaphor for life than *adaptation* because organisms construct their own environments:

> (1) organisms determine what is relevant. . . . (2) organisms alter the external world as it becomes part of their environment. . . . (3) organisms transduce the physical signal of the external world. . . . (4) organisms create a statistical pattern of environment different from the pattern in the external world. (p. 64)

(Readers will notice parallels with Piaget's concept of *cognitive construction*.)

After identifying gaps and inconsistencies in modern evolutionary theory, West-Eberhard (2003) concludes, "The piece that is missing from a synthesis of development and neo-Darwinism is an adequate theory of phenotype organization that incorporates the influence of the environment" (p. 19). She responds by proposing a theory explaining how *adaptive novelty is generated by phenotypic plasticity occurring in response to environmental pressures during development*, as well as by mutation:

> In the origin of a novel phenotype, there are two related events. First, there is a new input in the form of a genetic mutation or an environmental change. Second, there is a developmental response that produces a new phenotype. For a novelty to have evolutionary potential, both of these events must be recurrent in a population and across generations. (West-Eberhard, 2003, p. 200)

She exhaustively catalogs how phenotypic plasticity arising from responses to environmental pressures generates new adaptive structures and behaviors in plants and animals. She also traces the evolutionary history of these adaptive structures.

Following Waddington, Lewontin, and others, West-Eberhard (2003) emphasized that (a) the rich genetic variation already present in natural

---

be regarded as a phenotypic tool by which that gene can potentially lever itself into the next generation" (Dawkins, 1982, p. 199). As the name implies, Dawkins' concept emphasizes the idea that artefacts and other extended phenotypes are tools of the genes.

populations underlies the pervasive plasticity of phenotypes; (b) that all characteristics arise from and respond to both genes and environment; and consequently (c) that phenotypes may show *interchangeable* responses to genes and environments, and/or to several different genes; and (d) that this interchangeability is the key to understanding the modification of development, which generates variation. She also notes that phenotypic continuity between connecting life cycles is the source of genotypic continuity of the germ line.

West-Eberhard (2003) emphasized that novelty always arises from transformation of ancestral phenotypes during development. She also emphasizes that novelty arises from reorganization of existing traits. She argues that switch points in development lead to discrete, semiindependent developmental subunits or modules, which are distinctive in their gene expression. This ubiquitous feature of development, which occurs at every level of organization from the genetic to the cellular to the behavioral, leads to the *plasticity* and *modularity* of biological systems. Landmark examples of modules include (a) exons within genes, which can be recombined to make new proteins; (b) membrane bound organelles in eukaryotic cells, which can differentiate and aggregate to form new tissues and organs; (c) body segments, which can duplicate and specialize; (d) metamorphic stages in the life cycles of insects and amphibians, which can be omitted or truncated; (e) alternative phenotypes such as queens and workers in social insects; and (f) discrete behaviors, which can be recombined into new functional patterns. (She cautions against the use of the *modularity* concept without a clear developmental basis as has occurred in the so-called mental modules of evolutionary psychologists.)

West-Eberhard (2003) argued that plasticity and modularity facilitate the origin of developmental and behavioral novelties through the following forms of reorganization: duplication, deletion, recombination, reversion, correlated change, and cross-sexual transfer of morphological, developmental, and behavioral modules. As she notes, many of these phenomena have been described by ethologists tracing the origins of communication patterns through recombination and "ritualization" of preexisting behaviors in new contexts as in the re-use of infantile behaviors in appeasement signals (e.g., Eibl-Eibesfeldt, 1970).

West-Eberhard (2003) argued that *learning*—which she defines as ". . . a change in the nervous system manifested as altered behavior due to experience" (p. 337)—is a form of phenotypic recombination that can influence the rate and direction of evolution by influencing the frequency and contexts in which behaviors are expressed. She views learning as one among many "environmentally responsive regulatory mechanisms that coordinate trait expression and determine the circumstances in which they are exposed to selection" (West-Eberhard, 2003, p. 338). Contrary to the common belief that learning is unaffected by selection, West-Eberhard (2003) argued that its recurrence gives it greater evolutionary potential than mutation: "since many individuals in the same population may simultaneously and suddenly learn the same things in the same circumstances or through mimicry of other individuals" (p. 339). She notes, for example, that accidentally-learned behaviors in feeding such as kleptoparasitism in birds can contribute to individual or population specializations or even ecological specialization and ultimately species differences.

She notes that although behavior and physiology are the most responsive aspects of the phenotype in their speed and reversibility of response, they do not always lead morphology in the evolution of new characters: "Probably the best general approach to the relationship between behavior and morphology in evolution is to acknowledge that they will often evolve in concert, and that the aspect that takes the lead will be the one most flexible in producing a recurrent adaptive response" (West-Eberhard, 2003, p. 182). She continues by noting that flexibility of organization of development contributes to the *evolvability* of organisms. (Readers will note some similarity to Piaget's ideas, but they should also note that West-Eberhard [2003] explicitly denies any influence on the germ cells; see chap. 2, this volume.)

The concept of *niche construction*, also inspired by Lewontin's (2001) concept of construction as a metaphor for evolution, has been proposed as a paradigm that places artefacts (and other behavioral products) outside the genes (Griffiths, 2003).[7]

---

[7]It should be noted that "niche" is an ecological construct that emphasizes species actively create and carry out particular roles in their ecological communities. Ecologists define the habitat as the address, and niche as the profession of a species (Odum, 1971). From this traditional perspective, a species niche entails its life history, its intraspecific and interspecific relations of competition, predation, parasitism, and mutualism with other biotic community members, as well as its response to such abiotic features as water, soil, rocks, and climate. All of these biotic and abiotic interactions exert selection pressures resulting in co-evolution of species within a community, and transformations of abiotic environments as well. This co-evolution has led to primary succession of seral stages culminating in such mature ecosystems as tropical rainforests, tundras, and deserts. When small patches of these mature ecosystems are disturbed, they regenerate through a series of seral stages in a secondary succession (Odum, 1971). Note that the occurrence of ecological succession through a series of seral stages has implicit in it a notion of interactional transformation.

Niche construction refers to the activities, choices, and metabolic processes of organisms, through which they define, choose, modify, and partly create their own niches. . . . In every case, however, the niche construction modifies one or more sources of natural selection in a population's environment and, in doing so generates a form of feedback in evolution that is not yet fully appreciated. (K. Laland et al., 2000, p. 132)

Niche construction results in substantially modified environments of offspring and more distant descendants, a phenomenon they call *ecological inheritance* (K. Laland et al., 2000, p. 133).
In their model,

. . . adaptation ceases to be a one-way process, exclusively a response to environmentally imposed problems; it becomes instead a two-way process, with populations of organisms setting as well as solving problems. Evolution consists of mutual and simultaneous processes of natural selection and niche construction. (K. Laland et al., 2000, p. 133)

In other words, niche construction places artefacts and other behavioral products outside the phenotype and genotype, thereby conferring selective power on them especially as they persist through generations. (This contrasts with Richard Dawkins' [1982] *extended phenotype*, which places artefacts and other behavioral products within the phenotype and genotype, thereby conferring selective value.)

Odling-Smee (1988) called his two-way version of the evolutionary relationship between organisms and environments, *organism-environment co-evolution*. He notes that whereas it is impossible to describe an internal environmental inheritance system, it is possible to describe a succession of environments relative to successive generations of organisms. He proposes a model of

. . . two complementary systems of descent working via two transmission mechanisms under the influence of two reciprocal modifying, forces. Thus a *genetic inheritance* is transmitted from ancestral organisms (O) at time t_0 to successor organisms at time t_1 via the mechanisms of reproduction and under the direction of natural selection. Also, an ecological inheritance is transmitted to successor organisms at time t_i via the external environment (E). One component of this *ecological inheritance* is directed by the niche-constructing outputs of ancestral organisms at time t_0, however, a second component is independent. . . . (Odling-Smee, 1988, p. 80)

Odling-Smee and his colleagues (1996) argued that, whereas evolutionary theory traditionally assumes that only genes are transmitted from generation to generation (excepting human cultural traditions), niche construction entails inheritance of modified environments, that is, *extragenetic inheritance*: ". . . ancestral niche constructing organisms effectively transmit legacies of modified natural selection pressures in their environments" (Odling-Smee et al., 1996, p. 642). Therefore, they argue that natural selection and niche construction are two parallel interacting processes. As they phrase it, ". . . evolution proceeds in reciprocal and simultaneous cycles of selection and niche construction. Evolution is characterized by these cycles of contingency" (K. Laland et al., 2001, p. 118).

While noting that human culture is a powerful means for niche construction, these authors emphasize that niche construction is general and pervasive in the organic world. They cite many examples of invertebrate and vertebrate organisms including ants, bees, wasps, badgers, gophers, ground squirrels, hedgehogs, moles, molerats, and prairie dogs that construct nests or burrow systems. These artefacts in turn have led to the evolution of defensive and regulatory behaviors.

Odling-Smee (1988) described three kinds of niche constructing behavior: *phenotypic selection, phenotypic perturbation, and prediction.* The phenotypic selection refers to habitat selection, migration, and dispersal. Phenotypic perturbation refers to qualitative changes organisms inflict on environments, including depleting of resources, hoarding, nest and burrow construction, dumping detritus, and dam building (some of the same examples Dawkins [1982] uses). Prediction involves learning and anticipation, which can redefine organisms' problem space. (Human technology falls in this last category.) As Laland et al. (2001) noted, niche construction of both past and present generations can influence the current and future environment.

In a complementary formulation, Laland et al. (2001) characterized the processes by which organisms acquire information about their environments as follows: (a) population genetics processes (that is, mutation, recombination and selection); (b) individual learning processes (conditioning, trial and error); and (c) social learning processes, including imitation and teaching (which ultimately lead to cognitive prediction). They note that the concept of niche construction allows us to explain how acquired characteristics can influence evolution without resorting to Lamarckian explanations of (genetic) inheritance of acquired characteristics.

As Laland et al. (2000) emphasized, hominid evolution has been driven in part by such cultural traits as use of tools, weapons, fire, cooking, and symbols. Human language and culture are particularly powerful social learning mechanisms of niche construction because their products are not only long-lasting and pervasive, but are also cumulative and progressive. Indeed, in many societies, humans live in largely constructed habitats (cultural landscapes) and social institutions dominated by material (architectural and technological) and ideational (artistic, religious, political, and scientific) culture, all transmitted symbolically.

Niche Construction by Other Names. Without using the term, niche construction, anthropologists and other scholars have emphasized the central role these feedback relationships have played in hominid evolution. Fredrich Engels (1896), for example, prefigured modern anthropological formulations in his argument that labor, particularly tool use, "created man himself" (p. 279). Later, echoing Charles Darwin (1871), Sherwood L. Washburn (1960) made the feedback mechanism more explicit when he said "The success of the new way of life based on the use of tools changed the selection pressures on many parts of the body, notably the teeth, the hands, and brain as well as the pelvis" (p. 63). As Laland et al. (2000) noted, several anthropologists and biologists have proposed models for culture-gene interaction (Boyd & Richerson, 1985; Durham, 1991; Lumsden & Wilson, 1981).

Anthropological archaeologists study of remains of the material culture of earlier peoples provides important clues to the history of technology and settlement patterns (McBrearty & Brooks, 2000; Mellors & Stringer, 1989). Very recently, anthropologists have begun to trace archeological evidence from abandoned nut-cracking sites of West African chimpanzees, which should help future researchers identify the earliest hominid technologies (Mercader, Panger, & Boesch, 2002). This and other recent work on telltale remains of foraging by orangutans illustrates the information-carrying potential that phenotypic perturbations offer intelligent conspecifics (Russon, in press).

As Durham (1991) chronicled, anthropologists have noted that the spread of slash-and-burn agriculture created increased habitats for malaria-carrying mosquitoes in densely populated areas is sub-Saharan Africa, resulting in selection of malaria resistant hemoglobins (Livingstone, 1958). Similar feedback loops probably exist in the history of many genetic diseases, including other hemoglobin pathologies, cystic fibrosis, and diabetes.

Durham (1991) also emphasized that human culture depends on language and other forms of symbolizing. As a feedback mechanism, language vastly increases the content, accuracy, and efficiency of (inter- and intragroup and intergenerational) social transmission. Language confers a "constitutive function" by creating realities, organizing relations, and producing systems of knowledge. Moreover, language creates a social history of shared procedures, instructions, plans, values, and beliefs.

There is a vast literature focusing on the evolution of the capacity for language. Based on comparative studies of great apes, many investigators believe that the earliest form of symbolic communication in our lineage may have been gestural. In any case, most believe that the capacity for fully modern culture and language only evolved with modern *Homo sapiens* (Gibson & Ingold, 1993; Mellors & Stringer, 1989; Mithen, 1996; Noble & Davidson, 1996). Increasingly, biological anthropologists have emphasized the feedback relationship between habitual use of the vocal apparatus for symbolic communication and selection of greater control of organs of articulation, breathing, and neurological processing of sound comprehension and production in humans. More important, symbolic communication creates selection pressures on memory and symbolizing capacity (Deacon, 1997). Once attained, language creates, modifies, maintains, and transmits the routines of daily life, embodies the social institutions of production, reproduction, and political and religious life, and performs transformations of status during the life cycle (Austin, 1975).

Deconstructing the Global Concept of Environment in Niche Construction. Ecologists conventionally distinguish biotic and abiotic environmental elements, and different ecological roles among biotic elements; that is, predators, prey, competitors, parasites, and so on. Brandon (1988) extended this pattern by distinguishing the following three notions of environment: (a) external environment, that is, all biotic and abiotic elements external to an evolving population; (b) ecological environment, that is, features of the external environment that impact the demography of a populations; and (c) selective environment, that is, features of the external environment that affect differential fitnesses or contributions to succeeding generations among members of a population. Brandon goes on to note that the selective environment depends on the differing sensitivities of organisms to features of the external environment, and their abilities to damp out their effects.

Lewontin's (2001a, 2001b) and Waddington's (1975) co-evolutionary models and their derivative niche construction also imply the need to deconstruct the global concept of *the environment*. The critical distinction implied by constructivist notions is the distinction between environments that are directly or indirectly constructed and/or elicited by organisms themselves versus those that are independent of their activities. Self-generated and self-elicited environments are important for thinking about feedback loops that occur in environments at many levels: ecological communities, populations and societies, individual life cycles, and embryogenesis. Table 1.1 summarizes various hierarchical levels of environment.

Zooming Down from Niche Construction to Individual Development

As suggested by the hierarchical nature of these levels, the ontogenetic niches co-constructed by individuals during their life histories are embedded in niche construction by populations at the community level. These co-constructed ontogenetic niches are stage-typical physical and social environments in which immature individuals develop, and on which they depend for their subsistence, survival, and training (Parker & McKinney, 1999). In humans, these constructed environments include such physical aspects as shelters, clothes, tools, and such activities as transportation, prolonged provisioning, and extensive social and technological apprenticeships.

Human offspring and their caretakers co-construct cascading relationships and environments, which through feedback progressively transform their participants in predicable ways. Typically, these feedback loops result in a sequence of developmental stages. The predictable co-generated trajectory of these stages is reminiscent of the sequentially dependent (seral) stages of ecosystem succession during which each ecological community creates the conditions necessary for the emergence of the next (Odum, 1971). As noted in Table 1.1, both of these feedback loops are similar to feedback loops occurring in populations and their habitats through niche construction and natural selection.

Ontogenetic niches are *co-constructed* through routine daily activities and interactions of the developing organisms and their interacting caretakers, siblings, and playmates (Farver, 1999; Rogoff, 1993). As Bronfenbrenner (1979) emphasized, children develop in a nested set of progressively broader environments. Analysis of these nested environments and activities reveals that social and cultural knowledge is never embodied in a single individual but always *distributed* among interactants and even within the constructed environment (Hutchins, 1995; Strum, Forster, & Hutchins, 1997).

TABLE 1.1

Contingent Cycles of Feedback From Niche Construction and Selection Operating at Various Hierarchically-Organized Environmental Levels

Levels:	Components	Processes	Larger Phenomena	Time Frames
Abiotic elements in ecological communities	Atmosphere, climate, soil, water, and so on	Transformation through bio- and geochemical cycles	Primary or secondary ecological succession of seral stages	Geologic time or generations
Biotic elements in ecological communities	Interactions among species in roles of producers and consumers, predators, prey, parasite, competitor, and so on	Transformation of demography, biogeography, and community interactions		
Populations including social groups	Interactions among conspecifics of differing generations, roles and statuses	Transformation of demography and population structure	Niche construction and selection	Generations
Constructed environments	Interactions with cultural landscapes and technologies	Transformation of habitat and niche		
Ontogenetic development	Interactions with caretakers, siblings, peers, teachers, and so on	Species typical interactions and transformation of phenotype	Cascading phenotype transformation	Lifespan
Embryological and fetal development	Epigenetic interactions among genes, proteins, and tissues	Species typical phenotype construction	Cascading phenotype construction	Prenatal period

Piaget's model of cognitive development constitutes a paradigm for the child's active ontogenetic construction. Like niche construction and phenotype plasticity, Piaget's constructivist model emphasizes the self-transforming nature of feedback from the organism's actions on itself and other objects. The feedback occurs through assimilation of objects to existing schemes and accommodation of those schemes to the qualities of the objects. The resulting epigenesis pulls the child along predictable developmental trajectories.

Piaget (1952) proposed an epigenetic model based on the notion of assimilation of experience to existing schemes and accommodation of those schemes to the qualities of objects. These reciprocal processes result in new coordinations and differentiations of schemes, followed by reflection on the observable outcomes of these actions, resulting in cycles of disequilibration and re-equilibration (Piaget, 1985).

Piaget's model is also consistent with the perspective of niche construction and phenotype plasticity in focusing on what I call "the self-transforming phenotype" (Parker, 2001). As Lewontin (2001b) said, phenotypes have histories: "The phenotype at any instant is not simply the consequence of its genotype and current environment, but also of its phenotype at the previous instant" (p. 63). As he also says, "The final step in the integration of developmental biology into evolution is the incorporation of the organism itself as a *cause* of its own development, as a mediating mechanism by which external and internal factors influence its future" (Lewontin, 2001b, p. 62).

Comparative studies of development are a necessary component of this integration into evolutionary biology. As will be indicated in chapter 2, studies of distantly related model organisms (including fruit flies, zebra fish, and mice) have revealed much about such developmental features as the genetic basis of formation of body axes *common to metazoans*. It is important to emphasize, however, that the *evolution of developmental differences* can only be reconstructed through comparative studies of closely related species (Raff, 1996).

Evolutionary reconstruction of human developmental patterns rests on phylogenetic analysis (Brooks & McLennan, 1991) of the characteristics of great apes, our closest living primate kin, relative to those of lesser apes and Old World monkeys, their next closest outgroup (Parker & McKinney, 1999). Systematic comparative studies of this kind have revealed species differences in the rate, pattern, and extent of cognitive development, which apparently result from differences in the organization of, and interrelations among, schemes (Langer, 1996, 2000).

These studies also reveal that our human developmental pattern generates species-typical interactions with animate and inanimate objects. For example, circular reactions and vocal and facial imitation generate species-typical socializing interactions between human infants and their caretakers, that is, "the game" (Watson, 1972). These interactions stimulate development of centers in the brain. In other words, these schemes function as self-teaching and teaching-eliciting mechanisms, which, via brain development, transform the phenotype (Parker, 1993).

Self-Transforming Phenotype Fields. Consistent with Lewontin's (2001a, 2001b) constructivist view, West-Eberhard's (2003) phenotype plasticity, Dawkins' (1982) extended phenotype, and a Piagetian (1952, 1954) constructivist view of comparative data on primate cognitive development, I suggest that through time, developing organisms generate *phenotype fields*. These phenotypes and their fields successively transform themselves through their own activities and the feedback they generate. This self-generated feedback canalizes species-typical development by stimulating activity and experience-contingent brain development (Edelman, 1987; Trevarthan, 1973, 1987).

Interacting phenotype fields induce reciprocal and/or complementary changes in interacting organisms. This can be seen, for example, as human mothers react to their babies growing abilities—their extended "zone of proximal development" (Vygotsky, 1978) by progressively handing over more elements of the game of "peek-a-boo" to them (Bruner, 1983). In addition, feedback from these more-or-less predictable interactions among phenotype fields can cause developmental divergences through amplification of small biases as seems to occur in the development of gender differences (McCoby, 1998).

My self-transforming phenotype field construct differs from some popular evolutionary psychological models of cognitive modules (Cosmides, Tooby, & Barkow, 1992; Pinker, 1994) in emphasizing the constructivist nature of development. Specifically, it emphasizes the progressive and experience-contingent, yet canalized, nature of epigenesis. As Lewontin (2001a, 2001b) emphasized, the phenotype rather than the genotype is the self-transforming entity. As niche construction theory emphasizes, during development, phenotypes progressively transform themselves by constructing and transforming a series of environments in which they participate. Ultimately they act on, and are acted on by these environments, which also represent the legacy of prior generations.

CONCLUSIONS

Virtually every textbook in the field testifies to the fact that Piaget's constructivist model of cognitive development initiated an ongoing research program in developmental psychology. Moreover, as I have tried to show, Piaget's emphasis on phenotype construction and the creative role of behavior resonates with recent concepts of phenotype centered evolution and niche construction. In contrast, as is discussed in chapter 2, his model of behavioral evolution, particularly his phenocopy model, has not fared so well. Nevertheless, his core idea that behavior is a driving force of evolution seems to be vindicated in the models of phenotype plasticity and niche construction.

Finally, I would like to return to Piaget's questioning of the adequacy of neo-Darwinian theory to explain the origins of logical mathematical reasoning which is so admirably suited to predicting and explaining the laws of nature. In my view, modern genetics (see chap. 2, this volume) and emerging models in the constructivist tradition go far toward explaining the origins of this ability. Specifically, they converge to support the following logic:

IF evolution of sensory and locomotor organs allowed organisms to seek food and mates and avoid dangers,

IF evolution of neurons capable of learning allowed organisms to recognize and remember which actions and environments were fruitful and safe,

IF evolution of nervous systems capable to intelligence expanded their knowledge through mental maps and tool use, and allowed them to transmit their knowledge through imitation and teaching, and,

IF, as Piaget argues, logical-mathematical knowledge, like physical knowledge, arises from interiorization of and reflective abstraction on causal and logical outcomes of our actions on objects in space and time,

THEN, logical-mathematical knowledge, like physical knowledge, is based in action and therefore, the evolution of the capacity for logical-mathematical thought in our human ancestors can be understood historically as the culmination of a trend that shows both evolutionary and developmental continuities with simpler action-based forms of knowledge in related ancestral species (Parker & McKinney, 1999). Moreover, it can be understood as the underpinning of human language and culture, and therefore of the complex niche we have constructed.

ACKNOWLEDGMENTS

The 2001 Meeting of the Jean Piaget Society introduced the innovation of a pre-announced seminar discussion of Piaget's (1971) *Biology and Knowledge* by interested participants. One of the most useful elements in this seminar was a discussion of the historical context in which Piaget developed the idea expressed in these related works. Inspired by these remarks and my own reading of relevant literature, I have tried to present some of this context in chapter 1 and chapter 2.

I am grateful to the participants in the informal seminar discussion of Piaget's (1971) book, *Biology and Knowledge*, which took place at the 2001 Jean Piaget Society Meeting in Berkeley, California. I particularly thank the following participants who sent me comments and or suggestions: Joe Becker, Terrance Brown, Elinor Duckworth, Jeannette Gallagher, Jonas Langer, and Dan Slobin. I am also grateful to participants in the ongoing discussion group of the Michael Ghiselin Center for the History and Philosophy of Science at the California Academy of Sciences, particularly Rasmas Winther, who commented extensively on an earlier draft of my chapters. Likewise, I thank the members of my psychology/anthropology reading group—Phyllis Dolhinow, Diana Divecha, Judith Fitzpatrick, and Constance Milbrath—for the suggestion to divide my original chapter into two. I particularly thank Elbert Branscomb and Felix Friedberg for their careful reading and correction of errors in the molecular biology section. Likewise, I particularly thank Terrence Brown for his corrections of the Piagetian concepts, and critique of my argument. The errors and interpretations in both of these chapters are my own.

REFERENCES

Austin, J. (1975). *How to do things with words* (2nd ed.). Cambridge, MA: Harvard University Press.

Baldwin, J. M. (1896). A new factor in evolution. *American Naturalist, 30*, 441–451.

Baldwin, J. M. (1902). *Development and evolution*. New York: Macmillan.

Bickerton, D. (1981). *Roots of language*. Ann Arbor, MI: Karoma.

Bickerton, D. (1990). *Language and species*. Chicago: University of Chicago Press.

Boyd, R., & Richerson, P. (1985). *Culture and the evolutionary process.* Chicago: University of Chicago Press.

Brandon, R. N. (1988). The levels of selection: A hierarchy of interactors. In H. C. Plotkin (Ed.), *The role of behavior in evolution* (pp. 51–71). Cambridge, MA: MIT Press.

Bronfenbrenner, U. (1979). *The ecology of human development.* Cambridge, MA: Harvard University Press.

Brooks, D., & McLennan, D. (1991). *Phylogeny, ecology, and behavior.* Chicago: University of Chicago Press.

Bruner, J. (1983). *Child's talk.* New York: Norton.

Cosmides, L., Tooby, J., & Barkow, J. (1992). Introduction: Evolutionary psychology and conceptual integration. In J. Barkow, L. Cosmides, & J. Tooby (Eds.), *The adapted mind* (pp. 3–18). New York: Oxford University Press.

Darwin, C. (1859). *The origin of species.* New York: The Modern Library, Random House.

Darwin, C. (1868). *The variation of animals and plants under domestication* (1st ed.). London: John Murray.

Darwin, C. (1871). *The descent of man.* New York: The Modern Library, Random House.

Darwin, C. (1965). *The expression of the emotions in man and animals.* Chicago: University of Chicago Press.

Dawkins, R. (1982). *The extended phenotype.* Oxford, England: W. H. Freeman.

Deacon, T. (1997). *The symbolic species.* New York: W. W. Norton.

Deacon, T. (2003). Multilevel selection in a complex adaptive system; the problem of language origins. In B. Weber & D. Depew (Eds.), *Evolution and learning: The Baldwin effect reconsidered* (pp. 81–106). Cambridge, MA: MIT Press.

Durham, W. (1991). *Coevolution: Genes, culture and human diversity.* Stanford, CA: Stanford University Press.

Edelman, G. (1987). *Neural Darwinism.* New York: Basic Books.

Eibl-Eibesfeldt, I. (1970). *Ethology, the biology of behavior.* New York: Holt, Rinehart & Winston.

Engels, F. (1896). *The dialectics of nature.* Moscow: Progress Publisher.

Farver, J. A. M. (1999). Activity setting analysis: A model for examining the role of culture in development. In A. Goncu (Ed.), *Children's engagement in the world: Sociocultural perspectives* (pp. 99–124). New York: Cambridge University Press.

Gesell, A. (1945). *The embryology of behavior.* New Haven, CT: Greenwood Press.

Gibson, K. R., & Ingold, T. (Eds.). (1993). *Tools, language and cognition in human evolution.* Cambridge, England: Cambridge University Press.

Gibson, K. R., & Peterson, A. C. (Eds.). (1991). *Brain maturation and cognitive development*. New York: Aldine.

Griffiths, P. (2003). Beyond the Baldwin Effect: James Mark Baldwin's 'social heredity,' epigenetic inheritance and niche-construction. In B. Weber & D. Depew (Eds.), *Learning, meaning, and emergence: Possible Baldwinian mechanisms of evolution* (pp. 193–215). Cambridge, MA: MIT Press.

Hamburger, V. (1964). Some aspects of the embryology of behavior. *Quarterly Review of Biology, 38*, 342–365.

Hamilton, W. D. (1964). The genetical evolution of social behavior. I, II. *Journal of Theoretical Biology, 7*, 1–52.

Hutchins, E. (1995). *Cognition in the wild*. Cambridge, MA: MIT Press.

Inhelder, B., & Piaget, J. (1964). *The early growth of logical thinking*. New York: Routledge, Kegan Paul.

Laland, K., Odling-Smee, J., & Feldman, M. (2000). Niche construction, biological evolution and cultural change. *Behavioral and Brain Sciences, 23*(1), 131–146.

Laland, K., Odling-Smee, J., & Feldman, M. (2001). Niche construction, ecological inheritance, and cycles of contingency in evolution. In S. Oyama, P. Griffiths, & R. Gray (Eds.), *Cycles of contingency: Developmental systems and evolution* (pp. 117–126). Cambridge, MA: MIT Press.

Laland, K. N., Odling-Smee, F. J., & Feldman, M. W. (1999). Evolutionary consequences of niche construction and their implications for ecology. *Proceedings National Academy Sciences USA, 96*(18), 10242–10247.

Lamarck, J.-B. (1984). *Philosophical zoology*. Chicago: University of Chicago Press.

Langer, J. (1996). Heterochrony and the evolution of primate cognitive development. In A. Russon, K. Bard, & S. T. Parker (Eds.), *Reaching into thought: The minds of great apes* (pp. 257–277). Cambridge, England: Cambridge University Press.

Langer, J. (2000). The descent of cognitive development. *Developmental Science, 3*(4), 361–379, 385–389.

Langer, J., & Killen, M. (Eds.). (1998). *Piaget, evolution, and development*. Mahwah, NJ: Lawrence Erlbaum Associates.

Lewontin, R. (1974). *The genetic basis of evolutionary change*. New York: Columbia University Press.

Lewontin, R. (2001a). Gene, organism and environment: A new introduction. In S. Oyama, P. Griffiths, & R. Gray (Eds.), *Cycles of contingency: Developmental systems and evolution* (pp. 54–57). Cambridge, MA: MIT Press.

Lewontin, R. (2001b). Genes, organism and environment. In S. Oyama, P. Griffiths, & R. Gray (Eds.), *Cycles of contingency: Developmental systems and evolution* (pp. 58–66). Cambridge, MA: MIT Press.

Livingstone, F. B. (1958). Anthropological implications of sickle cell gene distribution in West Africa. *American Anthropologist, 60*(3), 533–562.

Lorenz, K. (1965). *The evolution and modification of behavior.* Chicago: University of Chicago Press.

Lorenz, K. (1981). *The foundations of ethology.* New York: Simon & Schuster.

Lumsden, C., & Wilson, E. O. (1981). *Genes, mind, and culture.* Cambridge, MA: Harvard University Press.

Mayr, E. (1998). Prologue: Some thoughts on the history of the evolutionary synthesis. In E. Mayr & W. Provine (Eds.), *The evolutionary synthesis* (pp. 1–48). Cambridge, MA: Harvard University Press.

Mayr, E., & Provine, W. (Eds.). (1998). *The evolutionary synthesis* (4th ed.). Cambridge, MA: Harvard University Press.

McBrearty, S., & Brooks, A. (2000). The revolution that wasn't: A new interpretation of the origin of modern human behavior. *Journal of Human Evolution, 39*, 453–563.

McCoby, E. E. (1998). *The two sexes: Growing up apart coming together.* Cambridge, MA: Belknap Press of the Harvard University Press.

Mellors, P., & Stringer, C. (Eds.). (1989). *The human revolution.* Princeton, NJ: Princeton University Press.

Mercader, J., Panger, M., & Boesch, C. (2002). Excavation of a chimpanzee stone tool site in the African rainforest. *Science, 296*(5572), 1452–1455.

Mithen, S. (1996). *The prehistory of the mind.* New York: Thames and Hudson.

Noble, W., & Davidson, I. (1996). *Human evolution, language and mind.* Cambridge, England: Cambridge University Press.

Odling-Smee, J. (1988). Niche-constructing phenotypes. In H. C. Plotkin (Ed.), *The role of behavior in evolution* (pp. 73–132). Cambridge, MA: MIT Press.

Odling-Smee, J., Laland, K., & Feldman, M. (1996). Niche construction. *American Naturalist, 147*(4), 641–648.

Odling-Smee, J., Laland, K., & Feldman, M. (2003). *Niche construction: The neglected process in evolution* (Vol. 37). Princeton: Princeton University Press.

Odum, E. B. (1971). *Fundamentals of ecology* (3rd ed.). Philadelphia: W. W. Saunders Co.

Parker, S. T. (1993). Imitation and circular reactions as evolved mechanisms for cognitive construction. *Human Development, 36*, 309–323.

Parker, S. T. (2001, June). *The nature in nurture.* Paper delivered at Jean Piaget Society Meeting, Berkeley, CA.

Parker, S. T., & McKinney, M. L. (1999). *Origins of intelligence: The evolution of cognitive development in monkeys, apes, and humans.* Baltimore, MD: Johns Hopkins University Press.

Piaget, J. (1952). *The origins of intelligence in children.* New York: International Universities Press.

Piaget, J. (1954). *The construction of reality in the child.* New York: Basic Books.

Piaget, J. (1962). *Play, dreams, and imitation.* New York: W. W. Norton.

Piaget, J. (1971). *Biology and knowledge: An essay on the relations between organic regulations and cognitive processes* (B. Walsh, Trans.). Chicago: University of Chicago Press.

Piaget, J. (1978). *Behavior and evolution.* New York: Pantheon Books.

Piaget, J. (1980). *Adaptation and intelligence, organic selection and phenocopy.* Chicago: University of Chicago Press.

Piaget, J. (1985). *The equilibration of cognitive structures, the central problem of intellectual development* (T. Brown & K. J. Thampy, Trans.). Chicago: University of Chicago Press.

Piaget, J., & Inhelder, B. (1967). *The child's conception of space.* New York: W. W. Norton.

Pinker, S. (1994). *The language instinct.* New York: HarperCollins.

Plotkin, H. C. (1988a). Behavior and evolution. In H. C. Plotkin (Ed.), *The role of behavior in evolution* (pp. 1–17). Cambridge, MA: MIT Press.

Plotkin, H. C. (1988b). Learning and evolution. In H. C. Plotkin (Ed.), *The role of behavior in evolution* (pp. 133–164). Cambridge, MA: MIT Press.

Raff, R. A. (1996). *The shape of life.* Chicago: University of Chicago Press.

Richards, R. J. (1987). *Darwin and the emergence of evolutionary theories of mind and behavior.* Chicago: University of Chicago Press.

Roe, A., & Simpson, G. G. (Eds.). (1959). *Behavior and evolution.* New Haven: Yale University Press.

Rogoff, B. (1993). Children's guided participation and participatory appropriation in sociocultural activity. In R. Wozniak & K. Fischer (Eds.), *Development in context* (pp. 121–153). Hillsdale, NJ: Lawrence Erlbaum Associates.

Russon, A. E. (in press). Developmental perspectives on great apes traditions. In D. Fragaszy & S. Perry (Eds.), *Towards a biology of traditions: Models and evidence.* Cambridge, England: Cambridge University Press.

Schmalhausen, I. I. (1986). *Factors of evolution: The theory of stabilizing selection.* Chicago: University of Chicago Press.

Spencer, H. (1872). *Principles of psychology* (2 Vols., 2nd ed.). London: Williams and Norgate.

Strum, S. C., Forster, D., & Hutchins, E. (1997). Why Machiavellian intelligence may not be Machiavellian. In A. Whiten & R. Byrne (Eds.), *Machiavellian intelligence II* (pp. 50–85). Cambridge, England: Cambridge University Press.

Tinbergen, N. (1963). On aims and methods of ethology. *Zeitschrift für Tierpsychologie, 20,* 410–429.

Trevarthan, C. (1973). Behavioral embryology. In E. C. Carterette & M. P. Friedman (Eds.), *Handbook of perception: Biology of perceptual systems* (Vol. 3, pp. 89–117). New York: Academic Press.

Trevathan, W. (1987). *Human birth: An evolutionary perspective.* Hawthorne, NY: Aldine de Gruyter.

Vygotsky, L. S. (1978). *Mind in society: The development of higher psychological processes.* Cambridge, MA: Harvard University Press.

Waddington, C. H. (1975). *The evolution of an evolutionist.* London: W & J Macay Limited.

Washburn, S. L. (1960). Tools and human evolution. *Scientific American, 203*(3), 62–75.

Watson, J. S. (1972). Smiling, cooing, and "the game." *Merrill Palmer Quarterly, 18*, 323–339.

West-Eberhard, M. J. (2003). *Developmental plasticity and evolution.* New York: Oxford University Press.

2

Piaget's Phenocopy Model Revisited: A Brief History of Ideas About the Origins of Adaptive Genetic Variations

Sue Taylor Parker

Professor Emeritus, Sonoma State University

PART I: PIAGET'S DEVELOPMENTAL EVOLUTIONARY MODEL

In his three books on evolution, *Biology and Knowledge* (*B&K*; Piaget, 1971), *Behavior and Evolution* (*B&E*; Piaget, 1978), and *Adaptation and Intelligence* (*A&I*; Piaget, 1980), Piaget describes what he regards as a nexus of related evolutionary problems expressed at different levels of generality. These include (a) how humans evolved the capacity for constructing logical-mathematical structures; (b) how these structures can so accurately describe the world; (c) how organic and cognitive regulations (functions) correspond; and the most general question (d) how behavior and morphology could have co-evolved.

The core of Piaget's argument is that none of these phenomena can be explained by orthodox, that is, neo-Darwinian evolutionary models. The structure of Piaget's argument involves an analysis of two alternatives, Lamarckism versus orthodox neo-Darwinism (mutationism and selectionism), counterposed to his own "cybernetic" synthesis, a *tertium quid*, based on his interpretation of Waddington's (1975) notions of phenocopy and genetic assimilation. His analysis and synthesis is based on the idea that there is a functional equivalence between the mechanisms generating variation in cognitive development and in behavioral and cognitive evolution. Readers will recognize Piaget's dialectical method of analysis from *The Origins of Intelligence in Children* (Piaget, 1952) and other works.

This chapter focuses on Piaget's thinking about the phylogenetic origins of adaptive behaviors and intelligence, particularly his idea that hereditary adaptations are passed from the soma to the genome, that is, his *phenocopy model*, for the origins of adaptive variations. Piaget's primary interest was in the idea of a functional equivalence between the origins of intelligence in phylogeny and ontogeny rather than in the posited mechanism of phenocopy (T. Brown, personal communication, August 2003). Despite this, Piaget devoted significant portions of his evolution books to this idea. Moreover, the mechanism of his phenocopy model has been taken seriously by some Piagetians and others (e.g., Gottlieb, 1992, 2002).[1] Piaget's idea that developmental adaptations play a role in the origin of new adaptive traits, however, gains support from the recently proposed phenotype-centered model of evolution (West-Eberhard, 2003), which will be described.

In this chapter, I attempt to place Piaget's thinking about these matters in the context of the history of related thought, and in the context of modern biological knowledge. This analysis leads to me to conclude that Piaget's phenocopy model is incompatible with what is now known about the processes of heredity and evolution. In particular, it suggests that the central misinterpretation in his phenocopy model and similar formulations lies in the notion that the feedback loops operating between genes

---

[1]Gottlieb (1992, 2002) attempted to describe a developmental mechanism for the origin of adaptive variations somewhat similar to Piaget's. Gottlieb presents a scheme for a hierarchically organized system of "reciprocal influences" or feedback to genes from chromosomal, nuclear, cytoplasmic, tissue, organismic, and environmental levels. Specifically, he argues

> While the feed forward or feed upward nature of the genes has always been appreciated from the time of Weismann and Mendel on, the feed backward or feed downward influences have usually been thought to stop at the level of the cell membrane. The newer conception is one of a totally interrelated, fully coactional system in which the activity of the genes themselves can be affected through the cytoplasm of the cell by events originating at any other level in the system. . . . (Gottlieb, 1992, pp. 143–144)

Gottlieb's model fails to discuss genetic transmission (he is apparently unaware that what he calls the feed-backward model has long been the dominant molecular biological model as it applies to *somatic genes as opposed to germ cell genes*). Other recent work on epigenetic inheritance suggests some routes by which environmental effects are transmitted to progeny (Jablonka & Lamb, 1995).

and environment in somatic cells also occur in germ line cells, allowing organisms to inherit adaptive environmental responses of their parents. As described, Piaget's notion of hereditary adaptive modification of genes during development has not been supported. As previously mentioned, his larger notion of the evolutionary importance of behavioral adaptations during development, however, has found support in a new theory of phenotype-centered evolution (West-Eberhard, 2003).

Piaget's Evolutionary Model of Auto-Regulation

The concept of *auto-regulation* and the posited analogy between organic and psychosocial evolution, that is, phylogenetic and ontogenetic processes lie at the heart of Piaget's alternative explanation for the origins of adaptive behavior and knowledge. A few pages into *B&K*, after discussing analogies between cognitive regulations and physiological regulation (homeostasis), dynamic regulation (homeorhesis) in ontogeny, and regulatory genes, he asserts that the central problem the book will deal with is ". . . the relationships between cognitive and organic regulations at all levels" (Piaget, 1971, p. 12).

Early in *B&K*, Piaget (1971) said his guiding hypothesis is that "Life is essentially auto-regulation" and that "Cognitive processes seem, then to be at one and the same time the outcome of organic auto-regulation, reflecting its essential mechanisms, and the most highly differentiated organs of the regulation at the core of interaction with the environment . . ." (Piaget, 1971, p. 26). He elaborates on this later, referring to some functional analogies between ". . . coordination of schemata on the genetic or epigenetic plane of the organization pertaining to instinct and the individual coordination of schemata in the domain of intelligence, at least sensorimotor intelligence . . ." (Piaget, 1971, pp. 228–229).[2]

In the second chapter of *B&K*, Piaget discusses the parallels between problems raised by what he saw as two forms of epigenesis: embryogenesis and *"mental embryology,"* both of which are characterized by auto-regulation. Table 2.1 lists some of the parallels Piaget sees between embryogenesis and cognitive development.

---

[2]Contrary to this translation of the term, Piaget (Piaget & Inhelder, 1971) used the term *schemes* to refer to actions and *schemas* to refer to imagery or figurative instruments of knowledge.

TABLE 2.1
Piaget's (1971) Parallels Between Embryogenesis and Psychogenesis
in His Model for "Construction of the Phenotype"

Epigenetic Processes in Embryology	Epigenetic Processes in "Cognitive Embryology"
Stages of embryogenesis:	Stages of cognitive development:
Stage specific embryonic competence to respond to inducers	Stage specific assimilation and accommodation by cognitive schemes
Homeorhesis (dynamic auto-regulation)	Process of equilibration
Homeostasis	Final product of equilibration
Regulatory mechanisms	Operational reversibility
	Convergent reconstructions and overtakings
Differentiation and integration	Reflective abstraction
	Reorganization
Independent chreods	Displacement/decalage among domains

This stimulates him to introduce some concepts derived from Waddington's (1962, 1975) studies of embryology including chreods, competence, homeorhesis, and homeostasis (see discussion in Part II). Piaget sees embryonic competence, the "physiological state of a tissue which permits it to react in a specific way to given stimuli" (Piaget, 1971, p. 22) at a specific time, as analogous to stage-specific cognitive schema of assimilation, which allows a child to acquire knowledge when presented with specific situations. He sees independent *chreods*, or "necessary routes" along canalized developmental pathways in the embryo, as analogous to displacements or *décalages* among developmental pathways within related domains of cognitive development.

Piaget sees *homeorhesis*, a form of dynamic equilibrium in embryonic development, as analogous to the process of dynamic equilibration during cognitive development:

> The various channelings as well as the auto-corrections which assure their homeorhetical equilibrium are under the control of a "time tally" which might as well be described as a speed control for the processes of assimilation and organization. It is, then, only at the completion of each structural achievement that, homeorhesis gives place to homeostasis or functional equilibrium. (Piaget, 1971, p. 19)

Following this discussion of parallel auto-regulatory processes in embryogenesis and cognitive development, Piaget proclaims: "The explanation of evolutionary mechanisms, for so long shackled to the inescapable alternatives offered by Lamarckism and classical neo-Darwinism, seems set in the direction of a third solution, which is cybernetic and is, in effect, biased toward the theory of auto-regulation" (Piaget, 1971, p. 26). Piaget makes it clear that he does not consider his solution to be Lamarckian even though, as discussed, it entails incorporation of somatic adaptations into the germ line.

Piaget returns to his model for the origins of adaptive variations in the latter part of chapter 6 and in 7 at the end of *B&K*. He begins this discussion with a reprise of his view of the alternative Lamarckian and neo-Darwinian approaches to the origins of variations. He characterizes Lamarckians as saying

> . . . instinct is only a habit fixed in heredity . . . [it] consists of a series of associations imposed on the subject by the environment and all that is inherited is the memory of it, which is handed down to the descendents. As a result, the adaptation of the instinct to the environment simply consists of anticipations based upon previous information transmitted from the environment to the germinative system. (Piaget, 1971, pp. 271–272)

Piaget (1971) criticized Lamarckians for failing to recognize that the organism "instead of passively accepting pressures from the environment (Lamarck did go so far as to admit there was some active part taken by the living creature in the actual choice of environment), assimilates them into structures that are endowed with the power of auto conservation" (p. 273). He argues that Lamarckians simply confirm the existence of the problem without solving it: "We still need to understand how, among all the details of the causal mechanism, the genome can acquire any information about the environment . . ." (Piaget, 1971, p. 273).

He contrasts the Lamarckian solution to the neo-Darwinist "mutationist" solution: "According to this theory, instinct owes its origin, just as morphological, anatomical, and physiological characteristics (including the human brain) do, to chance mutational variations, progressively sorted out and thus refined in the process of selection" (Piaget, 1971, p. 273).

He asserts that whereas such a mechanism of differential survival might fashion crude intellectual adaptations, it could never have produced refined intellectual instruments of logical-mathematical thought. Therefore he attaches no credence to this solution. He also dismisses the vitalist solution. (Also see Piaget's [1978] discussion of Lamarckism in *B&E*.)

After a brief discussion of population genetics and of probabilistic stabilizing ("channelizing") selection, Piaget makes an analogy between individual level genomic reorganization with population level reorganization of the gene pool, which, he says ". . . it makes possible a complete integration of Lamarck's two principle processes—information transmitted from the environment and the heredity of the characteristics acquired in this way—but by putting a new interpretation upon their causal mechanism . . ." (Piaget, 1971, p. 281).

Regarding the purported analogy between the gene pool and the genome, Piaget (1971) said:

> Our hypothesis is that every process taking place within the population and involving the fundamental relations between the genetic pool and the environments (variation and selection) may correspond to a parallel qualitative process involving the relations between the individual genome . . . and the individual environment. (pp. 283–284)

Specifically, he argues,

> . . . information is continuously transmitted from the environment to the genetic pool, and variations are fixed by "*genetic assimilation*," with heredity of acquired information again proceeding by means of selective stabilization, or, in other words, by modification—now irreversible of the proportions of the collective genome. . . . every stable genotype variation is a "response" made by the genotype to the tensions set up by the environment. (Piaget, 1971, p. 281, italics added)

In a key statement, Piaget (1971) elaborates the proposed mechanism:

> To put it more precisely, the hereditary fixation of a new behavior *seems to imply some transmission from the soma to the genome*, whereas, following the neo-Darwinian tradition, the general opinion of population geneticists (except for Waddington . . .) is that there is a radical isolation of the genome. . . . (p. 282, italics added)

He elaborates this point saying

> The great difficulty here, however, is that in the perspective of present-day geneticists there is no reciprocal relationship between the structure of the genome and its morphogenetic activities. The structure is supposed to be the source of its activities,

whereas these activities bring about no reaction in the structure. (Piaget, 1971, p. 290)

Piaget invokes his interpretation of Waddington's (1975) *genetic assimilation* to explain the assimilation of final re-equilibrations of phenotypic adaptations above the norm of reaction into the genes. He conceives of the resulting new phenotype as a *phenocopy*:

> Any biologist reading this summary of analysis is bound to think of situations in which *phenotypic variation precedes the appearance of a genotype that seems to be an imitation of it*, which is sometimes called a phenocopy precisely in order to show that an active and endogenous imitation has taken place, not a mere transmission of external causal influences. (Piaget, 1971, p. 344, italics added)[3]

Piaget also relates phenocopies and genetic assimilation to Baldwin's "rather ambiguous concept" of *organic selection*, using Horvasse's (1943) definition: ". . . replacing an accomodat by a mutation" (Piaget, 1971, p. 298). (See the discussion of organic selection in chapter 1.) He illustrates these concepts with a discussion of his own early research on phenotypic adaptations of the pond snail, *Limnaea*, in different lake environments.

John Messerly (1996) summarized Piaget's phenocopy model as follows:

> First the organism responds to a change in the external environment with a somatic (pertaining to the body) modification. If this modification does not cause disequilibrium, the phenotypic adaptation does not become fixed. Second, if there is disequilibrium between the exogenous modification and the endogenous hereditary program, then disequilibrium is "transmitted" to the internal environment. Third, if epigenetic development cannot reestablish equilibrium, then disequilibrium may descend all the way to the genome. Fourth, at the level of the genome, mutations respond to disequilibrium. The response of the genome is random in the sense that mutations do not necessarily restore equilibrium, but they are directed towards the needs of the organism. Fifth, the endogenous variations are then selected by both the internal and the external environments until stability is restored. (p. 95)

---

[3]In both *B&K* and *B&E*, Piaget's concept of the *phenocopy* is a new phenotype that is subsequently "imitated" by the genotype. However, in *A&E*, he notes that "The phenocopy is capable of three kinds of interpretations" (Piaget, 1980, p. 47) which include this interpretation, and its converse, that the phenocopy is an environmentally induced copy of a phenotype produced by a gene.

Biographical Context of Piaget's Evolutionary Model

Piaget's ideas should be seen within the context of his intellectual history. His formal education was in biology; both his baccalaureate degree (1915) and his doctorate (1918) were conferred by the University of Neuchâtel in the natural sciences (Piaget, 1952). His thesis was on the mollusks of Valais, which he had studied since he was 10 years of age when he asked the director of Neuchâtel's Museum of Natural History, his godfather, Paul Godet, for permission to work there. Soon he was apprenticed in collecting, cataloguing, describing, and classifying pond snails. When Piaget was 13, his early intellectual interests were further stimulated by his initiation into the Friends of Nature, a local naturalists club, in which he was actively involved in presenting and attending lectures and discussions for several years (Vidal, 1994).

According to Vidal's biography, the young Piaget was strongly influenced by Henri Bergson's (1941) book, *Creative Evolution*, which opposes a mechanistic theory of evolution. Frequent references to Bergson's ideas are scattered through Piaget's books (see Messerly, 1996, for a discussion of Bergson's influence). Another early influence was neo-Lamarckism, which was prevalent in the United States, Germany, and France from the 1880s through the 1920s (Mayr, 1989), and in France until recent times (Boesiger, 1980). Consequently, it was influential among young natural historians in the club Piaget frequented, the Friends of Nature. This was especially true of Piaget's friend, Juvet, who was influenced by the French biologist, LeDantec. Like some other French biologists at the time, LeDantec thought that Darwinism was useful and synthetic, but believed that natural selection could not explain the origins of new adaptations (Vidal, 1994).

Piaget's biological education preceded the modern synthesis of evolutionary biology, which settled many of the disputes that were alive during his youthful initiation into malacology. His mentor, Paul Godet, was a turn-of-the-century taxonomist devoted to the *morphological/typological species* concept, according to which names of species, varieties, subvarieties, and so on, were based on combinations and proportions of characteristics. According to this essentialist view, the type specimen was supposed to show the fixed defining characteristics of the species. In contrast, according to Darwin's evolutionary species concept, which was beginning to influence taxonomy at this time in Switzerland, species were characterized by variability and varieties were incipient species (Vidal, 1994).

In 1918, after finishing his doctorate, Piaget went to Zurich, where in his words, he "attended" two psychological laboratories. After returning briefly to study snails, in 1919 he left for Paris where he studied psychology for 2 years. It was there that he met Dr. Simon, who invited him to analyze Burt's data on reasoning tests on Parisian children. This led him to focus on age-related errors children made in classification and number and cause and effect. Consequently, as he says, "At last I had found my field of research" (Piaget, 1952, p. 245).

In 1921, he began experimental work on cognitive development at the *Maison des Petits Infants Jean Jacques Rousseau* (which later became part of the University of Geneva). From 1925 to 1929, he worked as a teacher of philosophy and psychology at the University of Neuchâtel (during which time he continued to work with students at the University of Geneva and continued his research on adaptation in pond snails). Piaget returned to Geneva in 1929 as Professor of Scientific Thought and Assistant Director of the *Institut J. J. Rousseau,* of which he became co-director in 1932 (Piaget, 1952).

Piaget's last publications on *Limnaea* were in 1929, however, he discusses the implications of this work in *Origins of Intelligence* (1936/1952), and later in *B&K* (1971), *B&E* (1978), and *A&I* (1980). In an interview with a French journalist in 1962 (Bringuier, 1980) regarding the phenocopy concept, Piaget says that his interest in *Limnaea* was reawakened by Waddington's (1975) work. Piaget also discusses the phenocopy concept with the French geneticist, Jacob, in *Language and Learning: The Debate Between Jean Piaget and Noam Chomsky* (Piatelli-Palmarini, 1980). (This book also contains "A Critical Note on the Phenocopy" by Antoine Danchin.)

Commentary on Piaget's Phenocopy Model

As Piaget professes, he has been strongly influenced by Waddington's ideas. It is interesting in this regard to compare Piaget's discussion of the third way between Lamarckism and neo-Darwinism to Waddington's (1942) similar formulation in his paper "Canalization of Development and the Inheritance of Acquired Characteristics." Piaget refers to Waddington's ideas throughout *B&K*, citing two sources (Waddington, 1957) and a talk Waddington gave at a symposium in Geneva in 1964. He also discusses Waddington's ideas through *B&E* and cites Waddington's autobiography (Waddington, 1975). Significantly, however, he only briefly mentions Waddington in *I&A*, which elaborates his phenocopy model.

Reciprocally, Waddington's (1975) autobiography includes a chapter on Piaget's work on pond snails. In his discussion of Piaget's work on *Limnaea*, Waddington (1975) concluded that Piaget's explanation for the origin of genetic variations in varieties of *L. laucustris* and *L. bodamica* ". . . seems to overlook the fact that there will be genetic variability in the capacity to perform the physiological adaptation which brings about the contraction [of the *Limnaea's* shell]" (Waddington, 1975, p. 94). In other words, Piaget ignores the pre-existence of genetic variation, which Waddington believes is the basis for selection resulting in genetic assimilation (see discussion of Waddington's work to come).

Waddington (1975) noted that his definition of phenocopy as a phenotype that mimics one produced by a recognized mutant allele differs from Piaget's definition of a phenocopy as a new characteristic produced in response to the environment and later assimilated into the genome. Waddington (1975) also criticized Piaget's model for the vagueness of the mechanisms of "progressive reorganization" he proposes. Piaget devotes several pages in *B&E* to a defense against Waddington's criticism. He does not, however, substantially change his formulation.

In fairness to Piaget, however, it should be noted that Waddington sometimes described phenocopy ambiguously in a way that may have misled Piaget, as for example,

> The process of genetic assimilation is one by which a phenotypic character, which initially is produced only in response to some environmental influence, becomes, through the process of selection, taken over by the genotypes, so that it is formed even in the absence of the environmental influence which had at first been necessary. (Waddington, 1975, p. 91)

This definition omits the crucial factor of the preexistence of genes involved in the phenomenon. More significant, Waddington (1957) did discuss the possibility of genetic assimilation of a new gene that did not already appear in the population, featured in Piaget's scenario. He also addressed the question of "directed mutation" and whether if such a gene occurred, it could have been triggered by environmentally influenced mutation. He concluded that there is very little evidence for this. As we will see, there is some evidence for limited epigenetic inheritance in some organisms (Jablonka & Lamb, 1995).

Piaget's evolutionary developmental model presented in the three books, *B&K*, *B&E*, and *A&I* is a hybrid between ideas common in the premodern synthesis of evolutionary biology and his interpretation of post DNA ideas from developmental biology of the 1950s and 1960s. On one hand, although Piaget cites several figures in the modern synthesis, including T. Dobzhansky, Julian Huxley, and George G. Simpson, he seems not to have fully understood or integrated their ideas. On the other hand, Piaget raises knotty issues regarding the relationship between evolution and development that also troubled Waddington. Many of these issues only began to be elucidated with the emergence of modern developmental biology in the 1970s, 1980s, 1990s, and the early 21st century.

This interpretation may explain Piaget's introduction of ideas that were common before the synthesis into the discussion of genetics. His presynthesis training, for example, may explain Piaget's failure to clearly distinguish (a) between the individual's genes (the genome) and the gene pool of the population, (b) between individual adaptation and population changes, and (c) between feedback loops in DNA functioning in somatic cells from those in germ cells (see text to come). This combined with an inverted definition of phenocopy as proceeding rather than mimicking an existing genotype results in an anachronistic model for the origin of heritable variations.

Piaget was a synthetic thinker, whose ambition was to explicate a genetic epistemology based on an integration of biology, philosophy, and developmental psychology. He focuses on the *origins of new heritable adaptations*, which as we shall see, was a major problem in 20th-century biology. His model of biology and knowledge addresses age-old questions of the relationship between development and evolution that have yet to be answered fully. Today these questions are again at the forefront of biology in the new research programs known as developmental biology and evolutionary developmental biology. Likewise, the related problem of the evolution and development of brain function and its relationship to the evolution and development of language and intelligence is in the forefront of both developmental biology and developmental psychology.

Whether or not he was aware of it, Piaget's search for the origins of new adaptive variations was in the tradition of ongoing efforts to delineate and explain this phenomenon. As we will see, this tradition goes back at least to Lamarck (1809/1984); Darwin (1859, 1868); Haeckel (1866); Weismann (1892); Baldwin (1902); Schmalhausen (1986); and Waddington (1962, 1975). Very recently, West-Eberhard (2003) presented a synthesis of these and other efforts.

Interestingly, Piaget's phenocopy model of genomic regulation through feedback from regulations of the developing organism is similar to Darwin's model of *pangenesis* to be described. Therefore, the same criticism could be made of Piaget's model that Ghiselin (1975) made of Darwin's:

> The idea that evolution involves changes in developmental mechanisms is an enduring truth. . . . Yet the provisional hypothesis does seem to me to have one major fault: it was anachronistic. It gave a nineteenth century answer to an eighteenth century question that needed to be dealt with in twentieth century terms. (p. 55)

(Except I would say that Piaget's question needs to be dealt with in 21st-century terms.)

The basis for this judgment is developed in the following section, which briefly addresses some relevant themes in evolutionary and developmental biology from 1700 to 2000: Lamarckism, pangenesis, Wiesmannism, recapitulationism, the modern synthesis, molecular biology, developmental biology, and recent work on epigenetic inheritance. I hope this will allow readers to judge for themselves the strengths and weaknesses of Piaget's evolutionary model, but most of all to see the great historical tradition in which it falls.

A BRIEF HISTORY OF SOME ISSUES IN EVOLUTIONARY AND DEVELOPMENTAL BIOLOGY

Although Lamarck and other French biologists at the turn of the 19th century espoused the idea of evolution and transformation of species, Darwin was the first to explicate and demonstrate the major mechanism of evolutionary change, that is, natural selection (Darwin, 1859). Although Darwin discussed the origins of the variations on which selection acts in the *Origin of Species* (Darwin, 1859) and *The Variation of Animals and Plants Under Domestication* (Darwin, 1868), these are only now coming to be fully understood. Darwin was prescient, however, in investigating the relationship between development and evolution.

Sources of hereditary variation, that is, mutation and recombination, and its transmission, began to be elucidated after the rediscovery of Mendel's (1966) work at the turn of the last century. The "modern synthesis" of natural selection and mutation and recombination took at least 40 years from that time. But embryology was excluded from that synthesis. Today, 50 years after the discovery of the genetic code, integration of developmental biology into evolutionary biology is still in its infancy. A brief history of evolutionary ideas may help explain this time lag as well as provide background for Piaget's ideas.

Lamarck and Lamarckism and Neo-Lamarckism

Jean-Baptiste Lamarck published his ideas about evolution and development in *Philosophie Zoologique* in 1809 (Lamarck, 1984). A product of the French Enlightenment, Lamarck believed in natural, mechanistic explanations for species transformation. He believed that life arose though spontaneous generation and that species evolved toward greater complexity rarely suffering extinction. He believed that the environment changed gradually over time and that animals adjusted their habits to these changes. He also believed that changes in behavior preceded changes in structure (Jablonka & Lamb, 1995; Richards, 1987). His theory of transformation is summarized in two laws:

First law: In every animal which has not passed the limit of its development, a more frequent and continuous use of any organ gradually strengthens, develops and enlarges that organ, and gives it a power proportional to the length of time it has been so used; while the permanent disuse of any organ imperceptibly weakens and deteriorates it, and progressively diminishes its functional capacity, until it finally disappears.

Second law: All the acquisitions or losses wrought by nature on individuals, through the influence of the environment in which their race has long been placed, and hence through the influence of the predominant use or permanent disuse of any organ; all these are preserved by reproduction to the new individuals which arise, provided that the acquired modifications are common to both sexes, or at least to the individuals which produce the young. (Lamarck, 1984, p. 113)

Although the idea did not originate with Lamarck, he is remembered chiefly for the notion of the inheritance of acquired characteristics expressed in these two laws. Lamarck's ideas have been widely misrepresented and caricatured by his detractors. This began with the notoriously damaging "eulogy" delivered to the French Academy of Sciences on November 26, 1832, 3 years after Lamarck's death, by his chief opponent, the creationist paleontologist, Baron Georges Cuvier (1984). Ever since then, Lamarck's ideas have been chronically misrepresented (Burkhardt, 1984).

Nevertheless, neo-Lamarckism was a popular movement among American paleontologists, especially Cope, Hyatt, and Packard at the turn of the last century in the United States (Burkhardt, 1998). It had a strong following in the United States, Europe, and Russia in the 1920s and 1930s (Mayr, 1998), and in France until the 1960s (Boesiger, 1980). Its fall from grace was one of the factors that facilitated the modern synthesis. (As we see later, its fall was also due to the predominance of zoologists as opposed to botanists and specialists in invertebrate animals in this movement.)

Pangenesis: Darwin's Theory of the Origins of Variation

In his three great books on evolution, Charles Darwin (1859, 1930, 1965) argued for the common descent of all life forms, gradual change in species through descent with modification, and natural and sexual selection of favorable variants through differential survival and reproductive success (Mayr, 1998). Despite his understanding of the key role variation plays in evolution and his attempts to understand it (Darwin, 1868), Darwin was unable to explain its origins. It is well known that he accepted the inheritance of acquired characteristics. It is less well known that he proposed a Lamarckian mechanism (involving emission and movement of "gemmules") to explain its origin and transmission in his theory of pangenesis in *The Variation of Animals and Plants Under Domestication* (Darwin, 1868, 1896; Ghiselin, 1969, 1975). (This mechanism is similar to the model of genetic assimilation Piaget proposes.)

Moreover, this theory, summarized by Winther (2000), ties variation, development, and heredity together ". . . into a coherent developmental, theory of heredity" (pp. 447–448). As Winther (2000) described it:

> . . . when changes in the conditions of life caused somatically-mediated variation, the body was affected . . . and emitted modified gemmules. When changes in the conditions of life caused germinally-mediated variation, the reproductive organs were affected . . . and collected gemmules irregularly. Changes in the conditions of life could also alter the development of the offspring . . . , which led either to reversions or to another change in the reproductive organs or body. These variations might, in turn, be inherited by the offspring of that offspring. (pp. 447–448)

Whereas Darwinism was compatible with Lamarckism, neo-Darwinism, the revised version of Darwinism after the modern synthesis, excised Lamarckism (Mayr, 1989).

Weismann and Weismannism

August Weismann, another great synthesizer, was a German biologist who tried to integrate the five biological fields of evolution, heredity, cytology, physiology, and development into a single conceptual model (Allen, 1989). Between 1875 and 1913, he published books and papers on a variety of topics including, prominently, the nature and significance of sexual reproduction, particularly the mechanics of meiosis (Winther, 2001). He devised a model for the behavior of units conceptually similar to chromosome, genes, and alleles (idants, ids, and determinants). He is best known for explicating the concept of the *germ-plasm* in 1883 as opposed to the soma and for his thesis regarding the intra- and intergenerational continuity of the germ plasm. He is widely credited with experimental work disproving the inheritance of acquired characteristics, by which he meant acquired somatic characteristics (Winther, 2001).

Ironically, given his reputation as an anti-Lamarckian, Weismann believed in the inheritance of variations acquired by the germ-plasm through a hierarchy of external conditions ranging from cytoplasm, somatic cells, bodily substances, and the extra-organismic environment (Winther, 2001). Significantly, over the years, Weismann changed his views regarding the directed or adaptive nature of germ-plasm variations (Winther, 2001). Despite these transformations, however, the always argued "changes in external conditions were necessary causes of variation in hereditary material" (Winther, 2001, p. 522).

Winther (2001) described and renamed two key concepts pioneered by Weismann. The first concept is *variational sequestration* of the germ-plasm, which occurred when external conditions caused no changes in the germ-plasm. In relation to this, Winther documents Weismann's changing views regarding whether meiosis simply reshuffled variants or produced them, that is, regarding the degree to which variational sequestration of the germ-plasm occurred. The second concept is *morphological sequestration* of the germ-plasm, that is, spatial separation from both germ-cell cytoplasm and somatic cells. This is perhaps Weismann's best known concept because it contributed to the demise of Lamarckism and hence to the development of the modern synthesis.

Mendelianism and Mutationism: Origins of Hereditary Variations in Mutations and Genetic Recombination

Gregor Mendel's three great discoveries (the particulate nature of traits, the laws of segregation and random assortment of those particles during seed formation, and subsequent recombinations of traits during fertilization) provided the first systematic knowledge of the mechanisms of inheritance. (They did not, however, explain the origins of new hereditary variants.) His work, which came out of a long tradition of research by the hybridists (Olby, 1966), remained obscure until it was rediscovered at the turn of the 20th century by three scientists (Correns, deVries, and Tschermak) after they replicated his work. A long controversy ensued over the generality of that discovery, and the differences between so-called continuously and discontinuously varying traits.

Meanwhile, Hugo deVries (1906), a Dutch hybridist studying *Oenothera* (evening primrose), developed a saltatory theory of speciation based on occasional occurrence of extreme differences between parent and offspring. He therefore introduced the concept of *mutation*, which he defined as a drastic change leading to formation of a new species. He also concluded that small variations had a negligible role in evolution because their effects could be swamped before they accumulated. This usage continued among naturalists into the 1920s and 1930s (deVries' "mutations" later proved to be results of hybridization; Adams, 1998).

The meaning of the term *mutation* changed under the influence of Thomas Henry Morgan, who—although he initially agreed with deVries' anti-Darwinian theory—gradually realized that small mutations are heritable and can contribute to evolutionary change (Adams, 1998). Other geneticists followed Morgan's revision of the term mutation to mean any kind of genetic change, however small (Mayr, 1998). A long-running debate over the cause of evolution ensued between saltationists, who believed that macromutationists caused evolution and the neo-Darwinians, who argued that selection was the primary cause.

The Modern Synthesis of Evolutionary Biology

Ultimately, the various contending schools of thought agreed on an integrated model of evolution according to which (a) mutation is the ultimate source of genetic variation producing new chromosomes, genes, and alleles; (b) recombination during meiotic generation of haploid sex cells and reformation of diploid cells through fertilization is the process that produces recombined chromosomal variants; and (c) selection (differential reproduction and survival); and (d) genetic drift or sampling error in fertilization and survival are the means by which the proportions of competing variants change through generations. Selection leads to the proliferation of adaptations, whereas genetic drift does not. Speciation is thought to occur primarily through the evolution of reproductive isolating mechanisms (generally during geographic isolation of a small population from its larger mass), which ultimately prevents fertile mating between individuals of different species (Mayr, 1998; Mayr & Provine, 1998).

The time lag between the discovery of particular parts of the picture and the emergence of the synthesis was approximately 40 years. Mayr (1989) explained this lag as the consequence of disciplinary differences between naturalists and geneticists working at different hierarchical levels, and conceptual confusions. Specifically, he notes (a) a failure to distinguish between proximate (physiological mechanisms) and ultimate (evolutionary causes) levels of analysis; (b) confounding individual change and population change; (c) confounding of genotype and phenotype; (d) wide acceptance of "soft inheritance," or Lamarckism; (e) seeing selection as solely a negative factor; (f) falsely distinguishing between large "saltatory" mutations and small mutations; and most of all (g) typological or essentialist thinking versus populational thinking about species.

The term, *modern synthesis*, which takes its name from Julian Huxley's book (1942), arose first out of an integration of new concepts (e.g., *the polytypic species*, the biological species concept, and *isolating mechanisms*, the gene pool and population genetics), and, second, out of bridge-building among naturalists, geneticists, and paleontologists who began to read one another's literature (Mayr, 1998). Key figures included systematists like Mayr, naturalist/biogeographers like Bernard Rensch, paleontologists like George Gaylord Simpson, and geneticists like Theodosis Dobzhansky (Mayr, 1998), but particularly generalists like Julian Huxley (Churchill, 1998).

On the other hand, and more central to our story, the modern synthesis bypassed embryologists, botanists, mycologists, and zoologists of colonial invertebrates (Buss, 1987). Because of this, the synthesizers were most familiar with organisms like humans, fruit flies, and mice that sequester their germ lines before embryogenesis. Therefore they were unaware that plants, fungi, and colonial invertebrates do not separate their germ plasm from their somatic tissue and therefore may be more likely to inherit adaptations to local conditions (Buss, 1987; Jablonka & Lamb, 1995).

In summary, it is important to emphasize, as Mayr and Provine (1998) noted, that the evolutionary synthesis was a complex, multifaceted, historical process that occurred on many levels and in different fields. It also progressed at different rates and to different degrees in different countries.

The response to Darwinism and neo-Darwinism in France provides some insight into the context of Piaget's biological training and concepts. When *The Origin of Species* was published in French, it generated little interest among naturalists (there was one favorable review by Edouard Claparède, the Swiss naturalist, in 1861). Darwin's name was placed before the French Academy of Sciences six times before he was elected to the botanical section after being rejected by the zoological section (Corsi, 1985).

Based on the few available studies of this subject, Corsi (1985) concluded that most French naturalists viewed Darwin's work as a reformulation of Jean-Baptiste Lamarck and Etienne Geoffroy Saint-Hilaire's theories of species transformism. They preferred alternative Lamarckian and embryological recapitulation explanations to Darwin's theory of natural selection.

Transcendental Morphology, Recapitulationism, and Evolution

In contrast to French biologists, German biologists displayed mixed reactions to Darwin's ideas, reflecting the nation's political fragmentation and cultural diversity. After noting a cool response to Darwin's work in Berlin, in 1863, Ernst Haeckel began a campaign for acceptance of natural selection from his post at the University of Jena, arguing that Darwin's work continued the traditions of Goethe's morphology and *Naturphilosophie* (Weindling, 1985). Haeckel's defense of Darwinism was part of a larger social political argument (also based on Herbert Spencer's ideas) about society and education. The consequent decline in his reputation influenced responses to his ideas.

Haeckel was another great synthesizer who sought to integrate comparative anatomy, cytology, and embryology (Haeckel, 1866). Darwin's book, *Origins*, ". . . inspired Haeckel to fundamentally reinterpret comparative anatomy in the evolutionary terms of the genealogy of organisms" (Weindling, 1985, p. 698). Haeckel is best known for his "biogenetic law," according to which ontogeny recapitulates phylogeny: "Ontogeny (embryology or development of the individual) is a concise and compressed recapitulation of phylogeny (the paleontological or genealogical series) conditioned by laws of heredity and adaptation" (cited in Ghiselin, 1969, p. 122).

Although Haeckel placed it in the context of genealogy and descent with modification, the idea of *recapitulation*, which originated in *Naturphilosophie*, long preceded theories of evolution. *Naturphilosophie* was devoted to the notions that all animals are built on a single structural plan or archetype, that there is a great chain of being, and that there was recapitulation. Early proponents of nonevolutionary recapitulationism included Enlightenment embryologists, Oken, Mekle, and Serres (Gould, 1977).

Recapitulationism was first and famously attacked by Karl Ernst von Baer, the Estonian embryologist and author of *Die Entwickelungsgeschichte de Thiere* (Baer, 1828), who was himself an opponent of evolution. In contrast to the idea that ontogeny repeats a series of lower forms, von Baer argued that development is differentiation, that features general to a group of animals appear earlier than the special features do, that each species differentiates more and more as it develops, and that embryos are never like the adults of lower animals but only like their embryos (Gould, 1977).

In any case, despite its bad name among psychologists, terminal addition of new features resulting in recapitulation is recognized by evolutionary biologists as one of several forms of *heterochrony* (a term meaning changes in developmental timing, coined by Haeckel; Gould, 1977; McNamara, 1997; McKinney & McNamara, 1991). As Gould (1977) said,

> ... recapitulation was not "disproved"; It could not be, for too many well-established cases fit its expectations. It was, instead, abandoned as a universal proposition and displayed as but one possible result of a more general process—evolutionary alteration of times and rates to produce acceleration and retardation in ontogenetic development of specific characters. (p. 206)

Like Haeckel, Darwin was vitally interested in development and evolution. Comparative anatomy and especially comparative embryology provided Darwin's first and best evidence for "descent with modification," that is, evolution. Therefore, it is striking that 100 years later, for a variety of reasons, morphologists were largely outside the modern synthesis. This omission is particularly interesting considering that Julian Huxley, one of the architects of the modern synthesis, worked with Gavin de Beer in embryology (Churchill, 1998; Huxley & de Beer, 1934).

The embryologist, Viktor Hamburger (1998), argued that the neglected developmental chapter in modern synthesis has not been written owing to both disciplinary and national/linguistic barriers. He argues that disagreements over recapitulation and Weismannism turned Haeckel's student, William Roux, away from ultimate (evolutionary) concerns and toward the more tractable proximate mechanisms. In 1894, Roux called for a new science of developmental mechanics (*Entwicklungsmechanik*) based on experimentation on proximate causes of embryo formation.

As a consequence of this new focus, embryologists generally ignored advances in genetics. So whereas this new approach freed European and American morphologists from the baggage of Haeckel's ideas, it deepened the division between geneticists, who focused on the nucleus, and embryologists, who focused on the cytoplasm (Hamburger, 1998). Hamburger (1998) suggested that the integration of embryology and genetics,

> . . . would have taken a biologist with very broad interests, who would be familiar with genetics, speciation, evolution and at the same time knowledgeable in experimental embryology and the intricacies of epigenetic development to write it. The only person of this rank at the time was Schmalhausen. (p. 108)

(As we shall see, Waddington was another.)

Schmalhausen's synthetic approach came from his teacher, Alexei Nikolaevich Severtsov, the best known Russian morphologist of his day, a devoted Darwinian who established the Laboratory of Evolutionary Morphology of the USSR in 1934. Severtsov conceived the task of *evolutionary morphology* to be to study evolution and its causes. He used Haeckel's three methods of comparative anatomy, comparative embryology, and paleontology, complemented by the study of physiology, heredity, and experimental sciences. As a consequence, the evolutionary synthesis was more advanced in Russia than anywhere else during the period from 1925 to 1948 (when all its proponents were killed or lost their positions during the Lysenko period; Adams, 1998).

PUTTING HUMPTY-DUMPTY TOGETHER AGAIN:
THE BEGINNING OF DEVELOPMENTAL BIOLOGY

Ivan Ivanovich Schmalhausen's (1986) *Factors of Evolution: The Theory of Stabilizing Selection* was published in English in 1946, 10 years after he died. Schmalhausen's contribution was not widely recognized by other evolutionary biologists of the time. He is known in the West primarily for the concepts of *norm of reactions* during development and *stabilizing selection* (as opposed to dynamic or directional selection) of patterns of morphogenetic development, auto-regulation of morphogenetic development, and for the notion of hierarchical levels in biological systems. He also focused on individual variability built up through invisible neutral and minor mutations that remain hidden by stabilizing selection (Wake, 1986). This is vital to understanding the hidden variation underlying genetic assimilation.

Norm of reactions in development refers to genetically determined modifications or range of expression in the phenotype in response to various environments. These reactions maybe highly auto-regulated or canalized so that above a certain threshold, the reaction occurs independent of stimulus intensity. As Schmalhausen (1986) noted,

> In the process of evolution, the entire morphogenesis (i.e., the entire reaction apparatus) together with all its adaptive reactions is endowed with regulating mechanisms which protect the processes of individual development against possible disturbances by changing and accidental influences of the external environment. Auto-regulation is characteristic of all adaptive modifications. (p. 10)

Auto-regulation is maintained through stabilizing selection, which maintains the mean value of a character by selecting against both extremes (as opposed to *dynamic selection*, which progressively changes the mean of a character in a population). Schmalhausen was a pioneer in connecting individual development to population genetics, and in investigating interactions between the cytoplasm and the nucleus in development.

Like Schmalhausen, Waddington and Salome Gleucksohn-Schoenheimer, another pioneer in developmental biology, tried to unite embryology and genetics. After doing her doctoral research in Spemann's laboratory in Freiburg, Gleucksohn-Schoenheimer fled Germany in the mid-1930s. She went to Columbia University to study mouse genetics, and then mouse embryology. In 1938, she published a programmatic statement distinguishing developmental genetics from experimental embryology and arguing for the emergence of the former from the latter. And again in 1945, she wrote a programmatic statement on the nature of gene action in embryo development (Gilbert, 1991).

Chester Waddington, like Piaget, whose work he influenced, was an academic maverick strongly influenced by philosophy. Like Piaget, he was critical of the modern synthesis of evolution (Waddington, 1942, 1957). Waddington took a first-class degree in geology at Cambridge University and studied paleontology in graduate school without finishing his doctorate. In pursuit of his interest in embryology, he had visited Gleucksohn-Schenheimer in Spemann's laboratory in Frieberg, and re-encountered her when he went to study genetics with Leslie C. Dunn at Columbia University in the United States. From there he went on to work on *Drosophila* mutants with Morgan's fly group at the California Institute of Technology (Gilbert, 1991).[4]

As a fly geneticist trained in Morgan's laboratory, Waddington knew that wild type flies display phenotypic constancy despite their genetic diversity. He coined the term, *canalization*, to refer to the developmental processes that tend to produce similar phenotypes among a population of diverse genotypes in the wild: "The canalization, or perhaps it would be better to call it the buffering, of the genotype, is evidenced most clearly by constancy of the wild type" (Waddington, 1942, p. 563).

This buffering creates the same "non proportional outcome" irrespective of the magnitude of the environmental stimulus within a certain norm of reaction. These canalized systems contrast with uncanalized systems of proportional response to environmental stimulation. Either type of sys-

---

[4]According to Gilbert (1991), Waddington's work was guided by two beliefs radically different from those of Spemann and most other embryologists of the time: (a) cells are systems whose development depends on their *competence* to be transformed as well as on inducers that transform them, and (b) cytoplasm of cells activates genes that guide development (Waddington, 1942).

Moreover, unlike most of his colleagues, Waddington envisioned application of Jacob and Monod's *E. coli* operon model to embryology. Gilbert traces Waddington's belief in systems, competence, and canalization to his devotion of Alfred North Whitehead's philosophy:

Thus we see that Waddington's idiosyncratic approach to development—concrescence, canalization, and genetic assimilation—arose from his placing fundamental emphasis on competence (rather than on induction) and in his placing these observations in the context of a Whiteheadian philosophy of organismic change. (Gilbert, 1991, p. 199)

tem may be built up by natural selection (Waddington, 1942). He notes, however, that canalization, which arises through natural selection, breaks down in mutants and in pathological conditions.

Later, citing his own earlier (1940) publication, Waddington (1975) defines canalization as "The property of a developmental process, of being to some extent modifiable, but to some extent resistant to modification, has been referred to as its 'canalization' . . . This notion can be applied whether the agents which tend to modify the course of development arise from genetic changes or from changes in the environment" (p. 71).

Referring to the first study of canalization in wild type *Drosophila*, Waddington (1975) described the breeding of stock with differing patterns of posterior cross veins in the wings:

> . . . the conclusion clearly emerged that the wild type phenotype can conceal within it a much greater range of dosages of vein-producing genes than can any other phenotype. That implies that the canalization of the normal vein pattern is such that it is highly resistant to the disturbing effects of changes in the dosage of genes tending either to make more or to make less vein. (p. 81)

Waddington conceived of *embryo development* as the product of interacting "gene action systems," that is, series of biochemical processes leading from gene to phenotype. He described these systems as working together to create a stabilized or buffered pathway of change. He called these pathways *chreods* or canalized developmental trajectories, remarking that they are the basic elements of developmental theory just more complex than that of single gene-action systems (Waddington, 1962). Waddington defined a dynamic form of homeostasis during embryo development called *homeorhesis*, in which ". . . the thing that is being held constant is not a single parameter but is a time-extended course of change, that is to say, a trajectory" (Waddington, 1975, p. 221).

The *epigenetic landscape* is Waddington's diagrammatic representation of a normal developmental trajectory as a ball rolling down a curved surface, whose sides represent resistance to alternative courses around a valley: Chreods maintain the landscape's topology but selection may alter this landscape as alternative developmental patterns are favored. (West-Eberhard [2003] noted that even this model is incomplete because it fails to account for ongoing changes in developmental potentiality.)

As indicated earlier, in Waddington's usage, the term, *phenocopy*, refers to an environmentally induced phenotype that mimics the phenotype produced by a mutant gene. In such cases, environmental stresses act to change a developmental pathway to mimic a genetically altered developmental pathway (Waddington, 1975). The tendency of flies to form a bithorax under the influence of ether, for example, may mimic the effects of the bithorax mutation. He emphasizes, however, that "A phenocopy must result from the combined action of the environment and the genotype of the organism. That is to say, genes must be involved in the production of phenocopies, and *ex hypothesi* they must be sub-threshold genes, at least as regards the abnormal phenotype" (Waddington, 1975, p. 78). As we will see, this is consistent with West-Eberhard's (2003) principle of the complementary relationship between genetic and environmental influences on trait development, and her emphasis on genotype–phenotype equivalence.

Waddington's concept of *genetic assimilation*, which he first published in 1953, is the most controversial of his developmental ideas. Like Piaget, and many early biologists, Waddington (1942) questioned the adequacy of the modern synthetic paradigm of random mutation and selection to explain the origin of novel adaptations. Like Piaget, he proposed a third way, which he eventually called *genetic assimilation*.

According to Waddington, genetic assimilation is the process by which selection in a new environment favors a phenotype that mimics a preexisting (usually aberrant) phenotype produced by a mutant genotype in a normal environment. Through selection for a shift in canalization and the epigenetic landscape, the phenocopy evolves a stable genotype no longer dependent on the new environment for its expression. A different environment, especially a stressful environment, stimulates this phenomenon: "A greater amount of the variation [in the wild type] can be revealed if some way is found to push the process of development away from the canalized phenotype, so that they follow a path which is more susceptible to the influence of minor genetic variation" (Waddington, 1975, p. 82). Other biologists also argue that the phenocopy uncovers pre-existing hidden variation in the developing organism (Gilbert, 1994).

In his critique of neo-Darwinian theory, Waddington (1957) said that in light of the epigenetic landscape and genetic assimilation,

> We have been led to conclude that natural selection for the ability to develop adaptively in relation to the environment will build up an epigenetic landscape which in turn guides the phenotype effects of the mutations available. In light of this, the conventional statement that the raw materials of evolution are provided by random mutation appears hollow. (p. 189)

In his discussion of Waddington's contributions, Gilbert (1991) says, "The pathway forged by Waddington could not have been made by a person who was purely a geneticist or purely an embryologist. Waddington's goal of synthesizing genetics, embryology, and evolution was critical for his ability to connect the disparate studies" (p. 203).

As noted, developmental biology was also prefigured by work of the Russian biologist, Schmalhausen, in the 1930s as well as by that of Chester Waddington and Salome Gleucksohn-Schenheimer. As Piaget acknowledges, both Schmalhausen and Waddington focused on two issues central to his concerns: the origins of new phenotypes and regulatory mechanisms stabilizing embryo development.

The molecular mechanisms underlying the developmental phenomena discovered by Schmalhausen, Waddington, and Gleucksohn-Schenheimer are just beginning to be elucidated. Understanding of these mechanisms could only come after the rise of the field of developmental molecular biology, after a major detour into the wonderland of molecular biology. Recently, for example, Rutherford and Linquist (1998) explained in molecular terms how genetic or environmental stress can produce heritable changes in development, that is, through what Waddington called genetic assimilation. Specifically, they showed

> that when the *Drosophila* heat shock protein Hsp90 was compromised, either genetically (by mutation) or environmentally (by high temperatures or drugs), genetic variation that had been silenced by the buffering action of Hsp90 became expressed and, after selection, continued to be expressed even when Hsp90 was restored to its normal state. (McLaren, 1999)

(See discussion to come on the class of proteins known as *chaperones* to which Hs90 belongs.)

EXPERIMENTAL EMBRYOLOGY TO GENETICS TO DEVELOPMENTAL BIOLOGY: OR FROM ORGANISM TO GENE AND BACK AGAIN

The field of genetics began at the turn of the 20th century when the particles or factors in Mendel's rediscovered laws were associated with the chromosomes cytologists had identified in cell nuclei. Genetics became distinct, and free from the hegemony of the older field of embryology at the same time the concepts of *evolution* and *heredity* became clearly differentiated into hereditary transmission as distinct from ontogenetic development (Keller, 1995).

As previously indicated, embryology already had began to free itself of the baggage of earlier recapitulationist speculations. Following Roux's call in 1894, it turned away from evolutionary questions and toward the experimental study of the mechanics of embryogenesis or *Entwicklungsmechanik* (Kaufman & Raff, 1983). Hans Spemann, the German embryologist who discovered the organizer region of the embryo, reinforced this in his 1924 declaration that embryologists should resist the invasion by geneticists (Gilbert, 1991). Two years later in *The Theory of the Gene*, T. H. Morgan (1926) decreed a division of labor between embryology and genetics when he declared that genetics dealt with the transmission of traits whereas embryology dealt with the expression of traits (Gilbert, 1991).

Ironically, modern genetics and modern developmental biology can both be traced back to Thomas Henry Morgan, who began to study neo-Mendelian heredity by experimenting on the development of such organ systems as eye color, wing shape, thorasic pattern, and body color in fruit flies (*Drosophila*). Kohler (1994) described how the large scale breeding of the fruit fly in Morgan's laboratory at Columbia University in the 1910s produced an "avalanche" of the previously elusive mutations Morgan had been seeking. Originally, as an embryologist, Morgan used these mutants to trace the pedigree of various developmental patterns.

Morgan and his students, Sturtevant, Bridges and Muller, discovered that genes occur in a linear order on chromosomes, that they mutate in forward and backward directions, that they occur in alternative forms or alleles, and that their functioning can be changed by changing their position along the chromosomes (Lewis, 1995).

Morgan (1917) is famous for three percepts about the nature of genes that arose from this early work on classical developmental genetics: (a) each gene has "manifold effects," that is, affects many different characters (i.e., displays *epistatic interactions*); (b) "all characters are variable" owing to the effects of external conditions on the embryo; (c) each character is the product of many genes (i.e., *polygenic effects*) (cited in Judson, 1996, pp. 608–609). As Judson (1996) notes, Morgan's percepts reveal the problem with the common phrase a "gene for X characteristic." This is misleading shorthand for a mutant allele that disrupts any one of the myriad steps in the normal process of development. The misleading locution arose as shorthand among those discussing the effects of deleterious mutations, and has been widely misconstrued to mean that one gene produces one character. Moreover, it ignores the well-known fact that, owing to canalization, there may be many genetic paths to the same traits (Waddington, 1975). This results in the phenomenon of genotype–genotype equivalence (West-Eberhard, 2003).

Gradually, the slow, cumbersome method for studying the effects of mutants on embryonic organ systems (which required systemic revision for each new allele or variant gene) was overshadowed by the ease, speed, and productivity of the new technology for constructing genetic linkage maps of new mutants along the chromosome (Kohler, 1994). This new mapping methodology developed by Morgan's students, was originally a sideline in the research. (It was based on the frequency of crossing over of alleles between homologous chromosomes, which is proportional to their distance from one another; Kohler, 1994.) This highly productive group worked together for many years, first at Columbia University and later at the California Institute of Technology.

Ironically, given the serendipitous nature of rise of genetic mapping (Kohler, 1994), Morgan, and his student, H. J. Muller, contributed key ideas that established the dominance of a new uniquely American school of genetics as distinct from the slightly older German school of experimental embryology (Keller, 1995). Investigators from many countries and data/techniques from least five disciplines including genetics, x-ray crystallography, physical chemistry, microbiology, and biochemistry contributed to this new field (Judson, 1996).

The Rise of the Molecular Gene

The success and productivity of the new genetic research program was clinched (and the discourse of gene action enhanced) by a series of stunning discoveries in the 1950s and 1960s (Keller, 1995). First, in 1944, Oswald Avery and his associates (Avery, MacLeod, & McCarty, 1944) demonstrated that DNA was the genetic material. Then Watson and Crick (1953) elucidated the structure of DNA. This was followed quickly by discovery of the *genetic code* (triplets of the four complementary DNA nucleotide bases, A&T and C&G, whose linear sequence specified a sequence of amino acids). Keller (1995) noted that the concept embodies the new cybernetic terminology of linear codes, programs, and information. Overnight, these discoveries changed the *gene* from a functional concept of unknown mechanism to one of known mechanism. Molecular biology was born.

Molecular biologists discovered that reading the DNA code was mediated by three forms of RNA, a close relative of DNA: (a) messenger RNA, which made a temporary complementary copy of a gene; (b) transfer or soluble RNA, a set of small RNA molecules that transport amino acids; and (c) ribosomal RNA, a large molecular machine that translates the RNA message into a protein. In 1958, Crick enunciated the "central dogma" that genetic information moves in one direction only, from DNA to RNA to protein (Keller, 1995). Soon after these announcements came Jacob and Monod's (1961) discovery of the mechanism of gene regulation in production of an enzyme in the bacteria *E. coli*. It took many molecular biologists until 1966 to unravel the mechanism of protein synthesis (*transcription* of the DNA sequence in the nucleus by the mRNA, and *translation* in the cytoplasm of the mRNA on the rRNA into a sequence of amino acids carried by tRNA; Judson, 1996).

Other critical discoveries in molecular biology relevant to our story include the discovery that the cells and the DNA of *eukaryotes* (organisms with nucleated cells) are more complex than those of bacteria (contrasts are summarized in Table 2.2). Specifically, in addition to the coding DNA (which is transcribed and translated into proteins and RNA), the genomes of eukaryotes are largely composed of noncoding DNA (about 80% to 90%), including many repeat sequences (comprising about 50% of the human genome). A recent review notes a genome is defined ". . . as the entire collection of genes encoded by a particular organism," but asks ". . . what is a gene?" (p. 258). The authors answer by pointing to historical changes in the concept, culminating in the current definition of a gene as "a complete chromosomal segment responsible for making a functional product" (Synder & Gerstein, 2003, p. 258). They note that this segment includes both coding and regulatory regions, and that the functional product can be either a protein subunit or an RNA.

Moreover, the protein coding sequences of most eukaryotic genes are dispersed along the DNA in fragments (exons) separated by many (typically 10 per gene) intervening sequences (introns), which must be edited or spliced out of the mRNA in order for translation from DNA to proteins to occur. The function, if any, of this apparently wasteful organization of eukaryotic genomes is unknown. It is clear, however, that processes of gene evolution, gene regulation, and genome evolution have exploited the existence of introns and repetitive elements. Large-scale comparisons of genome sequences tend not to support the idea that these elements in the genome owe their existence to any functional advantage

TABLE 2.2
Contrasts Between Prokaryotes and Eukaryotes

Taxa	Cellular Composition	DNA, and so on	Cell Division Mechanisms	Development
Prokaryotes: Kingdom Monera (bacteria, blue-green bacteria)	Lack nuclear membrane, mitochondria, chloroplasts, and other membrane-bound organelles	Circular single strand of DNA, RNAs for protein synthesis; all are coding and regulatory DNA	Binary fission	None
Eukaryotes: Kingdoms Protista (e.g., algae, protozoa) (includes both single celled and metazoan forms) Fungi Plantae Animalia (all metazoan)	Nuclear membrane, Endoplasmic reticulum, Mitochrondria, Chloroplasts (in plants), organelles including the Golgi complex, Lysosomes, Vacuoles, Cytoskeleton	Linear haploid and homologous diploid chromosomes with centromeres, nucleolus, and spindles in nucleus; haploid DNA in mitochondria; RNAs and endoplasmic reticulum for protein synthesis; DNA includes large segments of noncoding as well as coding DNA	Mitosis in somatic cells; Meiosis or reduction division (from haploid to diploid number of chromosomes) in sex cells	Metazoan life histories cycle between multicellular proliferation/differentiation and generative single-cell stages that undergo embryogenesis: Taxa differ in relationship between somatic and germline cells (see Table 2.4)

they may provide (E. Branscomb, personal communication, March 12, 2003).

Eukaryotic chromosomes are complex nuclear organelles composed of DNA and attached regulatory proteins—a complex known as *chromatin.* The smallest structural unit of the chromatin, the *nucleosome,* is a short segment of DNA wound around a core of *histone* proteins. These histone proteins are sometimes modified by the attachment of *acetyl, methyl or phosphate* groups, which are involved in the regulation of processes of DNA replication, gene expression, chromatin replication and cell division (Stewart, 2002).

Chromatin, the DNA and associated proteins, assumes different conformations depending on the phase of the cell cycle and the cell type. In its most relaxed state, the DNA molecule is maximally relaxed and accessible for protein synthesis; in its most compacted state, as seen in most photographs during mitosis (cell division), it is supercoiled. In supercoiling, the nucleosomes are packed into a coil of about 6 nucleosomes per turn called a fiber, which in turn is organized into loops, scaffolds and domains; at its maximum supercoiling, it is condensed about 10,000 times. Chromosomes also have specialized regions for replication, the centromeres and telomeres (McClean, 1997).

Gene expression, that is, the regulation of which genes are expressed in which tissues at which times, is controlled at both the transcriptional level in the nucleus and at the mRNA processing and stability, and translational levels in the cytoplasm. Transcription is regulated on a gene-by-gene basis by controlling the initiation of the process, which begins at a site near the start of a gene's first coding segment, the *promoter.* Close to the promoter and generally outside the coding segments of the gene (exons) are a number of small regulatory sequences. The regulatory sequences bind so-called transcription factor protein, which in turn influence the rate of transcription initiation (Kalthoff, 2001). The promoter sits immediately upstream from the start site of the transcribed region. The promoter site binds a large complex of proteins including an RNA polymerase molecule, which when freed initiates gene transcription by connecting RNA subunits into a mRNA string off the coding DNA template. The activity of the promoter is controlled both directly and indirectly by a dynamically changing complex of regulatory proteins (Kalthoff, 2001).

Regulatory proteins can either accelerate or inhibit transcription. (DNA regulatory sequences are bound by *transcription factors,* which interact with each other and with DNA to form a large 3-D transcription complex.) Transcriptional factors (proteins) are classified on the basis of their DNA binding domains (e.g., helix-turn-helix proteins, zinc finger proteins, leucine zipper proteins, helix-loop-helix proteins [including the homeodomain], and steroid receptor proteins), which allow them to recognize and bind to specific short sequences—the "cis-acting" domains of DNA.

> The salient aspect of transcription control is that a relatively small number of transcriptional activators and repressors can control, in combinatorial ways, the activities of many target genes. . . . Because the transcription factors themselves are gene products, their synthesis in turn is controlled by transcription factors. Thus the cells of an organism acquire different fates by establishing different hierarchies, feedback loops, and mutual inhibitions of transcription factors and their encoding genes. (Kalthoff, 2001, p. 413)

Many eukaryote genes occur in multigene families or sets of genes with closely related DNA sequences. These multigene families (e.g., the globin genes and histone genes, Human Lymphocyte Antigen (HLA) genes, and immunoglobin genes) apparently arose through gene duplication and subsequent differentiation of duplicates for new functions. They are often involved in related functions (Kaufman & Raff, 1983).

The history of genetics is one of an accelerating pace of (often surprising) discoveries of greater and greater complexity of genetic systems. One of the major surprises was the discovery of RNA *reverse transcriptase*, an enzyme that can copy an RNA transcript of a gene back into DNA (thereby violating the original form of the central dogma). This enzyme allows genes of viruses to insert themselves in the DNA of their hosts (Judson, 1996). As a result of sequencing of the genomes of mouse and human, it is known that a significant proportion of the DNA in a typical mammalian genome is comprised of leftover retroviral genes (E. Branscomb, personal communication, March 12, 2003).

Another surprise, was a recent string of discoveries of the regulatory role of two classes of small RNAs produced by genes: micro RNAs (miRNAs) and small interfering RNAs (siRNAs). These small (21 to 28 nucleotide) double-stranded RNA molecules can turn on or shut down genes or alter their levels of expression; siRNAs can identify and degrade messenger RNA with a complementary sequence; other small RNAs can alter the shape of chromatin, and permanently delete or shut down sections of DNA (Couzin, 2002).

Another surprise, was the discovery of transposable genetic elements (*transposons*) or insertion sequences or "jumping genes" in E. coli that can change location within a genome thereby changing their positions and hence their phenotypic effects, and even jump into the genomes of other species (Shapiro, 1969). (This phenomenon had been described earlier in maize by Barbara McClintock [1951].) At least 50% of the DNA in the human genome is now known to be the relics of transposable elements of several different types. As noted below, transposons have contributed significantly to the origin of variation.

Mutation and Variation at the Molecular Level

These and other discoveries in genetics and molecular biology contributed to better understanding of heredity and variation. Indeed, Shapiro (1983) argued "Now that the methods for directly characterizing genomic differences and similarities have become so powerful, it may prove useful to focus on the role of variation, rather than selection, in the origin of species" (p. 266). He also notes that variation is a complex phenomenon critical to understanding the process of evolution: "The term *variation* has multiple meanings, generally referring to the many processes that create new genomic configurations in cell lineages and thereby alter the characteristics of organisms" (Shapiro, 1983, p. 253). Among these processes, Shapiro includes (a) reassortment of Mendelian factors, (b) generation of recombinant chromosomes, (c) creation of new alleles at specific loci, (d) changes in chromosome number and structure, (e) alterations in structure of repeated genetic elements, and (f) "introduction of new hereditary determinants" such as transposable genetic elements or insertion sequences (Shapiro, 1983, p. 253).

Shapiro notes that transposable elements have profound effects on variation because they can insert into many different gene loci and their because their presence affects the expression of the new locus in characteristic ways frequently altering transcription patterns, He says that these elements are known to be major agents of spontaneous variation in *E. coli*, yeast, and corn. Moreover, he argues that, owing to their internal signals, "transposable elements can have both stimulatory and inhibitory effects on each other, in some cases serving as hot-spots for recombination with another element and in other cases blocking recombination over regions of tens or thousands of basepairs . . ." (Shapiro, 1983, p. 261). It should be noted that although transposable elements come from outside the organism, they do not constitute adaptations to environmental factors.

On the basis of work on the role of transposable elements in hybrid dysgenesis in fruit flies, Shapiro makes the radical argument that because chromosomal rearrangement requires cellular and populational interactions, rates of genomic variation are highly changeable: "Thus in a sexually reproducing metropolitan species like *D. melanogaster*, just as in *E. coli*, there can be no fundamental underlying rate of spontaneous variation independent of the history of any particular population" (Shapiro, 1983, p. 264). This is radical in that it suggests that some mutations involving jumping genes are under control of these transposable elements and therefore nonrandom. On the other hand, selection must have shaped the characteristics of these elements.

Another significant change in perspective on mutations came with the vindication by Richard Lewontin and other geneticists of Ohno's (1970) proposal regarding the predominance of *neutral mutations*, and of the consequently huge amount of silent genetic variation among copies of the same gene in natural populations (polymorphic gene loci) of fruit flies. Originally this was revealed by the technique of *electrophoresis* (producing a kind of protein fingerprint), which allows investigators to see that a single protein molecule in a population could have hundreds of variations, implying correlated variations in the genetic code (Ayala & Valentine, 1979). This discovery is consistent with the discovery of a large number of silent and neutral mutations, that is, mutations that have no discernible effect on the structure and function of the proteins they code (Kimura, 1979). This discovery can be understood as a consequence of the nature of the genetic code and protein structure: Because the genetic code is redundant, that is, the three-base codon for many amino acids is specified by the first two bases but can have any base in the third position, mutations affecting the third base do not change the amino acid specified by these codons. In addition, many amino acids in a typical protein can be changed with little or no effect because they share chemical properties, which make them functionally interchangeable.

Most mutations occur during DNA replication in cell division. For some time, geneticists have recognized that mutation frequency is not entirely random at the molecular level. Because certain base sequences or hotspots are subject to copying errors, mutation rates vary across the genome depending in part on the nucleotide sequence (Li & Grauer, 1991). It is important to note, however, that although they are somewhat nonrandom at the molecular level, mutations are random with respect to their possible selective value. Therefore it falls to selection to decide which mutations fade away and which persist. (The random or directedness of mutations has been an issue for some critics of neo-Darwinism, including Piaget.)

THE NEW DEVELOPMENTAL BIOLOGY: CELLULAR AND
PHENOTYPIC CONTINUITY, PARAGENETIC
INFORMATION AND EPIGENETIC INHERITANCE

The shift in resources from experimental embryology to molecular biology came in association with the refocus of the problem of development from the locus of the egg to that of gene action. Concomitant with this came a shift in the choice of model organism from complex multicellular organisms to the simplest forms of life. Morgan and Muller and their group worked on the fruit fly in the 1910s and 1920s, Beadle left corn to work on flies, and left flies to work on bread mold in the 1940s. Beginning in the 1950s, the new molecular biologists under the influence of the physicist, Max Delbrück, used the bacteria (*E. coli*) and bacteriophages or viruses as model organisms (Keller, 1995).

By eliminating the complexity of multicellular organisms, the newly emerging field of molecular biology narrowed and simplified the problem. Ultimately, however, these and other great discoveries outran the utility of the discourse of gene action that helped spawn them. As Keller (1995) put it ". . . the information yielded by those techniques is now radically subverting the doctrine of the gene as sole (or even primary) agent" (p. 22). Specifically, Jacob and Monod's discovery that protein synthesis by the lactose gene was controlled by feedback from the amount of lactose sugar in the *E. coli* cell (*lac operon*) showed the role of the cytoplasm in gene regulation.

The modern molecular version of the principle of *cellular continuity* helped to bridge the historic gap between geneticists, who focused primarily on genes, and embryologists, who focused primarily on the cytoplasm. Developmental biologists emphasize the dual inheritance of genes and cytoplasm (*genetic information* and *paragenetic information*):

> All cells of currently living organisms are the temporary ends of *uninterrupted* cell lineages extending back through their ancestors' germ lines to primordial cells billions of years ago. Two types of information have been passed on through these lineages. One is *genetic information*, which is coded in the DNA or RNA. Equally important is the unbroken chain of *structural organization* that is passed on directly, without being encoded in the genes, from each cell to its daughter cells. . . . it is called *paragenetic information*. (Kalthoff, 2001, p. 477)

West-Eberhard (2003) called this *phenotypic continuity* between generations.

Cell division entails not only replication of chromosomes, but replication of cytoplasm and its *organelles* including mitochrondia, endoplasmic reticulum, cell membranes, ribosomes, Golgi bodies, microtubules, and so on. Self-replication of the proteins and other molecules comprising these structures occurs through one of the following three processes of increasing complexity: self-assembly, aided assembly, or directed assembly. *Self-assembly* occurs from seed structures without molecular scaffolds or catalysts. *Aided assembly* requires help from auxiliary proteins, for example, chaperones, which ensure proper folding of the polypeptide chains into their active 3-D forms. *Directed assembly*, the most complex form, involves structures for assembling alternative forms from the same building blocks according to the phase of the cell cycle (Kalthoff, 2001).

As we have seen, long before geneticists began to shift their focus from bacterial and phage genetics to eukaryote genetics, some German, Russian, and English embryologists were thinking in terms of whole cells and organisms, seeking to reunite what Morgan had put asunder. Indeed, up through the 1930s under Morgan's fly group, geneticists had done pioneering work combining embryology and neo-Mendelian genetics in the study of the role of genes in the development of organ systems. Workers in Morgan's laboratory had invented methods for studying chromosome deficiencies in such fly embryo mutants as aristopedia, proboscopedia, and bithorax.

In the 1950s and 1960s, however, the novelties and productivity of molecular genetics soon overshadowed the slow painstaking work on *Drosophila* embryogenesis. As Kohler (1994) said, "Lines of work in developmental genetics, begun in the late 1930s and 1940s . . . only became really fruitful in the 1970s, when new techniques were devised for studying the master genes that regulate differentiation. Then *Drosophila* came once again into the scientific limelight" (p. 246).

For all these reasons, then, it was only in the 1970s that molecular genetics and embryology were solidly reunited in the new field of developmental biology that the focus turned from the gene to the organism and its development. The arrival of this field was institutionalized by the formation of the *Journal of Developmental Biology* and by the formation of an increasing number of university departments of molecular biology (Keller, 1995).

Classical experimental embryology had invented a variety of microscope-aided techniques for studying embryos. These included dyeing cells to allow their cell lines to be traced, and "controlled interference" through isolation, removal, and transplantation of cells at various developmental stages. Using these techniques, they identified the stages of embryogenesis (fertilization, cleavage, gastrulation, organogenesis, histogenesis), fate maps of the location of organs to be, and discovered organizers that induced tissue determination. They also tracked the formation of the three body axes typical of metazoans: anterior–posterior, dorsal–ventral, and left–right at the tissue and cellular levels (Kalthoff, 2001; Kohler, 1994).

The Molecular Biology of Embryogenesis

A variety of new molecular techniques from molecular biology contributed to the developing field of developmental biology. These include using the enzyme reverse transcriptase to go backward from a protein to RNA to DNA, thereby forming complementary DNA, robotic synthesis of new DNA and RNA sequences, mass cloning of existing DNA segments to form libraries of known genes, and cutting and splicing existing DNA into new combinations. These new DNA segments can be labeled and inserted into host genomes using viruses as vectors. The development of directed and insertional mutagenesis methods was also critical. If foreign genes are integrated into the host's genome, these inserted transgenes are replicated during cell division, potentially creating a genetically-transformed organism. Other techniques allow individual genes to be over expressed, suppressed, or knocked out (Kohler, 2001).

Classical experimental embryologists had used fruit flies, frogs, sea urchins, and mice. Developmental biologists have focused primarily on model animal organisms: nematodes, sea urchins, sea-squirts, fruit flies, frogs, zebrafish, and mice. Long-term studies of these organisms have yielded large numbers of mutant lines in each species that can reveal mechanisms of control of genetic development. They make good laboratory animals because they are hardy, mature rapidly, breed prolifically, and have small genomes. Beyond their general advantages, each offers specific advantages; for example, zebrafish are vertebrate models that develop rapidly and are easy to study developmentally because they are transparent. Because they can be tricked into developing haploid embryos, the effects of recessive alleles can be observed without recourse to breeding pure lines. Mice are mammalian models that develop rapidly. More important, they share many developmental patterns with humans and other mammals (Kohler, 2001). Table 2.3 summarizes genetic characteristics of these model organisms.

TABLE 2.3
Genetic Characteristics of Model Organisms Used
in Developmental Biology (Kalthoff, 2001)

Model Organism	Genome Size (m BP of DNA)	Haploid Number Chromosomes	Generation Time	History of Study	Known
Caenorhabditis elegans (nematode)	97 (19,000 genes)	6	3 days	S. Brenner, 1950s	Entire genome and all developmental paths
Drosophila melanogaster (fruit fly)	180 (14,000 genes)	4	3 weeks	H. L. Morgan, 1910s	Entire genome
Danio rerio (zebrafish)	1700	25	2 months	Nusselin-Vollard, 1980s	
Mus musculas (mice)	3000	20	2 months	1900	Entire genome

The new generation of developmental biologists in the United States, England, Spain, and Germany experimenting with these and other model species have discovered many of the genes involved in development including, for example, those involved in formation of the anterior–posterior, ventral–dorsal, and right–left axis, and also those involved in limb formation in vertebrates. Originally, most of this work was done on fruit fly larvae by Christiane Nüsslein-Volhard and her colleagues in Tübingen, Germany and in the United States.

Their work revealed a cascade of developmental gene regulation beginning with the localized RNA transcripts and other products of maternal effect genes, followed by expression of embryonic genes. Transitional between maternal effect and embryonic genes are the so-called *gap genes*. A famous set of genes involved in the formation of the anterior–posterior axis is the so-called *homeotic genes*, the mutant forms of which transform one body region into the likeness of another body region (e.g., antennapedia mutation in fruit flies produces a foot where an antenna should be). Subsequent development of the ventral–dorsal axis relies on another gene complex, as does the formation of body compartments.

Many other genes involved in development are also known now. The conservatism of developmental genes has amazed biologists.

Molecular developmental research on embryos of nematodes, mice, and other organisms has revealed many conserved homologies including those in the homeotic genes. These included common homeobox sequences and the resulting homeodomains in the proteins coded by these genes. (These proteins are all transcription factors that bind to particular DNA sequences, thereby regulating gene expression.) "Not only do *Drosophila* homeoic genes and mouse Hox genes share their chromosomal order and sequence information in crucial domains, they also have similar functions in specifying the anterioposterior pattern" (Kalthoff, 2001, p. 617).

Comparative data suggest that an ancestral *Hox gene*, which occurs in all bilaterally symmetrical animals, arose about 500 million years ago, duplicating many times and forming a tandem array of related genes called the *Hox complex*, which itself has duplicated several times during the evolution of vertebrates. Four copies of the Hox gene cluster persist on different chromosomes in mice and humans (Kalthoff, 2001). Intriguingly, "[T]he physical order of the Hox genes within each complex is related to the order of their expression domains along the anterior posterior axis of the embryo" (Kalthoff, 2001, p. 617). The majority of genes involved in development were directly inherited from the common ancestor of bilateral organisms that lived hundreds of millions of years ago.

Genetics and Epigenetics. From an epigenetic/phenotypic perspective, it is important to emphasize that, beginning with the maternal RNA transcripts in the egg and zygote, all of these processes depend on signals to the genes from molecules in regionally-differentiated cytoplasm, and reciprocal messages from the genes to the cytoplasm in the form of cell-specific proteins and RNAs. Subsequent to blastula formation, signals from other cells play a parallel role in sending and receiving messages to their neighboring cells.

Later stages of embryogenesis, beginning with blastulation but particularly during organogenesis and histogenesis or cellular differentiation, depend upon a variety of mechanisms including (a) induction of protein synthesis by molecules from within or outside the cell, (b) mitotic division, (c) cell movement aided by differential cell attraction and adhesion mechanisms, and (d) selective cell death or apoptosis. *Histogenesis*, differentiation of cells into their tissue-specific forms, involves switching on or off cell-specific genes as well as the general housekeeping genes shared by all cells (Kalthoff, 2001).

All genes are controlled by transcription factors (themselves gene products), which are activated by signals such as hormones that have been recognized by their target cells.[5] After organogenesis, development also depends on the action of various agents of cellular communication, including neurons, short-range chemical signals (e.g., inducers and growth factors), and long-range chemical signals or hormones, which act selectively on target organs with specialized receptors (Kalthoff, 2001). Clearly, many of these factors, particularly hormones, have been selected to respond to various external signals such as patterns of light and dark, sounds, tactile stimulation, and even social stimuli.

Genomic imprinting is an intriguing counterintuitive discovery made originally in maize and mouse embryology that in some organisms, gene expression can differ, depending on the parental origin of the chromosome. This is seen, for example, in the preferential inactivation of paternally derived insulin growth factor II: "Ig/2r and H19 . . . are active only when inherited from the mother; a third, called Ig/2, is turned on only when inherited from the father" (Penisi, 2001, p. 66). More than 40 genes of this kind have been identified including some implicated in disease. Imprinting seems to be responsible for differences between the mule and the hinny—hybrids between horses and mules, which differ depending on the sex of the parent. Imprinting seems to be mediated by methalytion, which generally suppresses gene expression (Jablonka & Lamb, 1995; Penisi, 2001). (Recall that histone proteins can be modified by the attachment of *acetyl, methyl or phosphate* groups, which are involved in the regulation of replication gene expression [Stewart, 2002].) It is also intriguing that

---

[5]Testosterone from the fetal gonad in humans, for example, acts on target cells in the Wolffian ducts to cause them to differentiate into the vas deferens and epididymis, and on the proto-genitals to differentiate into the penis and scrotum. Likewise, the fetal gonads produce the Müllerian inhibiting hormone, which cause the parallel Müllerian ducts to regress rather than forming internal female structures. Sexual development includes three basic aspects: dosage compensation, somatic sex differentiation, and germ-line sex differentiation. Dosage compensation disarms one sex chromosome in cells of the sex with two homologous sex chromosomes (XX in mammals) to equalize the expression of nonsex-related genes on that chromosome. Formation of the internal and external male genitalia, just described, is an example of somatic sex differentiation. Gonad differentiation into testis in humans occurs earlier under the influence of the SRY gene and the testis determining factor on the Y chromosome of the germ cells after they migrate into the proto gonad in the embryo from outside the embryo in the yolk sac. Germ-line sex determination in turn is triggered by interactions between genes in the germ-line cells and inductive signals from somatic cells in the gonads, which determine whether they initiate öogenesis or spermatogenesis (Kalthoff, 2001).

parental gene imprints are erased and reset before or during gameto-genesis when sperm and eggs are differentially packaged. Some imprints are subsequently erased and reset at other stages during embryogenesis (Jablonka & Lamb, 1995).

Modes of Embryogenesis and Epigenetic Inheritance

Buss (1987) described three modes of embryo development: (a) *somatic embryogenesis* in organisms (e.g., plants and fungi) lacking a distinct germ line; (b) *epigenesis* in organisms (e.g., some chordata and arthropoda) with a germ line that separates after organogenesis; and (c) *preformation* in organisms (e.g., flies, mice, and humans) with a distinct germ line that separates before organogenesis. Buss (1987) argued that organisms with the first two modes of embryo development can pass on new variants that arise before germ-line segregation. If these epigenetic variants are adaptive and triggered by environmental factors, they may be considered at least the short-term inheritance of acquired characteristics (Jablonka & Lamb, 1995). Although Piaget's phenocopy model might apply to some organisms, as creatures characterized by embryological "preformation," humans are not candidates for epigenetic inheritance.

As noted earlier, some historians (Mayr & Provine, 1998) argued that the rejection of Lamarckism occurred because most of the biologists who participated in the modern synthesis worked with organisms characterized by the preformationist mode. This experiential bias is also true of most developmental biologists. Table 2.4 summarizes the three modes of embryogenesis.

In their book, *Epigenetic Inheritance and Evolution*, Jablonka and Lamb (1995) reviewed molecular evidence for various *epigenetic inheritance systems* (EIS; i.e., those involved in neo-Lamarckian inheritance of acquired characteristics). Citing Buss' (1987) distinction among various modes of germ-line segregation, Jablonka and Lamb (1995) noted that much of the evidence for inheritance of acquired characteristics has come from studies of plants and other organisms lacking a distinct germ line:

> . . . in plants and other organisms that lack a segregated germ line, new epigenetic variants occurring in somatic lineages may be inherited. In contrast, organisms such as mammals, in which the germ line segregates early in development, can transmit . . . only those new variations that occur either before germ-line segregation, or in the germ line itself. (p. 152)

TABLE 2.4
Three Modes of Embryo Development in Metazoans
(Buss, 1987) and Their Implications for Inheritance
of Acquired Characteristics (Jablonka & Lamb, 1995)

Somatic Embryogenesis	Epigenesis	"Preformation"
Occurs in organisms lacking a distinct germ line	Occurs in organisms with a distinct germ line that separates from the somatic cells after primordial organ systems have formed	Occurs in organisms with a distinct germ line that separates from somatic cells before embryogenesis
Kingdoms: Protista, Fungi, Plantae	Kingdom: Animalia including some members of Phylum Chordata and Arthropoda	Kingdom: Animalia including some members of Phylum Chordata, for example, mice and humans; and Arthropoda, for example, flies
Can transmit new epigenetic variants that arise in somatic cells before germ-line segregation (most common in plants)	Can transmit new epigenetic variants that arise in somatic cells before germ-line segregation	Can only transmit variants that arise in the germ line itself; the epigenetic state of the germ cells is reset during meiosis

Jablonka and Lamb define an EIS as ". . . a system that enables a particular functional state or structural element to be transmitted from one cell generation to the next, even when the stimulus that originally induced it is no longer present" (Jablonka & Lamb, 1995, p. 80). Following Nanney (1958), they argue for the following differences between epigenetic inheritance and genetic inheritance: (a) EIS is usually less stable (persisting only a few generations) than GI because they are sensitive to environmental changes; (b) induced changes in EIS are more often directed (that is, responses to environmental factors); (c) EIS show limited variation because they are restricted by DNA sequences.

Jablonka and Lamb (1995) argued that the distinction between genetic and epigenetic inheritance systems is not absolute, perhaps because information is carried not only in the primary structure of DNA but in its superstructure or typology, which can be modified by interactions with proteins (i.e., transcription factors previously described). They describe three kinds of EIS: steady-state systems, structural inheritance systems, and chromatin-marking systems. *Steady-state systems*, i.e., genes that maintain their activity in the absence of inducing stimuli, are transmitted to daughter cells via regulatory proteins. *Structural inheritance systems* (which Kalthoff [2001] calls *paragenetic systems*) work through directed assembly of organelles. *Chromatin-marking systems* are involved in regulating gene transcription.

They summarize studies showing that EIS sometimes can be transmitted through the germ line in eukaryotes, occasionally through a few generations, which they say, may explain some puzzling cases of non-Mendelian heredity including incomplete penetrance of certain genotypes. Specifically, they cite induced response to temperature in beans and nematodes, induced dwarfism in rice, penetrance of induced bithorax in fruit flies (relevant to Waddington's concept of *genetic assimilation*), and drug and hormone induced changes in endocrine function in rats, mice, and guinea pigs (Jablonka & Lamb, 1995).

In their discussion of epigenetics, Jenuwein and Allis (Jenuwein & David, 2001) proposed the existence of a *histone code* that can change chromatin structure and thereby cause inherited differences in on-and-off transcriptional states of DNA. They say

> . . . Chromatin, the physiological template of all eukaryotic genetic information, is subject to a diverse array of post translational modifications that largely impinge on histone amino termini, thereby regulating access to the underling DNA. Distinct histone amino-terminal modifications can generate synergistic or antagonistic interaction affinities between transciptionally active or . . . silent chromatin states. The combinatorial nature of histone amino-terminal modifications thus reveals a "histone code" that considerably extends the information potential of the genetic code. We propose that this epigenetic marking system represents a fundamental regulatory mechanism. (Jenuwein & David, 2001, p. 1074)

In a recent review of epigenetics, Rutherford and Henikoff (2003) noted that although "the mysterious Lamarckian flavor of epigenetic phenomena initially slowed their acceptance into mainstream biology" (p. 6), molecular mechanisms—including DNA methylation, chromatin remodeling, histone modification—have been discovered for many of these phenomena. They say that problem of the relationship between somatic and germ-line inheritance, however, is just being unraveled. They cite the recent research by Sollars and colleagues demonstrating *epialleles*, that is epigenetic variants. Specifically, these investigators showed that a certain allele of the Kr gene in the anterior eye of fruit flies acting in confluence altered chromatin inheritance states in the female germ line ". . .

stochastically switches epialleles on or off at a number of modifier loci upon which selection is able to act" (Rutherford & Henikoff, 2003, p. 7). This altered state was accompanied by reduction or inhibition of heat shock protein (Hsp90), which buffers environmental and genetic disruptions. The effect of this was to change the expression of the Kr gene, causing a reduction of eye tissue.

New Light on Piaget's Concern With Sources of Heritable Variations

As indicated previously, Piaget, in common with several other thinkers working in the 19th and 20th centuries, puzzled over the mechanisms involved in the origin of heritable adaptive variations. Troubled by their belief that random mutations could not explain the complexity and adaptive value of new behavioral characteristics, Piaget and others have proposed a variety of mechanisms for the inheritance of characteristics acquired during embryogenesis or later. In Piaget's case (and certainly those of Lamarck and Darwin), this explanation was posed without benefit of detailed knowledge about genetic and epigenetic inheritance from molecular and developmental biology. The decalage or time lag in Piaget's concerns reflects his rejection of the modern synthesis and his ignorance of molecular biology.

In fairness to Piaget, the mechanisms of genetic, paragenetic, and epigenetic inheritance that interested him are just beginning to be elucidated. Because some knowledge of these mechanisms is key to evaluating Piaget's phenocopy thesis, I have summarized information about the complexity of eukaryote DNA and chromosomes and their role in embryogenesis relevant to the origins of adaptations. These data suggest a much richer repertoire of mutational and epigenetic changes than was previously known.

Mutation has proved to be a much more complex and subtle force than earlier critics imagined. Mutations can change characters (including development and behavior) at several levels including (a) regulatory DNA sites, that is, promoters and enhancers (which accelerate or inhibit the promoter controlling transcription); (b) genes coding ribosomal and transfer RNAs; (c) genes coding for structural proteins; (d) genes coding for proteins regulating growth; and (e) genes coding for transcription factors that bind to the enhancers and to each other. Because transcription factor proteins can bind to DNA regulatory sites and to each other, this can exert multiple effects. In addition, changes in one system can have cascading effects on other systems. Moreover, as we have seen, transposons or jumping genes can regulate the location and frequency of mutations (Shapiro, 1983).

Mutations affecting the amount, duration, and timing of growth factors can produce *heterochrony*, that is, changes in the timing and duration of growth and development of whole organisms, their component body parts, or their life histories (Parker, Langer, & McKinney, 2000). Mutations of this kind have been identified, for example, in nematodes (Kalthoff, 2001). In addition to these kinds of mutations, gene duplications can lead to mutations generating new structures and functions, and duplication of noncoding regions can change this the topology of the chromosomes. Variation is also increased by the recombinable, modular nature of some exons, which allows the same gene to generate mRNAs having different sets of exons and thus coding for different proteins. Another level of control is exerted at the level of translation of mRNA into polypeptide strings and then at the level of protein assembly. Some of these levels are summarized in Table 2.5.

A NEW PHENOTYPE-CENTERED THEORY
OF DEVELOPMENT AND EVOLUTION

In a recent review of developmental literature, West-Eberhard (2003) chronicled continuing gaps and inconsistencies in concepts of evolutionary biology, including such static concepts in genetics as cohesive co-adapted gene pool and stabilizing selection, and such problematic metaphors as genetic programs or blueprints. She argues that despite these impediments, progress in developmental molecular biology and evolutionary biology have laid the ground work for *a new synthesis* of development and evolution, and concludes that such a synthesis requires an "... adequate theory of the phenotype organization that incorporates the influence of the environment" (West-Eberhard, 2003, p. 19). I focus here on her treatment of trait origin first because that is the major theme of her book, and second, because it is the focus of Piaget's phenocopy model. In contrast to many biologists, who argue that only mutations give rise to evolutionary novelties, she argues that *environmental induction* is the most significant factor in the origin of evolutionary novelties.

In her view, development is the key or the "missing link between genotype and phenotype, a place too often occupied by metaphors in the past" (West-Eberhard, 2003, p. 89). Therefore she makes it the center of her

TABLE 2.5

Varieties of Mutations and Genomic Evolution Adapted from Table 3.1 (Kaufmann & Raff, 1984)

Location	Event	Consequence in DNA	Consequence in Protein	Consequence in Phenotype
Structural genes	Nucleotide substitution (NS)—silent	Change in base sequence	No change in amino acid (AA)	None or little
	NS—conservative	Same	Substitution of AA	None or little
	Deletion	Loss of base	Nonsense sequence or termination of sequence	Varies from loss of function
	Insertion	Addition of base	New related AA sequence	New function with old function retained
	Gene Duplication (followed by NS in duplicate)	Duplication (change in NS in duplicate)		Varies from none to new, to loss of function
	Gene Fusion	Loss of sequence of bases	Fused protein	Changes in timing or level of expression
Promoters or other regulators	Changes in base sequence	Changes in base sequence	None	None
Noncoding sequences: Repetitive Satellite DNA, Introns	Change is base sequence	Changes in base sequence	None to insertion of AA	None to changes in function
Sequence frequency of satellite	Changes in number of copies	Changes in number of copies	None	None
Changes in ploidy (chromosome number)	Changes in number of copies	Most or all sequences multiplied equally	None	Varies from none to reproductive isolation
Movement of sequences to new location in genome:	Insertion of intron;	Relocation of sequence	None	Possible change in function
Regulator	Insertion of promoter	Relocation	None	Change in time or level of expression
Transfer of genes between species	Introduction of new sequence		Introduction of new protein	Varies from none to introduction of novel function

synthesis. Consistent with ideas of Baldwin (1902) and Piaget, this synthesis emphasizes the central role that development and behavior play in the origins of new adaptations. As she emphasizes, it does so without resort the Lamarckian mechanisms invoked in Piaget's (1971, 1978, 1980) phenocopy model.

West-Eberhard (2003) enumerated the following major transformations involved in the origins of evolutionary novelties: (a) duplication (of genes, or chromosome segments, body segments, behaviors); (b) deletions (of genes, chromosome parts, body segments, or behaviors); (c) reversion to earlier more primitive form (of genes, organs or behaviors); (d) heterochrony, that is, changes in the timing of development (of organs, sexual maturity, behaviors, or whole organisms); (e) heterotopy, that is, changes in the location of organs; (f) cross-sex transfer (of a gene, organ, or behavior); (g) quantitative shifts and correlated changes in development, that is, shifts in the magnitude of trait expression (e.g., use correlated changes in bone); and (h) recombinations (of genes, chromosome segments, organs, or behaviors, e.g., through learning). All of these transformations are made possible by the modularity of developmental subunits (genes, chromosomes, organs, behaviors), which, in turn, results from the developmental process of switching.

West-Eberhard (2003) emphasized that development is organized by a series of binary decision points or *switches* that regulate the timing and selection of pathways, for example, the nutritional switch that determines whether male horned beetles form the adult fighting morph or the cryptic female morph, or the temperature-mediated switch that determines the sex of turtles, or the switch that determines when puberty occurs in human females. Switches ". . . determine the modularity of phenotypic traits and mediate the variation in phenotypic plasticity that permits the interchangeability of genetic and environmental influence on trait expression . . ." (West-Eberhard, 2003, p. 129). The mosaic nature of the phenotype is the result of the fact that functional traits are "switch-organized modular subunits" that are triggered by specific conditions. Switches are the focal point of evolution because selection can adjust the regulation of switches, for example, by altering their response thresholds in neural and hormonal systems. The modularity or dissociability of subunits engendered by these switch points allows recombination during their evolution. This recombination is a major source of novel traits.

Intergenerational continuity of the phenotype is a key idea in West-Eberhard's theory. She notes that whereas cross-generational continuity in genetic material has been recognized since Weismann's time, the fact that this continuity depends on the continuity of phenotype is seldom appreciated. As she notes, "Individual development always begins with an *inherited bridging phenotype*—a responsive, organized cell provided by a parent in the form of an egg, a newly divided cell, or a set of cells that springs entirely from the previous generation . . ." (West-Eberhard, 2003, p. 91).

Another key idea in the theory is the *interchangeability* of genetic and environmental influences on the phenotype (as revealed, e.g., by the phenomena of *phenocopy–genotype* and *genotype–genotype* equivalence): "Interchangeability during evolution occurs because in a population, variation in trait expression . . . that is due to plasticity and variation that is due to genotype are continuously variable, complementary determinants of total variation—their sum equals one" (West-Eberhard, 2003, p. 118). Therefore, she argues, "Interchangeability is a key to understanding both the organization of development and its modification during evolution" (West-Eberhard, 2003, pp. 117–118), that is, ". . . the phenotypic structures are the units of reproduction" (West-Eberhard, 2003, p. 98) and the units of selection. The development of the phenotype requires both environmental information, as for example, that processed by sensory structures, and environmental substances such as amino acids and vitamins: ". . . environmental factors function in the same ways as do gene products during development. They serve as building blocks and cues, as specific and essential for development as the selection-honed products of genes, and their effects can be transmitted between generations" (West-Eberhard, 2003, p. 115).

West-Eberhard (2003) proposed that *adaptive evolution* involves the following events: (a) trait origin of a distinctive developmental variant; (b) phenotypic accommodation by individual phenotypes owing to their plasticity; (c) recurrence or initial spread of these accommodations; (d) genetic accommodation, that is, evolutionary change owing to selection on variation; (e) persistence of an alternative phenotype; (f) modification of the trait; and (g) phenotypic fixation or deletion. She notes that, as with genotypes, new phenotypes arise as modifications and/or reorganizations of old ones during development. She says "Evolutionary change starts with a phenotype that is responsive to new inputs, and new inputs cause developmental change" (West-Eberhard, 2003, p. 145); this can happen, for example, through new inputs on preexisting developmental switches. In other words, "Selection depends upon phenotypic variation and environmental contingencies only. . . . But genetic variation is required for selection to have a cross-generational effect—an effect on evolution" (West-Eberhard, 2003, p. 141). On the other hand, "Change in frequency of an adaptive trait . . . can occur without genetic change, if the environmental change consistently induces (fixes) the trait" (West-Eberhard, 2003, p. 153).

SUMMARY AND CONCLUSION

In summary, powerful new models are providing insight into the manifold ways that developmental changes can evolve. Moreover, a new phenotype model is clarifying the role development and behavior play in evolution. These emerging insights have important implications for developmental psychologists and biological anthropologists interested in the evolution of development. This is especially true of those interested in evaluating the ideas Piaget (1971, 1978, 1980) proposed about the origins of adaptations in his three evolution books.

In these three books, Piaget argues that cognitive development is an extension of organic regulations and that both development and evolution proceed through assimilation and accommodation and auto-regulation. His argument is based on the conviction that ontogeny and phylogeny must by necessity involve the same functions (see Deacon, chap. 3, this volume). The crux of Piaget's phenocopy model is that the genome in somatic cells in the embryo can auto-regulate in response to environmental disequilibration in such a manner that these regulations can be passed on hereditarily to germ line cells through genetic assimilation. As he states, ". . . the hypothesis now being put forward shows . . . the organism to be reacting in an endogenous and active way to the pressures of environment and to be plainly assimilating them into its structures . . . by means of the genetic instruments at its disposal" (Piaget, 1971, p. 289). A few pages earlier, he is more explicit about the mechanism: ". . . the hereditary fixation of a new behavior seems to imply some transmission from the soma to the genome, whereas following the neo-Darwinian tradition, the general opinion . . . is that there is a radical isolation of the genome . . ." (Piaget, 1971, p. 282).

Piaget's discussion of evolutionary concepts embodies many of the conceptual confusions Mayr (1989) believes delayed the modern synthesis. Piaget sees selection as a negative force, and mutations as saltatory. He confounds variation at the individual level of the genome with variation at the population level of the gene pool (1989, see pp. 174–175). He confounds phenotype and genotype, misinterpreting Waddington's (1975) concepts of genetic assimilation and phenocopy as a Lamarckian process. In addition, he misattributes the feedback loops that occur between genes and environment in somatic cells to germ cells. Although as Piaget (1971, 1978, 1980) noted, cytoplasmic proteins and other molecules feedback into the nucleus and regulate gene expression; in mammals, this occurs almost exclusively in somatic cells (excepting the hormonal triggers involved in meiosis). (As indicated, humans and other mammals, indeed most vertebrates, sequester their germ-line cells during early stages of embryogenesis, thereby precluding influences on embryogenesis from being transmitted to these cells.)

Some of Piaget's phenocopy speculations, however, find limited support. It is true that there is evidence that some environmentally-induced epigenetic changes that affect the chromatin can, in turn, change gene expression, may be inherited in some species (primarily plants). It is also true that these could, in principle, change gene expression for at least a few generations. It is important to note, however, that these chemical responses cannot be described either as adaptive responses or as auto-regulation as Piaget's model proposes.

On the other hand, two larger themes in Piaget's biology books are partially vindicated by a recent new synthesis of development and evolution. The first theme is Piaget's dissatisfaction with the exclusive role of mutation in the classical neo-Darwinian account of the origins of adaptive variations, which partially parallels West-Eberhard's critique. The second theme that finds support is his emphasis on the role of development, especially behavioral adaptation, in the origins of adaptations. This theme finds parallels in West-Eberhard's phenotype-centered model of the origin of novel traits. (Unlike Piaget's phenocopy model, West-Eberhard's model avoids the mistake of Lamarckian inheritance.) As we have already seen, chapter 1 of this volume elaborates on the behavioral aspects of the phenotype-centered model, and introduces the complementary niche construction model, both of which find resonance in Piaget's constructivist model of intellectual development.

REFERENCES

Adams, G. (1998). The evolutionary synthesis: Morgan and natural selection revisited. In E. Mayr & W. Provine (Eds.), *The evolutionary synthesis* (pp. 356–382). Cambridge, MA: Harvard University Press.

Adams, M. B. (1998). Severtsov and Schmalhausen: Russian morphology and the modern synthesis. In E. Mayr & W. Provine (Eds.), *Evolutionary synthesis* (pp. 193–227). Cambridge, MA: Harvard University Press.

Allen, G. (1998). Evolutionary synthesis: Morgan and natural selection revisited. In E. Mayr & W. Provine (Eds.), *The evolutionary synthesis* (pp. 356–382). Cambridge, MA: Harvard University Press.

Avery, O., MacLeod, C., & McCarty, M. (1944). Studies on the chemical nature of the substance inducing transformation of Pneumococcal types. *Journal of Experimental Medicine, 79*, 137–158.

Ayala, F., & Valentine, J. (1979). *Evolving: The theory and processes of organic evolution.* Menlo Park, CA: Benjamin, Cummings.

Baer, K. E. v. (1828). *Die Entwickelungsgeschichte de Thiere.* Konigsburg, Germany: Borntrager.

Baldwin, J. M. (1902). *Development and evolution.* New York: The Macmillan Co.

Bergson, H. (1941). *Creative evolution.* New York: Random House.

Boesiger, E. (1989). Evolutionary biology in France at the time of the evolutionary synthesis. In E. Mayr & W. Provine (Eds.), *The evolutionary synthesis* (pp. 309–321). Cambridge, MA: Harvard University Press.

Bringuier, J. C. (1980). *Conversations with Jean Piaget.* Chicago: University of Chicago Press.

Burkhardt, R. W., Jr. (1984). The zoological philosophy of J. B. Lamarck. Introductory essay to *J. B. Lamarck, Zoological philosophy* (pp. xv–xxxix). Chicago: University of Chicago Press.

Burkhardt, R. W., Jr. (1998). Lamarckism in Britain and the United States. In E. Mayr & W. Provine (Eds.), *Evolutionary synthesis* (pp. 343–350). Cambridge, MA: Harvard University Press.

Buss, L. W. (1987). *The evolution of individuality.* Princeton: Princeton University Press.

Churchill, F. (1998). The modern evolutionary synthesis and the biogenetic law. In E. Mayr & W. Provine (Eds.), *Evolutionary synthesis* (pp. 112–122). Cambridge, MA: Harvard University Press.

Corsi, P. (1985). Recent studies on French reactions to Darwin. In D. Kohn (Ed.), *The Darwinian heritage* (pp. 698–711). Princeton: Princeton University Press.

Couzin, J. (2002, December 20). Small RNAs make big splash. *Science, 298.*

Cuvier, G. (1984). Biographical memoir of M. de Lamarck. In Appendix, *Lamarck, Jean-Baptiste, zoological philosophy* (pp. 434–453). Chicago: University of Chicago Press.

Darwin, C. (1859). *The origin of species.* London:

Darwin, C. (1868). *The variation of animals and plants under domestication.* London: John Murray.

Darwin, C. (1896). *The variation of animals and plants under domestication.* New York: Appleton.

Darwin, C. (1930). *The descent of man.* New York: Appleton and Co.

Darwin, C. (1965). *The expression of the emotions in man and animals.* Chicago: University of Chicago Press.

deVries, H. (1906). *Species and varieties: Their origin by mutation.* Chicago: Open Court.

Ghiselin, M. (1969). *The triumph of the Darwinian method.* Berkeley, CA: University of California Press.

Ghiselin, M. (1975). The rationale of pangenesis. *Genetics, 79,* 47–57.

Gilbert, S. F. (1991). Induction and the origin of developmental genetics. In S. Gilbert (Ed.), *Developmental biology: A comprehensive synthesis* (pp. 183–206). New York: Plenum.

Gottlieb, G. (1992). *Individual development and evolution.* New York: Oxford University Press. (Republished in 2002 by Lawrence Erlbaum Associates under same title)

Gould, S. J. (1977). *Ontogeny and phylogeny.* Cambridge, MA: Harvard University Press.

Haeckel, E. (1866). *Generelle morphologie der organism allgemeine Grundzuge der organischen Formen-Wissenschaft mechanisch begrundet durch die von Charles Darwin reformirte Descendenz-Theorie.* Berlin: Georg Reimer.

Hamburger, V. (1998). Embryology and the modern synthesis of evolutionary theory. In E. Mayr & W. Provine (Eds.), *Evolutionary synthesis* (pp. 97–110). Cambridge, MA: Harvard University Press.

Horvasse, R. (1943). *De l'adaptation a l'evolution par la selection.* Paris: Herman et Cie.

Huxley, J. (1942). *Evolution: The modern synthesis.* London: Allen & Unwin.

Huxley, J., & de Beer, G. (1934). *The elements of experimental embryology.* Cambridge, England: Cambridge University Press.

Jablonka, E. (2001). The systems of inheritance. In S. Oyama, P. Griffith, & R. Gray (Eds.), *Cycles of contingency: Developmental systems and evolution* (pp. 99–116). Cambridge, MA: MIT Press.

Jablonka, E., & Lamb, M. (1995). *Epigenetic inheritance and evolution: The Lamarckian dimension.* Oxford, England: Oxford University Press.

Jacob, F., & Monod, J. (1961). Genetical regulatory mechanisms in the synthesis of proteins. *Journal of Molecular Biology, 3,* 318–356.

Jenuwein, T., & David, A. C. (2001). Translating the histone code. *Science, 293,* 1074–1080.

Judson, H. (1996). *The eighth day of creation: Makers of the revolution in biology.* Plainview, NY: Cold Spring Harbor Press.

Kalthoff, K. (2001). *Analysis of biological development.* Boston, MA: McGraw-Hill.

Kaufman, T. C., & Raff, R. A. (1984). *Embryos, genes, and evolution.* Bloomington, IN: Indiana University Press.

Keller, E. F. (1995). *Refiguring life.* New York: Columbia University Press.

Kimura, M. (1979). The neutral theory of molecular evolution. *Scientific American, 24*(5), 98–130.

Kohler, R. E. (1994). *Lords of the fly.* Chicago: University of Chicago Press.

Lamarck, J.-B. (1809/1984). *Philosophical zoology.* Chicago: University of Chicago Press.

Lewis, E. B. (1995). The bithorax complex: The first fifty years. In N. Ringertz (Ed.), *Nobel lectures in physiology or medicine 1991–1995* (pp. 247–274). Singapore: World Scientific.

Lewontin, R. (2001). Genes, organism and environment. In S. Oyama, P. Griffith, & R. Gray (Eds.), *Cycles of consistency: Developmental systems and evolution* (pp. 58–66). Cambridge, MA: MIT Press.

Li, W.-H., & Grauer, D. (1991). *Fundamentals of molecular evolution.* Sunderland, MA: Sinauer.

Mayr, E. (1998). Prologue: Some thoughts on the history of the evolutionary synthesis. In E. Mayr & W. Provine (Eds.), *The evolutionary synthesis* (pp. 1–48). Cambridge, MA: Harvard University Press.

Mayr, E., & Provine, W. (Eds.). (1998). *The evolutionary synthesis.* Cambridge, MA: Harvard University Press.

McClean, P. (1997). Eurkaryotic chromosome structure (2002). Available on the World Wide Web.

McClintock, B. (1951). Chromosome organization and genic expression. *Cold Spring Harbor Symposium on Quantative Biology, 16,* 13–47.

McKinney, M., & McNamara, K. (1991). *Heterochrony: The evolution of ontogeny.* New York: Plenum.

McLaren, A. (1999). Too late for the midwife toad: Stress, variability and Hsp90. *Trends in Genetics, 15*(5), 169–171.

McNamara, K. (1997). *The shapes of time.* Baltimore: Johns Hopkins University Press.

Mendel, G. (1966). On Hieracium-hybrids obtained by artificial fertilization. In C. Stern & E. Sherwood (Eds.), *The origins of genetics: A Mendel source book.* San Francisco: W. H. Freeman.

Messerly, J. G. (1996). *Piaget's conception of evolution: Beyond Lamarck and Darwin.* Lanham, MD: Rowan & Littlefield.

Morgan, T. H. (1917). The theory of the gene. *American Naturalist, 51,* 517–520.

Morgan, T. H. (1926). *The theory of the gene.* New Haven, CT: Yale University Press.

Nanney, D. L. (1958). Epigenetic control systems. *Proceedings of the National Academy of Sciences, USA, 44,* 712–717.

Ohno, S. (1970). *Evolution by gene duplication.* Berlin: Springer.

Olby, R. (1966). *Origins of Mendelism*. New York: Schocken Press.

Parker, S. T., Langer, J., & McKinney, M. (Eds.). (2000). *Biology, brains, and behavior: The evolution of human development*. Santa Fe, NM: School of American Research Press.

Penisi, E. (2001). Behind the scenes of gene expression. *Science, 293*, 1064–1067.

Piaget, J. (1952). Jean Piaget. In R. M. Yerkes (Ed.), *History of psychology in autobiography* (Vol. 4, pp. 237–256). New York: Russell & Russell.

Piaget, J. (1952). *The origins of intelligence in children*. New York: International Universities Press.

Piaget, J. (1971). *Biology and knowledge: An essay on the relations between organic regulations and cognitive processes*. Chicago: University of Chicago Press.

Piaget, J. (1978). *Behavior and evolution*. New York: Pantheon Books.

Piaget, J. (1980). *Adaptation and intelligence, organic selection and phenocopy*. Chicago: University of Chicago Press.

Piaget, J., & Inhelder, B. (1971). *Mental imagery in the child: A study of the development of mental representation*. New York: Basic Books.

Piatelli-Palmarini, M. (Ed.). (1980). *Language and learning: The debate between Jean Piaget and Noam Chomsky*. Cambridge, MA: Harvard University Press.

Richards, R. J. (1987). *Darwin and the emergence of evolutionary theories of mind and behavior*. Chicago: University of Chicago Press.

Rutherford, S. L., & Henikoff, S. (2003). Quantitative epigenetics. *Nature Genetics, 33*, 6–8.

Rutherford, S. L., & Linquist, L. (1998). Hsp90 as a capacitor for morphological evolution. *Nature, 396*, 336–342.

Schmalhausen, I. I. (1986). *Factors of evolution: The theory of stabilizing selection*. Chicago: University of Chicago Press.

Shapiro, J. A. (1969). Mutations caused by the insertion of genetic material into the galactose operon of *Escherichera coli*. *Journal of Molecular Biology, 40*, 93–105.

Shapiro, J. A. (1983). Variation as a genetic engineering process. In D. S. Bendall (Ed.), *Evolution from molecules to men* (pp. 253–270). Cambridge, England: Cambridge University Press.

Stewart, R. (2002). A few words about DNA and chromatin, Pacific Northwest National Laboratory (2002). Available on the World Wide Web.

Synder, M., & Gerstein, M. (2003). Defining genes in the genomics era. *Science, 300*, 258–260.

Vidal, F. (1994). *Piaget before Piaget*. Cambridge, MA: Harvard University Press.

Waddington, C. (1942). Canalization of development and inheritance of acquired characteristics. *Nature, 150*, 563–565.

Waddington, C. (1957). *The strategy of the genes*. Bristol, England: George Allen & Unwin, Ltd.

Waddington, C. H. (1962). *New patterns in genetics and development.* New York: Columbia University Press.

Waddington, C. H. (1975). *The evolution of an evolutionist.* London: W. & J. Macay Limited.

Wake, D. (1986). Foreword to I. I. Schmalhausen, *Factors in evolution* (pp. v–xii). Chicago: University of Chicago Press.

Watson, J. B., & Crick, F. (1953). Genetic implications of the structure of deoxyribonucleic acid. *Nature, 171,* 964–967.

Weindling, P. (1985). Darwinism in Germany. In D. Kohn (Ed.), *The Darwinian heritage* (pp. 685–698). Princeton: Princeton University Press.

Weismann, A. (1892). *Das Keimplasma: Eine Theorie der Verebung.* Jenna, Germany: Gustav Fischer.

West-Eberhard, M. J. (2003). *Developmental plasticity and evolution.* New York: Oxford University Press.

Winther, R. (2000). Darwin on variation and heredity. *Journal of the History of Biology, 33,* 425–455.

Winther, R. (2001). August Weismann on germ-plasm variation. *Journal of the History of Biology, 34,* 517–555.

3

Beyond Piaget's Phenocopy: The Baby in the Lamarckian Bath

Terrence W. Deacon
University of California, Berkeley

> *I think that all structures are constructed and that the fundamental fea-*
> *ture is the course of this construction: Nothing is given at the start, ex-*
> *cept some limiting points on which all the rest is based. The structures are*
> *neither given in advance in the human mind nor in the external world, as*
> *we perceive or organize it.*
>
> —Jean Piaget (1977, p. 63)

In the beginning of his career, Jean Piaget was drawn to evolutionary biology. Although he was to subsequently turn most of his attention to
mental development, even when focused on childhood cognition, hints of
a broader evolutionary perspective seem implicit in his thoroughgoing
commitment to constructivism. This underlying metatheoretical perspective ultimately reexpressed itself in three probing and influential books
written toward the end of his career, the most substantial of which, *Biology and Knowledge* (*B&K*; Piaget, 1971) articulated a unified theory for
the construction of biological and cognitive information and outlined his
criticism of neo-Darwinism. The most recent summary of these ideas was
presented in his short, *Behavior and Evolution*, written in 1978. Characteristic of his eclectic and uncompromising intellectual enterprise, Piaget
sought to integrate his theoretical approach to cognitive development
with themes drawn from mathematical philosophy and evolutionary biology to the extent that these could be made compatible with his construc-

tivist vision. This remains an unfinished synthesis. Piaget's commitment to a unified "genetic epistemology" led him to struggle with dissonances between his own theoretical reconception of mental development and neo-Darwinian theories of evolution. The goal of this chapter is to examine the basis of this dissonance, and to ask whether we are now closer to a synthesis of evolutionary and developmental biology of the sort Piaget had envisioned. In this light, I consider to what extent neo-Darwinian theory, Piaget's developmental theory, and the dream of a unified theory need to be reconsidered.

Piaget's early research on the intergenerational transmission of acquired shell characters in freshwater mollusks (see summary by Parker, chap. 2, this volume) signaled an early dissatisfaction with standard neo-Darwinian evolutionary theory of the mid 20th century. His critical views reflected a decidedly Lamarckian perspective. At the very least, he was thoroughly dissatisfied with the simplicity of standard Darwinian models. Toward the end of his career, his interest in evolutionary issues was rekindled by his belief that a middle path between the Darwinian and Lamarckian alternatives could be articulated. This vision was informed by his model of the development of knowledge, by his studies of mathematical logic, systems thinking, and cybernetic theory, and by his reflections on the embryological research of his contemporary, Conrad Waddington.

The books written at this time, addressing the relation between cognition and evolution, show clearly that he was motivated by a belief in the possibility of formulating a single, overarching theory of the construction of knowledge in its most general sense. And he was convinced that the way forward would necessarily involve a synthesis of evolutionary theory with these infant sciences of complexity.

This quest for a unified theory appears to have been catalyzed by his encounter with C. H. Waddington's conception of the complex layering of epigenetic processes and how this might facilitate what Waddington had called "genetic assimilation" (Waddington, 1953). Waddington's evolutionary heresy was the suggestion that the evolutionary process could not be understood merely in terms of genes, traits, and populations undergoing natural selection. Instead, he posited that epigenesis itself played a crucial role, and could mediate the expression of traits in a way that mimicked Lamarckian evolution. In this respect, it was seen both by Waddington and many subsequent interpreters as related to (though not identical to) the concept of *organic selection*, as independently described by James Mark Baldwin (1896), Conway Lloyd Morgan (1896), and Henry Osborne (1896; see historical discussions in Richards, 1987). All had suggested means by which phenotypic flexibility—especially learned behaviors—might influence the future patterns of natural selection acting on a lineage, in a directional manner biased by the specific responses of the organism.

As I show, the similarities between Waddington's theory and these early predecessors are mostly superficial. Moreover, I argue that each include nearly fatal flaws, at least with respect to their supposed mode of action. Nevertheless, combining these insights and accounting for these theoretical difficulties will still show them to be complementary and conducive to including a significant role for organism agency in evolution, though not in quite the way originally conceived (nor as Piaget had presumed).

The notion that the developmental mediation of the expression of genes might be important for evolution was the key aspect of these theories that was resonant with Piaget's long mistrust of the completeness of orthodox neo-Darwinian theory and his belief that regulative processes in the development of cognition must also contribute to the evolution of mental abilities in phylogenesis. I think one can salvage the intent of Piaget's critique and show that a denial or at least an ignoring of the role of organism agency in evolution was, as he argued, overly restrictive, even though his attempted reformulation of evolutionary logic was flawed. His belief that the logic of cybernetic regulation could ultimately supplant, or at least augment, selectionist logic in evolutionary theory also contributed to his failure to abandon a crypto-Lamarckian paradigm. Nevertheless, his efforts to resolve the dissonance between neo-Darwinism and his constructivist view of knowledge development by way of appealing to systems theory and cybernetic theory foreshadowed current "evo-devo" (evolution mediated by development) approaches to evolution. In this sense, Piaget can be seen as recognizing the critically incomplete aspects of the neo-Darwinism of his time, even if he could not articulate an adequate alternative.

In what follows, I pay special attention to Piaget's focus on the concept of *phenocopy* (a term borrowed from Waddington with a slightly varied interpretation) as the bridge to his new approach, discussing both its implicit failings and how it points obliquely to a better alternative. The implicit Lamarckian logic of Piaget's conception of phenocopy is compared with Waddington's intermediate view, and both are compared to contemporary approaches to similar problems. Finally, I discuss the ultimate implications of the failure of Piaget's phenocopy-evolution theory and his conception of a unified constructivism applicable to all biological knowledge. Specifically, I argue that abandonment of Piaget's version of pheno-copy theory leaves two possible resolutions: Either we need to give up Piaget's dream of a unified constructivist mechanism in evolution and de-velopment, or alternatively sacrifice his conception of the cybernetic mechanisms underlying cognitive development in order to achieve it.

IS EVOLUTION CYBERNETIC?

> *Life is essentially autoregulation. The explanation of evolutionary mecha-nisms, for so long shackled to the inescapable alternatives offered by Lamarckism and classical neo-Darwinism, seems set in the direction of a third solution, which is cybernetic, and is, in effect, biased toward the theory of autoregulation.*
>
> —Piaget (1971, p. 26)

In the 1950s a newborn, cross-disciplinary approach to the study of com-plicated systems was taking its first steps. Variously developed in subdisciplines labeled *information theory* (e.g., Shannon & Weaver, 1963), *cybernetics* (e.g., Wiener, 1965), and *general systems theory* (e.g., Von Bertalanfy, 1968), this eclectic set of approaches began to suggest new ways of understanding biology and cognition. It was seen by some as a way of merely augmenting existing ideas about evolution and mental processes, but a growing number saw it instead as heralding an alterna-tive paradigm that might significantly revise prior conceptions of biologi-cal intelligence in all its forms. Piaget was among the latter group. In many ways he was predisposed to consider these new tools as compatible with his own critical assessment of the evolutionary and cognitive theo-ries of the time. This was reinforced by analogies he saw between pro-cesses of logico–mathematical knowledge construction and processes of cognitive development. Thus a mathematical reframing of control, sys-temic complexity, and goal-directed behavior seemed a natural extension of this approach. This was further supported by the central role played by circular causality and feedback mechanisms, both in Piaget's conception of knowledge construction and in cybernetic control theory.

To Piaget, the logic of recursive interaction with the environment promised a way to transcend the implicit exterior/interior dichotomy in the dominant theories of the origins of biological function and knowl-edge. Although the debates of the mid-20th century could no longer be strictly identified with classic empiricist/rationalist debates, what lingered was an opposition between those committed to what today would be called *eliminativism* and those committed to preserving a role for irreducible teleology or self-regulation. In evolutionary biology, the success of orthodox neo-Darwinism, bolstered by the "new synthesis" between population genetics and vertebrate paleontology, reinforced the view that functional adaptation can be achieved by entirely *a posteriori* selection processes. Strict Darwinian interpretations of evolution eliminate any role for purposive end-directed explanations, including any form of environmental "instruction" or active accommodation to environmental challenge. Evolution is thus reduced to antecedent chance mutation honed by *a posteriori* competitive elimination; or blind variation and selective retention (Campbell, 1960).

To the extent that Piaget envisioned a continuity between biological adaptations and acquired knowledge, this conception of evolution excluded what to him was the essential role of active construction. This likely motivated his attraction to aspects of Lamarckian theories of evolution. Although popular writing on evolutionary theory often caricatures the distinguishing (and most easily criticized) feature of Lamarck's theory to be the inheritance of acquired characters, this is an oversimplification. What ultimately distinguishes Lamarck from Darwin is not this genetic fallacy but the role that the purposeful striving and adaptive agency of the organism plays in evolution. Phylogenetic evolution was conceived by Lamarck to be a *consequence* of animal adaptation, and not the other way around, as was later argued by Darwin. This active role of facultative adaptation as an organizing factor in evolution was more consonant with Piaget's rejection of passive "copy" theories of knowledge (implicit in the empiricist tradition), and his view that Darwinian selection was also conceived as a passively imposed consequence. Piaget rejected the implicit vitalism and the simple inheritance of acquired characters inherent in Lamarck's theory but could not accept the passive reactive vision suggested by strict Darwinism. His search for a middle road between Darwin and Lamarck thus reflects his view that neither a strictly subjective (i.e., innate) nor a strictly environmental (i.e., perceptual) basis for knowledge could suffice (Piaget, 1971, p. 27). Both the evolution of biological adaptation and the construction of knowledge must, he thought, be explained as the progressive schematization of action (Piaget, 1971, pp. 6–7).

Piaget's dissatisfaction with *a posteriori* selection mechanisms can be usefully exemplified by his preference for instructional theories over selection theories of acquired immunity. Even while admitting that a mechanism of "selection within the genetic information already established" was the prevailing view of this system (now well-established), he opined that such a mechanism "must be more like learning by trial and error than an all or nothing process" (Piaget, 1971, p. 189). He could not accept that a noninteractive, nonreactive, blind variation and selection mechanism could be adequately flexible to serve in this regulative capacity.

This conviction was also the motivation for his rejection of the then new computational structuralist theories of language acquisition. Despite his rejection of behaviorist eliminativism and his attraction to mathematical analogies, he could not agree with the alternative approach suggested by nativist linguistics, such as that of Noam Chomsky. Though both were harsh critics of behaviorism, Piaget saw Chomsky's nativism as equally incompatible with his own views on development. The clash of these perspectives is well exemplified in Piatelli-Palmarini (1980), in which these two protagonists spell out their respective differences. For Piaget, Chomsky's nativism was equally passive and *ad hoc*. It made no more sense to him to posit the existence of innate knowledge than to posit that knowledge was of entirely environmental origin. For Piaget, knowledge was intrinsically relational and required co-construction via the interaction of organisms with their environment. In contrast, both behaviorism and Chomsky's new computation-inspired structuralism were based on passive conceptions of the role of the organism in the development of knowledge. Although Piaget is considered to be an icon of classic structuralism (even as he himself described his views), there is an important sense in which his approach depended on an implicit process metaphysics—an interactionism, where knowledge creation is a consequence of an organism actively establishing dynamic relationships, and in which knowledge is neither a structural feature of the world nor of the organism.

Piaget's views might today be understood as implicitly emergentist (in the modern sense; see Deacon, 2003a). His conception of both cognitive and evolutionary change were apparently inspired by classic Kantian and Hegelian dialectic approaches to the historical construction of knowledge, but he translated the classic thesis–antithesis–synthesis paradigm of conceptual development into a more physiologically realistic diachronic theory of active regulation. In Piaget's reformulation of this classic logic, the dialogue between organism and environment is one of physical prediction and concrete feedback as the organism acts and readjusts to what its actions accomplish or fail to accomplish with respect to the mental models from which the actions were generated. This is described in terms of an interactive circular dynamic of repeated assimilations and accommodations, as the organism's mental models of the world—in the form of action-perception schemas—are updated and retested. The goal is a kind of informational equilibration, a balance or matching achieved between one's mental models and the world they depict.

In this terminology, there is an implicit transference between teleological and energetic analogies. This implicit bivalence of the terms used to describe the developmental process reflects a commitment to emphasizing the dynamical and physical nature of cognition and to deny an energy/information dichotomy. Though wedded to a dialectic model of knowledge construction, implicit in neo-Kantian theory, Piaget saw in it a parallel to cybernetic regulative theories of signal processing and control that could be applied to biological processes. He thus appeals to both sides of this analogy in his effort to build a general theory of biological "knowledge" construction.

On one hand, Piaget's notion of "equilibration" epitomizes the role of energetic analogies in this theoretical juxtaposition. It captures both the connotation of a balancing of forces or an achievement of energetic stability and a sense of matching or fitting. It is the point before which there is a tendency for change, and after which there is a tendency to resist change and re-establish stability despite perturbation. On the other hand, his notions of assimilation and accommodation are essentially biological regulatory terms, with the implication of goal directed adaptive organization. And finally, the term, *reflective abstraction*, referring to the process by which prior constellations of knowledge schemes become consolidated into a higher order system (constituting a shift to a higher stage of cognitive processing) is an entirely cognitive notion with some relationship to the formal notion of logical types.

This juxtaposition of energetic, organismic, and logico–mathematical concepts reflects Piaget's somewhat ambiguous notion of agency. For Piaget, energetic analogies of disequilibrium and equilibration provide a sense of spontaneous dynamical tendency, while adaptive concepts like *assimilation* and *accommodation* provide a cybernetic goal-directedness. Describing agency in self-sufficient terms is required by Piaget's view that evolution is ultimately driven by the activity of the organism. The use of energetic analogies to describe adaptive relationships has a long history in both psychology and evolutionary theory. In their evolutionary theories, both Lamarck and Spencer had also invoked the concept of *equilibration* as a surrogate for organismic purposiveness (see Richards, 1987, for a discussion of the history of this metaphor). In these theories, as in Piaget's developmental theory, a natural tendency toward equilibration was conceived as the principal source of change. In contrast, Darwin took pains to avoid any such assumption of directed tendency, and appealed only to the essentially external contributions of "accident" and chance variation to account for the origins of evolutionary change. Even prior to conceiving the final details of natural selection, Darwin was deeply critical of Lamarck's appeal to a vital principle behind organisms' incessant "striving" in pursuit of food, comfort, and mates (Richards, 1987). To the extent that this striving ultimately determined evolutionary directionality toward improved adaptation, he felt that it reduced the explanation to a mystical principle.

Piaget's use of cybernetic theory provided a more sophisticated view of equilibration and adaptation that retained the Lamarckian role of adaptation while avoiding any hint of vitalism and intrinsic aim of evolution (also a goal of Lamarck). His conception of regulation can be partially analogized to simple servo-mechanism feedback function (though he diverges from this analogy as well). A homeostatic device, such as a thermostat, requires a flow of energy to carry the signals within its control circuit. Additionally it is coupled to some other energetic system (e.g., a heater coupled to room temperature), whose state it controls with respect to some reference value. Regulation is achieved by the "structure" of the circuit, which gates the flow of energy in the coupled physical systems in such a way that it opposes any deviation from this value. This is quite different from a system tending toward energetic equilibrium, in which the energy of a system or the distribution of its states exhibits a tendency toward minimal energy and homogeneous distribution. By analogy, Piaget uses "equilibration" to refer to this tendency of a cybernetic system (e.g., a thermostat) to produce compensatory actions that bring a perturbed system back to a "ground state," at which compensatory signals and actions (e.g., turning on heating/cooling devices) are not activated. *Equilibration*, in both senses, is the tendency of the system to develop toward a state where it will settle if not further perturbed. Piaget realized, however, that this analogy was overly simple and so he appealed to Waddington's notion of *homeorhesis* (i.e., homeostasis with respect to a systematically shifting set point) and *chreod* (i.e., the "trajectory" that system states follow in this process) to augment the simple energetic analogy (Piaget, 1971, pp. 18–21, 23–25). This was no mere tweak of an otherwise complete model, however, because progressive change of the whole dynamic of the equilibration process was perhaps its most important characteristic.

For Spencer, like Piaget, both life and mind could be conceived in the same terms, as the active adjustment of internal "organic relations" to external relations in the environment (Spencer, 1872, pp. 435–436, 486). For both it may be argued that their commitment to a Kantian conception of knowledge as intrinsically relational (between agent and environment) rendered a nondynamic nonteleological evolutionary theory incomplete. But where Spencer doggedly adhered to "the inheritance of functionally-wrought modifications" as the dominant factor in evolution to explain the source of this intrinsic developmental teleology (Spencer, quoted in Richards, 1987, p. 294), Piaget, writing with the benefit of another 75 years of debate on the issue, was more careful.

Piaget rejected Lamarckian theory on two counts: the lack of a mechanism for the simple inheritance of acquired characters and Lamarck's belief in the exogenous origins of the organization of animal behavior (via the progressive internalization of external conditions). By appealing to a cybernetic interpretation of both development and learning, he believed he could find a middle ground that neither begs the question of the origins of organism teleology nor reduces to a passive *a posteriori* conception of evolutionary adaptation and functional organization.

Trading on this ambiguity between energetic and informational conceptions of equilibration, Piaget derived a number of crucial concepts for his constructivist synthesis of evolution and learning. The organism's spontaneous actions are thus determined with respect to an internal structure that can be said to be "about" the world it acts on, yet conceived merely as structural organization as can the setting of a regulator. Knowledge is in this way conceived both structurally and dynamically, as a kind of schematic action or schema. The causal regularities of the external world become knowledge only to the extent they enter into some schema and only insofar as they lead to modification of its structure. Thus only a miniscule window on the world is opened, and this is determined by what the organism *does*. Here again, Piaget offers a more complex dynamical conception of regulation. Mental regulation, unlike the regulation of physiological parameters, is not for the production of homeostasis, but rather it operates at a higher level, by regulating the fit of these schemas to the world. Acquired knowledge is thus modification of a behavioral schema in a way that better predicts the consequences it produces. It is not merely a passive impression imposed on some mental substrate by sense data from without. Which aspects of things get assimilated to some mental structure depends on the extent to which they force accommodation of some action schema, and how that schema is changed, thus assimilating this mismatch. In this way, the organism is the engine of its own cognitive development.

His belief in the primacy of an internally-generated dynamic forced Piaget to reject both Lamarckian and Darwinian theories for the same reasons he rejected empiricism and behaviorism: They conceived of adaptive structure as passively acquired and structured only from without. But his focus on a central role for regulation led him to continually return to refrains found in both Lamarck and Spencer, whose evolutionary theories were also built around the centrality of an intrinsic tendency toward improvement of adaptation, and an explicit analogy between learning and biological evolution. Both evolutionists assumed that the directed activity by which the organism accommodates itself to its environment, whether via structural modification or via learning, is continuous with the means by which its lineage manages to explore the realm of possible adaptations to the environment and by which it perfects these over the generations. For Lamarck and Spencer this directed adaptation provided a degree of pretesting of the raw materials subject to selection, but this required that such "incarnated habits" themselves become heritable over time.

Where Darwin diverged from these other evolutionists was not in the addition of something more, but ultimately a simplification. Darwin argued that, although not implausible (as far as was known at the time), the inheritance of acquired traits was an unnecessary complication. Spontaneous variation of heritable organism traits, irrespective of any physiological fine-tuning, coupled with the selective reproduction and elimination of certain alternative variants of these would be enough, given sufficient time and variety of individuals. But his dissatisfaction with prior approaches went deeper than this. Darwin was critical of what he interpreted as Lamarck's invocation of vital energy as a driver of evolution because it took as given some implicit antecedent teleology. Despite Lamarck's expressed intent to provide an entirely mechanistic account of evolution, Darwin felt that this left the door open for mysterious nonphysical influences. The power of Darwin's vision was that it seemed to provide a more rigorous mechanistic account, in which the appearance of intelligent "striving" *itself* could be explained as a consequence of evolution and was not required *a priori*.

It was August Weismann (1892), however, who finally closed the door on Lamarck and Spencer, by providing evidence that the germ line of an animal lineage was ultimately sequestered and insulated from other cell lines in the body. As a result, the substrate of inheritance was unavailable to influences acquired by other somatic alterations during a lifetime. This soon became the orthodox view and was bolstered by the subsequent rediscovery of genetic inheritance and the effects of mutational change. The ultimate origin of evolutionary novelty, in this orthodox view, could only be attributed to sources of variation that were "blind" of any functional consequences, because they never could become directly involved in any physiological adaptation process. All appearance of "final causality" (e.g., anticipatory, end-directed functionality) had to be explained in terms of *a posteriori* coincidental utility born of this blind variation.

This more restrictive orthodoxy was what Piaget was unwilling to adopt, precisely because it denied any active role for the organism, and so was inconsistent with his constructivist vision (as was the parallel orthodoxy of strict behaviorism). Though wary of Lamarckism for its implicit empiricism (as well as for its questionable assumptions about inheritance), Piaget also had to reject orthodox Darwinism because it too conceived of functional adaptation as passively imposed exogenously. They were two sides of the same coin with merely a difference in mechanism. If constructivism was right, there must be a middle evolutionary mechanism that was more consistent with an organism's active engagement with the world playing a constitutive role in evolution.

This is why cybernetic theory seemed to offer a more congenial interpretation of the apparent teleology of evolution than did Darwinism. A cybernetic account of goal-directedness in mental development could exhibit the structure of goal-directedness without requiring external "intelligence" or any vital impetus to animate it. A cybernetic system was intrinsically dynamic and regulatory, and its apparently teleological behavior was derived from the circular causality of regulatory action, not from some antecedently designed purpose. The power of Darwin's theory was that it offered a nonteleological account of good biological "design." The power of cybernetic theory was that it appeared to do the same for regulative processes. Piaget felt that some combination of the cybernetic approach and the Darwinian approach might offer a way to reintegrate accounts of accommodation and assimilation of adaptive organization in both the biological and the mental realm.

Teleonomy was the term often used for cybernetic analogue to goal-directed behavior, implying that it was only nominally purposive. The apparent goal-directed behavior of a regulatory device merely reflects the circular logic of its energetic organization. So this structural notion of a goal-directed mechanism occupied a fruitful middle ground. Piaget was clearly attracted by the way it both preserved intrinsic dynamism and goal-orientation without invoking preformed design. It suggested a possible approach to a general theory of autonomously constructed biological information that might be applied equally well to evolution and to mental development. It promised to reserve a role for organism agency, embodied in the logic of the *structure* of a cybernetic circuit, and yet without the baggage of either acquired characters or implicit teleology.

Cybernetic models had another attraction for Piaget; they seemed applicable to the majority of physiological functions at all levels of biological organization. Besides the behavioral adaptations to the environment that were the focus of Piaget's psychology, the regulation of the internal milieu (e.g., via hormones) could also be described in terms of dynamically-changing homeostatic relations (i.e., "homeorhetic" processes). Piaget also interpreted many aspects of embryonic development in these terms as well. For him "the construction of a structure is inseparable from its regulation" (Piaget, 1971, p. 205). By invoking an ambiguity between epigenesis and regulation, Piaget could imagine a way that regulatory interaction with the environment might induce disequilibrium in the "internal environment" of the developing organism that would provide a kind of internal selection pressure and what he describes as "a sort of endo-adaptation" (Piaget, 1978, p. 21). For example, he argues that the genome's reaction to internal environmental disequilibrium is to "try out variations," and he describes the epigenetic consequences of this as "trials," that may then become subject to selection (Piaget, 1978, p. 80). So in Piaget's view, *a posteriori* selection plays only the role of adopting or rejecting what a regulatory process has already shaped.

Piaget's conception of the evolution–epigenesis relationship deviates from cybernetic models in one important sense. His theory is predicated on the additional assumption that complex cybernetic systems will also intrinsically exhibit a kind of hierarchically-constructed development as well as a regulatory function. This dynamic is presumed to be the result of the way higher order regulatory systems can develop on the foundation of lower order regulatory systems. This might be caricatured as blending aspects of Hegel's logic of dialectic idea construction with Norbert Wiener's (the mathematician who coined the term, *cybernetics*) logic of interactive error control. Piaget assumed that the interactions generated by intrinsic regulatory dynamics interacting with extrinsic environmental perturbations and constraints would follow a kind of dialectic logic, propelling the system to complexify and differentiate itself. Although it is clear that organisms differentiate and complexify as they develop, there are a variety of reasons to doubt that this is intrinsic to cybernetic regulation. But setting this hierarchical issue aside for the moment, we must first ask whether this is really an intermediate paradigm.

In many ways, the cybernetic paradigm is far more compatible with Lamarckian views than Darwinian views. The notion of physiological regulation may have been only vaguely grasped by Lamarck, yet it was the implicit prime mover in his theory. Animals were described as driven to improve their adaptation to the environment by the sensations of comfort and discomfort, pleasure and pain, which provided feedback concerning the degree of mismatch between their internal milieu and external conditions. Lamarck was also aware of the ability of organs, such as muscles and bones, to respond physiologically to habitual use and disuse by hypertrophy or atrophy, respectively. This active feedback could change the patterns of energy distribution that supported these structures with respect to their need. At a time when the logic of genetic inheritance was still a deep mystery, it was reasonable to further imagine that what became expressed in the structure of the organism might also be passed through physical reproduction to offspring. By such a mechanism, then, the goal-directed dynamic of physiology could play a constitutive role in shaping the adaptations of future generations. Habit formation served as a sort of evolutionary guidance mechanism.

But Piaget was acutely aware that a simple Lamarckian inheritance of acquired traits was theoretically questionable, unnecessary according to Darwinism, and unsupported by then-known mechanisms of animal inheritance (though some examples of nongenomic cytoplasmic inheritance mechanisms are now known in animals, e.g., see review by Jablonka & Lamb, 1995, these are generally considered far too limited in effect and in intergenerational preservation to serve any significant role in large-scale phyletic evolutionary trends). Is there some means by which cybernetic regulation might accomplish this Lamarckian function indirectly? Could there, for example, be a regulatory coupling between physiological plasticity and gene expression in the germ line, without a direct downward causal link? It was Piaget's belief that there must be.

THE PHENOCOPY CONCEPT
(FROM WADDINGTON TO PIAGET)

Waddington's notion of *phenocopy* appeared able to serve in this role. For Waddington, this term referred to a developmental phenomenon he hoped to account for by an elaboration of neo-Darwinian theory. According to Waddington, a phenocopy is a nongenetic, that is, epigenetic, simulacrum in one organism of what could be produced in another by gene expression. By virtue of the responsivity of organisms to environmental conditions, structural modifications can be induced in some individuals that are innate in others. For Piaget, this notion appeared to serve as a marker for something more; an intermediate mode of inheritance that might be able to bridge between generations carrying developmental adaptive information, even though not a part of the germ line itself. If such a mechanism existed, it might serve to influence germ-line information indirectly by way of development. Epigenesis could be influenced by extragenomic factors, and thus in some as yet unknown way, might pave the way for the evolution of progressively more gene-based surrogates. In other words, if the epigenetic process itself were an accumulator of structuring information, in quasiindependence of the genome, and if this too was indirectly inherited by virtue of continuance of these extragenomic influences, then this intermediate mode of inheritance would not require a further assimilation into the genome to be inherited.

Waddington's experimental results had suggested that something very close to this was indeed possible. In his experiments breeding fruit flies under abnormal conditions (e.g., in the presence of ether, high salinity, heat, and so on; Waddington, 1953, and reviewed in 1957 and 1962) he found that he could induce the developing flies to express abnormal body forms. Some of these were the morphological equivalents of traits exhibited by known genetic mutants in apparently normal strains, but that were not normal spontaneously occurring population variants. He labeled these features *phenocopies*, implying that they were parallel developmental outcomes produced by alternate epigenetic means. Key to Waddington's interpretation of what constituted a phenotype was the belief that all complex traits were produced by systemic interactions between a vast number of genes whose combined effects ultimately contributed to an epigenetic interaction that was ultimately responsible for the expressed phenotype. This suggested the possibility that there could be multiple epigenetic ways to produce the same expressed phenotype.

In experiments where he selectively bred individual flies expressing the same phenocopy produced under these unusual conditions, he found that he could eventually produce lineages in which these traits were ineluctably produced, irrespective of environment. This result had the superficial appearance of Lamarckian inheritance. Waddington, incautiously dubbed this transition, *genetic assimilation*, implicitly suggesting a Lamarckian-like transfer from phenotype to genotype. In fact, however, he did not envision something this straightforward, and it turns out not to be. Waddington's notion of the genotype–phenotype relationship was intrinsically systemic. Gene expression is inevitably mediated by a complex epigenetic process, in which genes contribute to combinatorial interactions whose outcome is more a function of the interaction dynamics than any specific gene's product. In Waddington's view, diverse physiological adjustments to environmental factors could play constitutive roles in the shaping of a phenotype, and diverse genetic backgrounds could likewise be biased to produce similar phenotypes. Different combinations of genetic background and epigenetic responses to environmental factors could thus lead to many ways of achieving a similar phenotypic end. He therefore did not assume that what he had demonstrated was produced by a Lamarckian mechanism. He described it only in terms of a kind of epigenetic convergence (which he called *canalization* on the analogy of a deep channel within an "epigenetic landscape"). If phenotypes were the results of multiple converging genomic and extragenomic influences, there might be multiple ways of increasing the probability of generating any given phenotype over the course of lineage evolution, including trading extragenomic for genomic influences.

Piaget followed this logic and often talks about this epigenetic pattern for producing a phenotype as a "matrix" into which different genetic and extragenetic factors can be exchanged and substituted, producing the same outcome. But in this regard, the concept appears to take on a more structural sense for Piaget, taking on the attributes of a more concrete sort of phenocopy. Whereas Waddington was cautious about making a leap to the inheritance of acquired characters, this reformulation of the phenocopy concept led Piaget to be less so. Although he too stopped short of invoking Lamarckian inheritance as the likely mechanism, he seems to have envisioned that phenocopies in his sense were subject to an intermediate mode of inheritance. For him, the possibility of a Lamarckian-like consequence was sufficient. As Piaget (1971) remarked: "to accept the second basic fact of the Lamarckian position [i.e., that acquired characters can become fixed in the lineage] does not involve acceptance of the Lamarckian explanation of it" (p. 108). But this meant he needed another way to conceive of the logic of inheritance of phenocopies.

It is now well accepted that a non-Lamarckian mechanism is involved in the phenomena that Waddington experimentally demonstrated. In an important recent reanalysis of one of Waddington's experimental models—the production of anomalous transverse wing veins in certain flies raised in a heated environment—it has been shown that the apparent genetic assimilation to uninduced expression of this trait is the result of the convergence of multiple independent "risk factors" for the expression of this trait. This phenocopy is due to some rare but naturally occurring variants of a class of "chaperone" proteins called *heat-shock proteins* (Rutherford & Lindquist, 1998). Individuals that bear one of these variants of heat-shock proteins will express the phenocopy (transverse wing vein) if exposed to heat during development, but not in normal temperatures. But there are many different and independently assorting variants of heat-shock protein that can produce the same effect. Breeding different individuals, each carrying one of these variants, ultimately inbreeds for these factors. Generation after generation, there will be an increased chance that both parents will pass a variant of one of these to their offspring. The result is that selective breeding among individuals expressing this heat-induced phenocopy will tend to co-assort these contributing genetic factors. Eventually, these variants will become more and more prevalent in the population and some individuals will inherit a sufficient number of them so that they potentiate each other's effects. It is not "assimilation" of new mutant genes, and not exactly the imposition of phenotypic information onto the genome, but a reorganization of systematic intergene relationships that has taken place in response to selection on a trait to which all these genes contribute.

Piaget appears to appreciate this possibility obliquely, and criticizes the comparatively passive conception of Lamarckian inheritance on these grounds, remarking that such changes may be conceived as "active reorganization [of the genome] in terms of selection" on what he calls a "pleuri-unit" (presumably the complex of factors producing a phenotype). But Piaget goes on to invoke "an internal organization which reacts actively," by which he appears to suggest a dynamical systemic analogue to Lamarck's mechanism (Piaget, 1971, p. 108), whereas the analysis of Waddington's result just presented is neither dynamic nor regulatory in the cybernetic sense, but entirely *post hoc* and Darwinian. Selection on the phenocopy *is* selection on these contributing genes because of their contribution to this effect. *The difference is that the genotype–phenotype link is only expressed, and thus made available for selection to act on it, because of the epigenetic contribution of an environmental factor.*

For Piaget, the question begged by Waddington's work was "How could this seeming transference from phenocopy to genocopy work?" It turns out, however, that the mechanism is quite different than he had imagined. It is not a kind of cybernetic feedback process, although it is not merely Darwinism in its classic sense either. Could this difference provide important insights into the questions Piaget was asking?

One important way that this process is incompatible with the core principles of Piaget's paradigm is that it does not involve an *active* reorganization of the genome by the organism. The pattern of gene–gene relationships within the genome is modified by *post hoc* selection on independently variant genes. But in a somewhat oblique way, organism action is indirectly implicated. To some extent, regulatory processes during epigenesis *are* involved in the expression of the phenocopy in response to heat in these individuals (though these processes are just as likely to involve *intraselection*; see next section). So although in a far more indirect manner than Piaget imagined, epigenetic regulatory processes *are* an essential ingredient. But such regulatory effects do not serve as the mechanism for transferring phenocopy to genocopy. This occurs via a systemic Darwinian mechanism involving the combinatorial relational structure of the genome. More importantly, this process is not confined within an individual lineage. It necessarily involves a kind of co-selection *across* lineages that include individuals expressing parallel phenocopies. Selection on these phenocopies results in what can be described as *parallel distributed selection* throughout the population that ends up reorganizing which gene–gene relationships tend to occur in individuals. The result is not a change of genes but a reorganization of genome structure with respect to the common effects of these phenocopies.

Equally important—whatever the mechanism—is whether this could reinstate an important role for organism agency and autoregulation in evolution? What this shows, however, is that although many physiological regulatory processes *can* be modeled in cybernetic terms—for example, body temperature, hormone levels, blood pressure, heart rate, appetite, control of movement, and so on—processes that have regulatory consequences need not be cybernetically organized. Feedback based correction mechanisms aren't the only viable solutions to all problems of regulation. What Piaget had predicted would involve a kind of physiological–genetic feedback mechanism turned out instead to be the consequence of a kind of blind, feed-forward selection process.

This has serious implications for Piaget's larger enterprise. Contrary to his speculation, in this case, the evidence suggests that a comparatively "passive" noncybernetic, *a posteriori* selection process is sufficient. Piaget did recognize that the relational complexity of gene expression in epigenesis contributes a significant organizing influence but not the possibility that there could be Darwinian means to achieve seemingly cybernetic effects. This opens the door to another more fundamental question that now must be considered. If the phenocopy assimilation mechanism is indeed fundamentally non-Lamarckian and non-cybernetic, what are the implications for Piaget's vision of the unification of biological knowledge processes? Even more to the point, does it even call into question his proposed mechanism of cognitive development?

INTRASELECTION AND PHENOCOPY

There are other options. The possibility of achieving Lamarckian-like evolution without invoking inheritance of acquired characters has been "in the air" since the 1880s. It was August Weismann, the architect of orthodox anti-Lamarckian Darwinism, who was one of the first to articulate a mechanism whereby epigenetic plasticity might contribute some structuring influence to the otherwise "blind" processes of variation and selection. He described this as *intraselection* because it involved sources of selection imposed on some phenotypes by other phenotypes within the same organism (Weismann, 1892; see also the discussion in Richards, 1987). It was his answer to critics who had suggested that selection that only acted at the level of whole organisms would be too crude to contribute to the evolution of traits of extreme perfection, to wit, the complex interdependencies of body systems and the reciprocal balance of their various functional demands and contributions. Though good design would be selectively favored over bad design, critics of Darwin pointed out that it would be astronomically too slow and vastly inefficient for evolution to achieve such exquisitely complex designs as manifested by the eye or brain in an all-or-none fashion. A Lamarckian mechanism could, on the other hand, quite effectively "discover" and fine-tune such complex synergies via the contributions of learning and physiological adaptation.

Weismann suggested, however, that physiological adaptation processes would be able to play a significant role by an entirely indirect process that did not need to rely on inheriting acquired responses. For example, consider a lineage of elk that by chance acquired the tendency to grow unusually large antlers. Members of future generations would be faced with a variety of correlated stresses, affecting the thickness of the skull, the strength of the muscles of the neck, the size and shape of cervical vertebrae, and so on. Generation after generation, the disproportionate weight of antlers would ultimately be compensated by acquired physiological adaptations in each individual to strengthen muscles and thicken bone in response to these demands. To see how this might contribute to evolution, we need to add that these regulatory responses will inevitably be produced at some metabolic and developmental costs due to trial-and-error fine-tuning, and costs due to reduced functionality during the nonequilibrium phase when the phenotypic compensation is still incomplete. Under these circumstances, if some serendipitous mutation appeared in some future member of the lineage that induced individuals to grow these compensatory structures in advance of (and irrespective of) the stressor, then such incidentally synergistic mutations would in effect find themselves in a selectively congenial context and so tend to be passed on down.

In other words, any distribution of stresses throughout the body that recruits compensatory physiological regulatory mechanisms will create a context that is selectively receptive to compensatory genetic changes. In this sense, Weismann had discovered a way that a physiologically produced phenocopy (e.g., a thickened skull generated in response to the stress of added weight) could provide a selectively favorable context for any analogous, genetically-"hard-wired" phenocopy that might arise spontaneously in future generations. The initial physiological accommodations might in hindsight be viewed as paving the way for the evolution of more ineluctable developmental counterparts with more determinate genetic bases—as though the phenocopy acted as a sort of template, eventually filled in by natural selection with corresponding genotypic changes. But again, although the physiological adaptations are active regulatory responses, the evolutionary mechanism involved is strictly Darwinian.

This turns out to be partially analogous to Waddington's examples of phenocopy assimilation. In Waddington's examples, the stressor is extrinsically imposed (e.g., heat), and yet like Weismann's example of a mutation that unbalances the body, it produces a distinctive phenotypic consequence that is subject to selection. Analogous to intraselection, the selective breeding of individuals expressing this environmentally induced trait creates a context that is congenial to selection favoring other ways of producing this trait. But as further analysis of Waddington's cases has shown, newly acquired mutations are not necessary to fill this role. Whether originating endogenously or exogenously, these "imbalances" create conditions that effectively "recruit" congenial genotype–phenotype combinations from a vast pool of previously neutral mutations already present in the population at large

And there is another difference besides this intrinsic/extrinsic source of stress. In the case of intraselection, the features that are selectively favored aid in the physiological compensation for a stressor, whereas in the case of genetic assimilation, the features favored augment the disturbing effects of the stressor. The outsized antlers in Weismann's example play an analogous role to Waddington's environmental stressor, but with inverse effects. In both processes a stressor "unmasks" some selective advantage of certain epigenetic and phenotypic factors that were previously "invisible" to selection. But because one is intrinsic to the body and the other is extrinsic, the nature of the synergy under selection is the inverse.

Both are opposite to what Piaget seems to have had in mind, and yet their consequences meet his more general goal of demonstrating ways that phenotypic plasticity might contribute to the trajectory of evolution.

ACTIVITY-DEPENDENT SELECTION
IN NEURAL DEVELOPMENT

Something like the process of intraselection is probably immensely important with respect to the development of the nervous system. This is because the mammalian central nervous system (CNS) appears to be highly sensitive during development to contingent structural variations as well as environmental input for determining structural features of neural circuitry. In many ways, this sensitivity is even more appropriately called intraselection than what Weismann proposed. It is now well established that neural development involves activity-dependent, selection-like processes in which populations of axons compete for synaptic contacts in a given brain region. Depending on the relative synchrony and asynchrony of converging signals conveyed down these many competing axons, a significant fraction will be eliminated, whereas the remainder will tend to organize in ways that are consistent with their signal correlation characteristics.

Classic cases include the formation of rodent vibrissae whisker maps and cat and primate binocular visual maps in the cerebral cortex. Manipulation of the inputs of these systems during early development (e.g., removing vibrissae or blocking input from one eye) can restructure the pattern of the central representation of these systems to match the induced disturbance. These modifications could quite appropriately be understood as phenocopies. They are functional connection patterns developed via phenotypic plasticity and environmental sensitivity that can be achieved by multiple genetic and nongenetic means. To be more accurate, they emerge from the interaction of multiple, incompletely-determinate genetic and extra-genetic influences.

In contrast to what might have been suggested by Weismann's logic, however, this sort of facultative fine-tuning of major neural maps during development is not merely a transient phase in the evolution of more completely prespecified developmental mechanisms. All mammals that have been studied in this respect demonstrate similar selection-like neural developmental mechanisms in one or more systems.

One advantage of maintaining some level of input-dependent neural development is exemplified by binocular visual maps in the visual cortex.

These are interdigitated zebralike projection maps, with projections from neighboring eyes occupying adjacent stripes. This allows signals from corresponding points in visual space, as seen by each eye, to arrive at neighboring points on the map and interact to allow binocular comparison (i.e., stereoscopic fusion of the images). There is considerable variation in the relative lateral placement of eyes in different mammal species, from nearly no overlap to nearly complete overlap of visual fields. A hard-wired map would require independent genetic respecification of numerous neural systems in order to be optimal for each such variant. A soft-wired map can approach optimal functional organization without coincidentally correlated genomic mutations of skull structure and the multiple neural targets of the eyes. Leaving the details genetically incomplete allows the brain to actively adapt to the pattern of inputs during development.

Before turning to some further variants of this logic, it is worth pausing to reflect on the significance of this for Piaget's enterprise. These examples of neural intraselection processes were mostly unknown at the time of Piaget's writing. Yet in many ways, this is consistent with Piaget's conception of the active regulation of neural development augmenting the more generic patterns provided by evolution, even though it is neither in the process of becoming "assimilated" nor exactly cybernetic in its logic. It also suggests that with respect to neural development, there may be considerable potential for phenocopy effects.

Probably the most remarkable example of a neural phenocopy produced by this kind of intraselection comes from an experiment by Law and Constantine-Paton (1981) involving manipulation of visual development in the frog brain. They transplanted a third eye between the existing eyes of a frog embryo and observed the growth of axons from this eye back to a primary visual processing center, the tectum (a bilateral, semispheric structure on the back of the midbrain). Normal projections to the frog tectum terminate in a retino-topic pattern (i.e., retaining retinal topography), with each retina projecting onto the opposite tectum. What the researchers observed as a result of the additional eye's projections was a zebra-stripe pattern analogous to that found in the mammalian visual cortex associated with binocular vision. In these frogs, as in mammals, this pattern was shaped by competition from axons from the two eyes alternately supporting and competing with each by virtue of the correlation and noncorrelation of signals from the two eyes. In these "triclops" frogs, parallel visual signaling was generated by eyes with overlapping visual fields and overlap of terminations. But frogs' eyes normally do not have any significant binocular overlap. And there were no binocular frogs in the evolutionary past, as far as we know. So this patterning cannot have been some vestigial potential nor could it have been anticipated in evolution. It is a purely emergent product of the activity-dependent competition between signal pathways conveying similar patterns of visual input. In this regard, this phenotypic pattern is a spontaneously produced cross-species phenocopy. It is the result, of the brain literally adapting to the body it finds itself in, like species adapting to their surroundings, by a selection-like process. In this case, there is not merely regulation of function but spontaneous "discovery" of function by an epigenetic process. It suggests that epigenesis could indeed play a role as a sort of evolutionary scout, as Piaget implied.

Piaget was not unaware of Weismann's concept of intraselection, but seems to have interpreted it as though it could serve as a sort of intra-organismic selection acting directly on genes with respect to somatic functions (Piaget, 1978), but neither as Weismann conceived of it nor in terms of intercellular competition (as already described with respect to neural development). Piaget cannot, of course, be faulted for not recognizing the latter effect. Neurobiological evidence for extensive developmental plasticity playing a major role in CNS pattern formation was still little appreciated when Piaget was writing. The significance of these mechanisms for Piaget's effort to bring evolutionary theory in line with his cognitive developmental theory derives from the fact that they are both consistent with this constructivist logic and yet somewhat the inverse of how he thought it was implemented. Rather than evolution exhibiting characteristics of cybernetic regulation, as he presumed, it now appears more likely that embryonic development at many levels exhibits features characteristic of Darwinian evolution!

The crucial point is that autoregulation can be achieved without cybernetic processes, at least as understood in classic feedback terms. *A posteriori* selectionlike mechanisms can mimic feedback effects to the extent that they produce adaptive consequences. In predictable contexts, selection may be slower, more wasteful, and perhaps less reliable. This may be why physiological regulation is often maintained by more direct feedback mechanisms. But unlike cybernetic regulation, selection-based autoregulation need not begin with a predetermined set point, trajectory, or optimization rule, nor even a clear correspondence with an environmental domain. So where conditions are not so predictable—and especially when many combining factors make things highly variable and flexible—selection-based autoregulation may be favored. Selection mechanisms can contribute to both the "search" for an optimal regulatory pattern as well as fine-tune existing systems to unpredictable environmental variations. Selection dynamics may thus be necessary to put the "auto" in autoregulation in the first place.

To put this in Piagetian terms: Darwinian-like processes enable CNS circuits to *assimilate* structural information from the rest of the body, its variable geometry, and the signals from these systems. Selection processes enable the developing brain to *accommodate* to these variations by virtue of the way that *active* signal processing is used to support or inhibit the stabilization of connections. This ironically inverts Piaget's logic of mechanism almost exactly while reproducing its functional logic and its constructive consequences, including an active role for organism-originated change. It is just not cybernetic in any standard sense. This new twist should matter to the application of Piagetian theory in more than just evolutionary domains.

OTHER END-RUNS AROUND ORTHODOX DARWINISM

Intraselection is only one way that a phenocopy may contribute a positive evolutionary influence, and so make an end-run around simple Darwinian processes. Shortly after Weismann introduced this concept, a second class of end-runs were suggested. Today these go by the name of the *Baldwin Effect* (named for James Mark Baldwin). In 1896, Baldwin and two other evolutionary theorists, Conway Lloyd Morgan and Henry Osborne, independently proposed variants of the same idea. Baldwin called it *organic selection* to emphasize the critical role played by the organism. In other words, like the Lamarckian conception, Baldwin saw the actions of a flexible, adapting organism as playing a constitutive role in evolution of its lineage, but not by virtue of physiologically internalizing these adaptations. Baldwin's effect was proposed as an augmentation of the Weismannian vision of strictly-sequestered, germ-line inheritance (Weismann, 1892). He believed that it could produce these Lamarckian-like consequences without invoking the Lamarckian inheritance mechanism. Despite this physiological "barrier" to the inheritance of acquired characters, there appeared to be indirect ways that animal behavior and intelligent agency might nevertheless be reintroduced as a factor in evolution. For this reason, and because of its superficial resemblance to Waddington's notion of genetic assimilation, Piaget saw in Baldwin's organic selection another plausible alternative to orthodox Darwinism, which could introduce a role for development in evolution.

Although often treated as variant accounts of the same assimilatory evolutionary process, the logic of Baldwin's and Waddington's "effects" differ in significant and important ways. Unlike Waddington's genetic assimilation or Weismann's intraselection, *organic selection theory* is predicated on the possibility that nongenetic inheritance can itself create the conditions whereby genetic change might be made more likely. This makes it even more relevant to Piaget's belief that learning plays a constitutive role in evolution. This was in fact Baldwin's contention as well, and ultimately motivated his attempt to unify evolution and mental development (Baldwin, 1902). Thus learning can also be considered a kind of phenocopy, developmentally acquired and assimilated, and potentially shaping the evolutionary process. And learned phenocopies have a clear form of extragenomic inheritance. Baldwin's contention was that Lamarck's transference of learned habit to inherited phenotype could be accomplished because of the way learned behaviors might hold selective elimination at bay and allow "space" for innate substitutes to accumulate by chance.

Before examining the plausibility of this mechanism, however, we should consider two other important differences that distinguish organic selection from both Weismann's and Waddington's theories. First, phenotypic plasticity is in this case functioning as a sort of shield against a lineage being eliminated by natural selection, whereas in these other processes, this plastic effect is what is subject to selection. Second, this effect is extrinsic to the organism. It is an *extended phenotype* to use Richard Dawkins' term (Dawkins, 1982). These differences are critical, and ultimately serve to undermine the logic of the proposed mechanism, as Baldwin envisioned it, though not eliminate its significance as a factor in evolution (see Table 3.1).

The problem is that this protection from the impact of selection will in most circumstances also block the very effects of selection that would be necessary to give advantage to any alternative ways to accomplish what is being produced by these acquired means. The Baldwin effect should actually accomplish the opposite of what Baldwin originally imagined; it should inhibit the evolution of innate substitutes for learning and other plastic responses in most circumstances. It should even degrade parallel genocopies of a regularly acquired phenocopy!

There are many parallels between this masking (or shielding) effect and a more common effect: the reduction of stabilizing selection (e.g.,

TABLE 3.1

The Three Major Theories of Epigenetically Mediated Systemic Selection Effects

	Intra-Selection	Organic Selection	Genetic Assimilation
System affected	Organ-organ interactions and inter-dependencies	Flexible adaptive responses (e.g., learning)	Epigenetic processes involving complex polygenic effects
Selection effect	Stress imposed by changes in one system imposes physiological regulatory demands on other linked systems causing selection to favor compensatory variants	Phenotypes are shielded from selective elimination by plasticity allowing variants to accumulate which may be selectively favored to substitute for shielded phenotypes	Stress imposed from environmental changes induces expression of previously unexpressed phenocopy variants exposing them to selection
Theoretically assumed functional outcome	Replacement of physiological mechanisms for internal regulation of organ interdependencies with genocopies of the optimally regulated state	Initially flexible environmentally responsive phenocopies get replaced by automatic, invariantly expressed phenocopies	Replacement of environmentally responsive epigenetic phenocopies with ineluctable genocopies
More likely evolutionary consequences	Selection-based epigenetic mechanisms evolve for adapting one organ system to others in the developing body	Reduced selection on genetic determinates of a phenocopy that is reliably produced with the aid of extragenomic influences allows genetic influences to progressively degrade	Progressive co-assortment of genes with overlapping or co-potentiating effects on phenocopy development reduces proportional role of environmental influences

(Continued)

TABLE 3.1
(*Continued*)

	Intra-Selection	Organic Selection	Genetic Assimilation
Empirical example	Non-specific overproduction followed by activity-dependent competitive neuronal and axonal elimination fine-tunes neural circuit development	Loss of function mutations accumulate in ascorbic acid synthetic pathway in primates due to behavioral adaptation to fruit eating in primates leading to dependency on fruit eating	Environmentally induced variants of fruit fly wing veins develop independent of environment in individuals inheriting multiple converging genetic predispositions
Significance	Epigenetic regulation is more often accomplished by selection processes than by cybernetic feedback mechanisms	Baldwin effect will more likely produce increased environment-dependency rather than increased innateness	Genetic assimilation is not mediated by cybernetic mechanisms in individuals but by self-organizing effects at the population genome level

Comparisons, theoretical predictions, most likely mechanisms and effects, examples, and significance for Piaget's theory of genetic epistemology and evolution. Empirical examples are described and referenced in the text.

Schmalhausen, 1986). Stabilizing selection is what maintains phenotypes (and their contributing genetic substrates) within a narrow domain of variation by selecting against deviant variants. So a reduction of stabilizing selection due to such shielding effects of environmentally influenced phenotypes ultimately should likewise tend to produce degenerative effects. For example, consider the effect of the facultative shift to fruit eating in early anthropoid primate evolution and its consequences for the necessity of obtaining ascorbic acid (vitamin C) in the primate diet. In most mammals, ascorbic acid is endogenously synthesized, but in anthropoid primates (including humans), the principal enzyme in this synthetic pathway is not produced, because the gene coding for it has been degraded to the point of nonfunctionality (Nishikimi, Fukuyama, Minoshoma, Shimizu, & Yagi, 1994). The early primate shift to frugivory provided these primates with a reliable extrinsic source of ascorbic acid. This, in turn, reduced selective pressure to maintain its endogenous production. Reduction of function mutations that spontaneously damaged this gene were no longer selected against and eventually spread throughout the lineage. In this way, a foraging option evolved to become a dietary necessity (Deacon, 2003b; see Table 3.1).

Likewise, something like Baldwinian masking of selection may contribute to the evolutionary persistence of intraselection mechanisms as major contributors to brain development, as in the case of ocular dominance columns already discussed. The ease of their epigenetic production via intraselection mechanisms may paradoxically have blocked selection that might otherwise have led to the evolution of a more hardwired phenotype. What is more, there may be greater potential for intraselection effects in larger, slower developing brains, such as mammal brains (due to greater statistical sampling possibilities during extended epigenesis). This suggests the counterintuitive possibility that there may have been an evolutionary tendency toward increasing the soft wiring of brains by intraselection mechanisms. Intraselection effects may have masked the natural selection maintaining hardwired brain circuitry, allowing systems to degrade to a point where there is relatively less prespecification. Thus, an incomplete specification of neural circuits and a permanent role for intraselection may be an evolutionary stable state for the brain.

A number of researchers have endeavored to model the Baldwin Effect by computer simulations using competing software agents. Most of the successful simulations of the Baldwin Effect appear to rely on large cost differentials between acquired and innate production (e.g., see Hinton & Noland, 1987; Mayley, 1996). In other words, if a facultatively-generated phenocopy must be acquired at some cost (e.g., a learned behavior in which errors can lead to harmful consequences) while an innate alternative avoids this and does not produce other costs (e.g., due to its inflexibility), then the innate variant could potentially be selected over the acquired variant. But even if there is such a cost-differential, the process will only evolve to an equilibrium between these costs and benefits, that is, to the point where acquisition is a bit easier and less costly because of innate aids (see critique by Simpson, 1953; see also Deacon, 2003b). But it turns out that even this is likely to be uncommon, because the acquired phenocopy (e.g., the learned adaptation) is not the most likely target for selection induced by the costs it imposes. There are innumerable ancillary and complementary systems within the organism that are impacted by these novel demands, and thus, we should expect a form of both Waddingtonian co-assortment and Weismannian intraselection to follow. Compensatory changes in these ancillary phenotypes will also be selectively favored to the extent that they assist the acquisition of the phenocopy in some way. Their multiplicity makes evolution toward phenocopy replacement less likely and evolution toward complementary support by flexible adaptations more likely. This tendency is again the inverse of Baldwin's prediction.

Because of the reduction of selection effects of an acquired phenocopy—that allows extrinsic factors to substitute for genetically-supported functionality in some way—the same selection dynamic will both tend to degrade this direct genetic contribution and shift selection to the suite of ancillary functions that supports acquisition of this external source of epigenetic influence. This suite of interdependent compensatory adaptive responses is then in some sense the converse or complement of a phenocopy.

This sheds a whole new light on the often-criticized concept of *preadaptation* and its modern counterpart, *exaptation* (Gould & Vrba, 1982). Both of these evolutionary concepts are based on a sort of hindsight projection of a phase space of potential untapped adaptive phenotypes from which epigenesis and natural selection have serendipitously drawn alternatives. Baldwin, Waddington, and Piaget each suggest that epigenetic plasticity may offer a means for a more or less active "exploration" and recruitment of this space of alternatives, which might be described as "adjacent" loci of functionality. Although I have suggested that each of these theoretical mechanisms provide individually flawed accounts of how this may be possible, it may, nevertheless, be afforded by a combination of Baldwinian masking and Waddingtonian unmasking of functional selection (Deacon, 2003b).

Consider again the example of the degeneration of the capacity to synthesize ascorbic acid that evolved due to the selection reducing influence of frugivory (as described). The parallel evolutionary development of three-color vision, via gene duplication, is an obvious candidate for a compensatory ancillary adaptation contributing an increased probability of acquiring this external phenocopy (ascorbic acid). The evolution of a capacity to more accurately assess color change, indicating ripeness, would help guarantee access to what had become an essential nutrient due to this "evolved addiction" to extrinsic ascorbic acid.

To restate this more generally, this combination of biasing effects on the distribution of selection over multiple loci and multiple phenotypes may account for the evolution of many different forms of ancillary synergistic adaptations (e.g., color vision with respect to ascorbic acid acquisition), which are otherwise difficult to explain under standard interpretations of Darwinian processes. For example, consider the way that numerous diverse learning biases often converge in evolution to aid species in learning complex species-typical skills. This probably applies to songbirds learning songs with only brief exposure to adult songs, to Oystercatchers (a species of shore bird) learning the trick of opening oysters safely by watching their mothers, and to human children learning language with little effort and no explicit training (Deacon, 2003b). Instinct does not likely replace learning in these cases, but learning is clearly streamlined by the combined influences of many independent biases affecting sensory attention, motor patterns, motivation, and memory consolidation, irrespective of whether these biases arise from genetic, epigenetic, or external sources.

Notice also that this inverse analogy to intraselection also invokes the logic of the co-assortment characteristic of what we might now call the Waddington Effect. Waddington's environmental manipulation and selective breeding with respect to a particular environment-sensitive facultative phenotype also served to unmask selection on diverse loci that in some previously subthreshold manner were supportive of the acquired phenotype under selection (e.g., the transverse wing vein pattern). This produced what I called *parallel distributed selection* because it selectively favored any variant genetic contributions distributed anywhere within the breeding population that were capable of decreasing the threshold of expression of this acquired phenotype. Correspondingly, with respect to the masking consequences of an acquired phenocopy, we should also expect parallel distributed selection at the population level. Here too it should provide a powerful tool for sampling the phase space of adjacent synergistic behavioral or physiological supports for supporting this dependency. So, in neither condition will the phenocopy be supplanted by a more genetically hardwired phenotype, and yet the expression of this phenocopy will be made more epigenetically stable. This is exactly what we see in the case of neural development; a highly reliable epigenetic production of complex phenotypes despite the fact that they are genetically underspecified and dependent on active signal processing and environmental input.

There may well be other variations on this theme of epigenetic autoregulation accomplished by indirect selection mechanisms. All are based on the fact that selection processes acting on complex animals inevitably get distributed nonrandomly because they are mediated by *a complex interdependent system that itself is subject to intraselection effects.*

A suspicion that epigenetic complexity should introduce a significant nonrandom element into evolution was the basis for the theories of Baldwin, Waddington, and Piaget. This suspicion has been vindicated, even if some of the mechanistic details are different. Selection processes can produce remarkably complex autoregulatory results, even involving highly distributed synergistic effects, because the blind variation that is its ultimate source is effectively filtered through this complex self-organizing system before being subject to selection. Although the production of the contributing variant forms of ancillary supportive adaptations may be diverse and "blind" to the ultimate synergistic function that they will be recruited to serve, their *expression* in the system and the ways they get exposed to selection are system dependent. To put it in terms that Piaget might have found entirely congenial, this sort of epigenetically-mediated selection is a kind of distributed systemic auto-regulation. It can thus achieve something that cybernetic auto-regulation, based on prespecified set points and feedback circuits, cannot. It can spontaneously "sample" (in an unbiased way) the vast space of possible functional interdependency relationships and feedback loops that are potentially available within the range of combinatorial effects of existing epigenetic mechanisms.

Piaget's appeals to both Baldwin and Waddington can now be seen to be insightful anticipations of a necessary complexification of evolutionary theory, though neither a repudiation of Darwinian mechanisms nor a return to Lamarckian paradigms. To explain the apparent autoregulatory power of biological evolution does, as he suspected, require incorporating the role of epigenetic processes as mediators between genotype and phenotype selection. Yet there turns out to be a far more prominent role for

Darwinian over cybernetic mechanisms of regulation in both development and evolution than Piaget could ever have imagined. Although opposite in mechanism from what he predicted, these processes are fully consistent with his constructivism.

SOME CONCLUSIONS

Dissatisfied with both Darwinian and Lamarckian logic and reaching for a constructivist intermediate, Piaget anticipated the contemporary convergence of developmental psychology and evolutionary biology. Even today, this new synthesis is only just beginning to bear fruit. But molecular and developmental biology were in their infancy when he was writing, so his effort to develop an evolutionary account of biological intelligence cannot be faulted by what he could not have anticipated. Piaget nevertheless did anticipate the importance of trying to integrate systems thinking with evolutionary theory, and epigenetic theory with evolutionary theory, even if he miscalculated where such a synthesis might eventually lead. Accepting this, however, we must still ask about the implications for the grand synthesis he ultimately envisioned, and what it might say about his cognitive theory.

- What aspects of his vision of the unification of biological intelligence can be preserved in light of the abandonment of cybernetic logic as an alternative for selection in evolution?
- How does the important role played by selectionlike processes in neural development impact his cybernetic conception of cognitive development? Can we safely assume that the logic of cognitive development is independent of the logic of neural development?
- What are the implications of finding that many if not all examples of apparently Lamarckian effects are best understood in Darwinian terms, though in far more systemic terms than any neo-Darwinian thinker, save Waddington, would have recognized during Piaget's time?

The cybernetic model as Piaget understood it, was not, as he believed, a distinctive third alternative to the Lamarckian versus neo-Darwinian dichotomy he was hoping to overcome. It leads ultimately, if cryptically, to a Lamarckian understanding of evolutionary processes. The regulatory logic that was characteristic of the cybernetic theory of Piaget's time—although able to be described in purely mechanistic terms, as teleonomic rather than teleological—nonetheless presupposes a design logic, with a causal architecture externally specified and antecedent to its structure. What evolution requires is a theory not just about striving, regulation, and goal-directedness, but also an account for the origins of the underlying teleonomic architecture itself. Piaget believed this could be achieved by simpler cybernetic autoregulatory systems building more complex autoregulatory systems. At present, there are no examples of this. Even in epigenesis, it appears that selectionlike processes play the critical roles in the interactive genesis of structural information (as in the circuits of the brain). Genes provide quite sketchy information for development, and functional structure depends on mechanisms that spontaneously generate structural information interactively, as Piaget surmised. Selection processes are coupled in complex ways with other self-organizing processes at the molecular and cellular levels to produce complexly-balanced functional synergies. Cybernetic regulatory mechanisms appear to be products of this, not the builders.

The regulatory conception that is at the heart of Piaget's developmental theory, with its circular reactions, accommodation, assimilation and equilibration, is more congenial with Lamarckian and Spencerian mechanisms for evolution than with Darwinism. Piaget, like these predecessors, envisioned a unified mechanism linking evolution to learning. Although his was a more sophisticated understanding of cybernetic processes in complex systems, he could not ultimately escape the attraction of this analogy to learning.

Although Piaget himself may not have succeeded in articulating a third evolutionary epistemology that could explain the origins of all forms of biological knowledge, this does not imply that such a unified theory is impossible. The problem I think lies in the way that Piaget (among many others) depicts the Lamarckian-cybernetic and Darwinian theories as simple alternatives to one another. From this framing of the problem, Piaget envisions a sort of Hegelian synthesis between them. It is my contention, however, that these supposed alternatives are not correctly portrayed as explanations for the same thing on the same level. There is a subtle but important difference between explaining complex, goal-oriented behavior and explaining the origins of systems with architectures able to give rise to such behaviors. His reflection on the analogies between the construction of logico-mathematical systems and the operation of cybernetic processes had suggested a physical-dynamic rendition of a Hegelian hierarchy of knowledge, in which tensions at earlier lower level stages would inevitably give way to higher order syntheses that re-equilibrated these tensions. Perhaps this made it all the more difficult to see that a blind, *post hoc*, selectionist approach to evolution and development need not be either purely *ad hoc* or passively empiricist.

Piaget was struggling with a problem for which the necessary tools were unavailable. Those that were most suggestive and most congenial to his theory of knowledge were misleading. But his turning to cybernetics and systems theory can be understood as marking a kind of promissory note for an information theory adequate to this challenge. In hindsight, one cannot hold Piaget responsible for not anticipating the ways in which more recent approaches to complex systemic processes would replace the simple feedback dynamics of cybernetic theory with stochastic, self-organizing dynamics or how developmental biology would demonstrate the importance of selection dynamics in embryogenesis, any more than we can hold Lamarck or Darwin responsible for not having an adequate theory of genetic transmission in their times. But although these offer advances over Piaget's model systems, a clear articulation of such a synthetic theory is not yet available. So despite these shortcomings, Piaget's analysis of the problem and his proposals for a solution offer a useful point of departure for reconsidering the problem that lies behind it.

But whereas Piaget underestimated the power of *post hoc* selection to serve an autoregulatory function and oversimplified the dichotomy between selection and regulation mechanisms, he nevertheless recognized that the missing element in both paradigms was systemic organization.

What Piaget envisioned as this third model of the evolutionary process has been at least partially answered by recognizing that selection processes are themselves systemic and self-organizing, if not strictly speaking auto-regulatory, in a cybernetic sense. Although this new understanding disconfirms Piaget's predictions concerning adaptive mechanism, these developments tend to strengthen the core tenets of his constructivist conception of development and evolution. His was a thoroughly emergentist theory, in which dynamical interactions themselves are responsible for the evolution and development of structures of "extreme perfection" (to use Darwin's phrase) in body and brain. Even if the predominant mechanisms are for the most part best characterized as *post hoc* selection, the generative processes are ultimately dynamic, interactive, and systemic, as Piaget anticipated. Piaget was half right. Systemic self-organizing dynamics do indeed provide the sought-after *tertium quid* in evolution and development, but this depends more on selection logic rather than cybernetic logic to achieve frugal and efficient auto-regulation.

But if Piaget underestimated the autoregulatory capacity of selection dynamics in development, might he also have done so in his cognitive theory as well? Insofar as cognitive development can be understood to be constructive in the strong sense of giving rise to emergent cognitive structure, it too might benefit by incorporating aspects of selection logic in the same ways. This suggests a much more radical critique than merely of Piaget's thoughts on evolution.

A unified theory of development should ultimately explain the phenomenological patterns of cognitive development in terms that are also consistent with the logic of neural development, and this logic also appears to be more selectionist than cybernetic. Yet it is interactive and constructive in Piaget's sense.

I have tried to show how much of what Piaget understood as epigenetic auto-regulation is now understood in terms that are more Darwinian than cybernetic. My belief is that Piaget's use of cybernetic terminology and logic was ultimately intended to be primarily analogical and heuristic. It was a surrogate for systemic causal accounts that he could only suggestively indicate in these ways. Whether a similar critical analysis and reassessment can be usefully applied to the essential details of Piaget's theory of cognitive development, with its cybernetic circularity, its logic of reflective abstraction, and its hierarchic system of cognitive stages, is unclear. Such an enterprise is beyond the scope of this chapter. What is clear is that auto-regulation and selection logic are compatible. So inasmuch as Piaget's ideas about evolution can be integrated with, rather than opposed to, Darwinian selection logic (as he believed), a parallel reassessment of his cognitive theory might benefit by a similar critique and synthesis.

We can only guess if Piaget would have been more wedded to his cybernetic logic or his constructivist metaphysics, but I would wager that he would have been willing to reject cybernetic logic and accept a systemic Darwinism in order to retain a constructive epistemology. Piaget's insistence on the importance of this philosophical stance was his primary motive for seeking a viable middle alternative to the Lamarckian and Darwinian extremes of evolutionary theory, not to mention his interactionist cognitive theory. Pursuing the spirit of the constructivist enterprise even at the expense of critiquing his efforts to formulate a non-Darwinian alternative to evolutionary theory can be understood as entirely within this vision. To the extent that this grander purpose can be realized, even if requiring some variant of selection logic, Piaget's vision of a constructive epistemology of biological knowledge will have been vindicated.

REFERENCES

Baldwin, J. M. (1896). A new factor in evolution. *American Naturalist, 30,* 441–451, 536–533.

Baldwin, J. M. (1902). *Development and evolution.* New York: The Macmillan Co.

Campbell, D. T. (1960). Blind variation and selective retention in creative thought as in other knowledge processes. *Psychological Review, 67,* 380–400.

Dawkins, R. (1982). *The extended phenotype.* Oxford: Oxford University Press.

Deacon, T. (2003a). The hierarchic logic of emergence: Untangling the interdependence of evolution and self-organization. In B. Weber & D. Depew (Eds.), *Evolution and learning: The Baldwin effect reconsidered* (pp. 273–308). Cambridge, MA: MIT Press.

Deacon, T. (2003b). Multilevel selection in a complex adaptive system: The problem of language origins. In B. Weber & D. Depew (Eds.), *Evolution and learning: The Baldwin effect reconsidered* (pp. 81–106). Cambridge, MA: MIT Press.

Gould, S. J., & Vrba, E. (1982). Exaptation—a missing term in the science of form. *Paleobiology, 8,* 4–15.

Hinton, G. E., & Noland, S. J. (1987). How learning can guide evolution. *Complex Systems, 1,* 495–502.

Jablonka, E., & Lamb, M. (1995). *Epigenetic inheritance and evolution: The Lamarkian dimension.* Oxford: Oxford University Press.

Law, M. I., & Constantine-Paton, M. (1981). Anatomy and physiology of experimentally produced striped tecta. *Journal of Neuroscience, 1,* 741–759.

Mayley, G. (1996). The evolutionary cost of learning. In P. Maes et al. (Eds.), *From animals to animats. 4. Proceedings of the Fourth International Congress on Simulation of Adaptive Behavior.* Cambridge, MA: MIT Press.

Morgan, C. L. (1896). On modification and variation. *Science, 99,* 733–740.

Nishikimi, M., Fukuyama, R., Minoshoma, S., Shimizu, N., & Yagi, K. (1994). Cloning and chromosomal mapping of the human nonfunctional gene for L-gulono-gamma-lactone oxidase, the enzyme for L-ascorbic acid biosynthesis missing in man. *Journal of Biological Chemistry, 269,* 13685–13688.

Piaget, J. (1971). *Biology and knowledge: An essay on the relations between organic regulations and cognitive processes.* Chicago: University of Chicago Press.

Piaget, J. (1977). Foreword. In J. -C. Bringuier, *Conversations libres avec Jean Piaget.* Paris: Editions Laffont.

Piaget, J. (1978). *Behavior and evolution* (D. Nicholson-Smith, Trans.). New York: Pantheon Books.

Piatelli-Palmarini, M. (Ed.). (1980). *Language and learning: The debate between Jean Piaget and Noam Chomsky.* Cambridge, MA: Harvard University Press.

Richards, R. J. (1987). *Darwin and the emergence of evolutionary theories of mind and behavior*. Chicago: University of Chicago Press.

Rutherford, S., & Lindquist, S. (1998). Hsp90 as a capacitor for morphological evolution. *Nature, 396*, 336–342.

Schmalhausen, I. I. (1986). *Factors of evolution: The theory of stabilizing selection*. Chicago: University of Chicago Press.

Shannon, C., & Weaver, W. (1963). *Mathematical theory of information*. Chicago: University of Illinois Press.

Simpson, G. G. (1953). The Baldwin effect. *Evolution, 7*, 110–117.

Spencer, H. (1872). *Principles of psychology* (2nd ed.). London: Williams & Norgate.

Von Bertalanfy, L. (1968). *General systems theory: Foundations, development, applications*. New York: D. Braziller.

Waddington, C. H. (1941). Canalization of development and inheritance of acquired characteristics. *Nature, 150*, 563–565.

Waddington, C. H. (1953). Genetic assimilation of an acquired character. *Evolution, 7*, 118–126.

Waddington, C. H. (1957). *The strategy of the genes*. Bristol, England: George Allen & Unwin, Ltd.

Waddington, C. H. (1962). *New patterns in genetics and development*. New York: Columbia University Press.

Weismann, A. (1892). *Das Keimplasma: Eine Theorie der Verebung*. Jenna: Gustav Fischer.

Wiener, N. (1965). *Cybernetics, second edition: Control and communication in the animal and machine*. Cambridge, MA: MIT Press.

4

Human Brain Evolution: Developmental Perspectives

Kathleen Rita Gibson
University of Texas–Houston

Charles Darwin stated that differences between animal and human minds are matters of degree, not of kind, thereby issuing a major challenge to the Cartesian philosophy that human minds are qualitatively different from and superior to those of other animals (Darwin, 1871, 1872). Darwin further hypothesized that human mental capacities evolved in a gradual stepwise fashion from those of other animals, and that the mental differences between "lower" animals, such as fish, and "higher" animals, such as apes, are greater than those between apes and humans.

Darwin's views had little immediate acceptance among behavioral scientists. In the late 20th century, many behaviorists were still selectively applying Morgan's Canon (that all behaviors should be explained by the simplest possible mechanism; Morgan, 1894) to animals but not to humans (Greenfield & Savage-Rumbaugh, 1990). Evolutionarily-oriented scientists were still proclaiming the qualitative uniqueness of varied human cognitive skills and behaviors, such as syntax, use of a tool to make a tool, imitation, deception and consciousness. Nonetheless, by the mid-20th century, primatologists were beginning to acquire data that bolstered Darwin's claim. The first major assault to human qualitative uniqueness paradigms was the discovery in the 1950s of socially-transmitted food-processing traditions in Japanese macaques (Kawai, 1965). The discovery of tool-making traditions in wild chimpanzees soon followed (Goodall, 1964); and we later learned that some great apes can learn to (a) use gestural or pictorial symbols in language-like ways (R. A. Gardner & B. T. Gardner, 1969; Greenfield & Savage-Rumbaugh, 1990; Rumbaugh, 1977); (b) that many apes can recognize themselves in mirrors (Parker, Mitchell, & Boccia, 1994); (c) that some monkeys and apes are capable of tactical deception (Whiten & Byrne, 1988); and (d) that some apes can imitate (Byrne & Russon, 1998). It now appears that most cognitive skills once thought to be uniquely human may exist in rudimentary form in our primate brethren (Gibson, 1996a, 2002; Gibson & Jessee, 1999).

These findings necessitate models that account in Darwinian terms for the evolution of human cognitive skills from similar, but more rudimentary, skills in the common great ape–human ancestor. Such models must also conform to accumulating evidence that great apes reared entirely or in part by human caretakers may more easily acquire human-like skills in language (Savage-Rumbaugh, Shanker, & Taylor, 1998) and mirror self-recognition (Parker et al., 1994) and possibly also in other tasks than those reared in the wild. Finally, acceptable evolutionary models must also accord with modern knowledge of minimal genetic differences between chimpanzees and humans and with the now rapidly-emerging data indicating considerable functional plasticity of mammalian brains. This chapter suggests that the constructionist views of cognitive development pioneered by Piaget and later elaborated by Case (1985) provided plausible steps in constructing models of human cognitive evolution, especially when considered in conjunction with modern understandings of environmental influences on brain development.

EPIGENESIS, NEURAL DEVELOPMENTAL PLASTICITY, AND THE CANALIZATION OF COGNITIVE DEVELOPMENT

Some evolutionary psychologists and linguists argue that adaptive cognitive skills, such as syntax or the ability to detect cheaters, are mediated by genetically-determined, functionally-dedicated neural modules (Cosmides & Tooby, 1992; Pinker, 1994; Tooby & Cosmides, 1992). Piaget (1974), in contrast, accepted Waddington's concepts that complex phenotypes reflect genetic and environmental interactions during development (epigenesis) and that similar phenotypes can reflect either environmental or genetic triggers (Piaget, 1974; Waddington, 1957) (see chap. 2, this volume, by Parker, for discussion of Waddington's concepts). The findings of the emerging sciences of ecological developmental biology (ecodevo) and evolutionary developmental biology strongly support and extend Waddington's hypotheses (Duscheck, 2002). According to Duscheck, complex phenotypes reflect developmental cascades of interacting genetic and environmental effects. For example, which fish becomes a male and which ant becomes a queen depends on environmental triggers.

Much vertebrate brain development is clearly of an epigenetic nature. Most vertebrate and some chordate central nervous systems, for example, possess anatomical and functional lateral asymmetries (Rogers & Andrew, 2002). Across a wide range of species ranging from the chordate, amphioxus, to fish, amphibians, birds, and mammals, the right eye and left brain focus on fine-grained visual analyses needed for obtaining food, whereas the left eye and right brain perceive holistic patterns useful for detecting predators, social stimuli, and other emotional events (Andrew, 2002). The cross-chordate consistency of these patterns could readily be interpreted to mean that brain lateralization is under strict genetic determination, but this is not so. Rather, brain lateralization develops via interactions between genetic and environmental input and can be prevented or reversed by environmental manipulations during early developmental stages.

In normal chickens and pigeons, for example, differential gene expression on the left and right embryonic sides precipitates embryonic folding patterns that result in the left eye resting on the embryo's body, and the right eye facing the egg shell (Deng & Rogers, 2002). Consequently, the right, but not the left eye, receives brief exposures to light when the mother hen or pigeon temporarily changes her position. Differential light exposure results in the excess visual connections from the right eye to the left brain as compared to those from the left eye to the right brain. These differences, in turn, stimulate additional functional, anatomical, and biochemical differentiation of the left and right brains. Experimental manipulations that deprive the right eye of light for the last few days prior to hatching prevent the normal development of lateralized brains and behaviors. Experimental manipulations that provide light to the left rather than right eye reverse normal lateralization patterns (Deng & Rogers, 2002; Güntürkün, 2002).

In mammalian species, behavioral and brain neural lateralization also reflects interactions between genes and environment. In rats, handling during infancy induces right hemisphere lateralization for spatial navigation during swimming. Male rats, not handled by humans during infancy, swim equally well with the right or left eye covered (Cowell, Waters, & Denenberg, 1997). Rats handled in infancy, however, swim mazes more effectively with the left eye opened and the right covered, than conversely. Developmental data on the genesis of lateralized brains and behaviors in primates is less extensive than that for chickens, pigeons, or rats, but what information exists suggests that primate brain/behavior lateralization also partially reflects environmental influences. Across a range of human and nonhuman primates, for example, mothers hold infants in their left arms, thereby reinforcing infants' tendencies to turn their heads to the right (Damerose & Vauclair, 2002). As a result, in primate infants, the left and right ears, eyes, and bodies receive somewhat different sensory input. This, in turn, provides differential stimulation to the right and left brains. Evidence that primate functional lateralization is partially under environmental control also derives from the experiences of brain-damaged human children. Adults who experience damage to classic language areas on the left side of the brain usually manifest severe language deficiencies, but when the same areas are damaged in young children, the right hemisphere often assumes language functions (Hallett, 2000).

Epigenetic processes impact many aspects of mammalian brain development in addition to brain lateralization. For one, environmental rearing conditions influence adult brain size. Thus, rats housed in laboratory environments enriched with social companions, toys, and exercise wheels experience neocortical expansion and enhanced learning abilities at whatever age the enrichment occurs. The effect, however, is strongest when the enrichment occurs in young animals (Diamond, 1998; Diamond, Krech, & Rosensweig, 1964). This neocortical expansion is accompanied by decreased neocortical neuronal density, increased neocortical connectivity, and increased numbers of neocortical glial cells, neuronal dendrites, and synapses (Greenough, Black, & Wallace, 1987). Indeed, a role for experience in determining the numbers of neurons and synapses is built into the basic mammalian neural developmental process, which is characterized by an overproduction of neurons and synapses during infancy followed by selective pruning of those that remain unused (Changeaux, 1985; Greenough et al., 1987; Rakic & Kornack, 2001).

Species-typical sensorimotor brain structures and functions also develop partially in response to environmental input. The removal of facial vibrissae in young mice and rats, for example, results in abnormal development of tactile areas of the thalamus and neocortex (Bates & Killackey, 1985; Killackey, 1979; Simons & Land, 1987). Similarly, visual input during critical developmental periods is essential for the normal maturation of the visual thalamus and visual cortex in cats and monkeys (Hubel, Wiesel, & LeVay, 1977). Although the effects of sensory or motor deprivation are most pronounced during the developmental period, neural plasticity in response to sensorimotor input is also evident in adult primates. Thus, ablation of fingers in adult owl monkeys leads to loss of cortical representation for the missing fingers and the expansion of cortical representation for adjacent fingers (Merzenich & Kaas, 1982; Pons, Garraghty, & Mishkin, 1988; Pons et al., 1991). Even temporary immobilization of the fingers and arms in adult squirrel monkeys results in altered neocortical representation of these structures (Nudo, Wise, Sifuentes, & Milliken, 1996).

Among the most dramatic examples of the influence of sensorimotor input on brain development derive from our own species. It is common knowledge that people who have no arms may learn to tie shoelaces and to draw or write with their feet and that blind people often have unusually acute hearing. We now know that cortical representation for the limbs and fingers may actually be altered in people who become paralyzed, lose a limb to amputation, or suddenly increase their finger usage, by assuming new behaviors such as piano playing (Grabowski & Damasio, 2000; Hallett, 2000). The human cortex and thalamus also require visual input for normal development (Von Noorden & Crawford, 1992) and may, sometimes, assume unusual functions when that input is absent. Thus, the primary visual cortex of blind individuals trained in Braille from a young age often assumes tactile functions (Sadato et al., 1996), whereas cortical areas that respond to auditory input in most humans respond, instead, to visual input in congenitally-deaf individuals trained in sign language from early childhood (Neville, 1991).

The realization that species-typical brain functions develop via gene–environmental interactions casts doubt on theories that functionally-specific neural modules develop under strict genetic determination. (Also see chap. 7 by Bates, chap. 10 by Karmiloff-Smith & Thomas, and chap. 8 by Slobin, this volume.) This realization does not, however, rule out hypotheses that species-typical brain functions and behaviors evolved under natural selection. When species-typical developmental conditions provide predictable sources of environmental input, selection need only act on genes that initiate the developmental cascade. In comparison to developmental processes under strict genetic control, environmentally-responsive developmental processes have the increased adaptive potential of providing potential neural and behavioral plasticity in response to new or unusual environmental conditions. Hence, the epigenetic and functionally plastic nature of brain development may partially explain certain behavioral phenomena such as the ability of apes reared in close human contact to develop rudimentary linguistic capacities not present in wild apes and not acquired by apes whose first human contacts occur in adulthood (Savage-Rumbaugh et al., 1998). Similarly, that modern humans can read, use computers and drive cars reflects, in part, the functionally plastic nature of the human brain, rather than natural selection for those specific capacities.

Despite the obvious importance of environmental and sensorimotor input to the developing human brain, some human capacities, such as language, appear in all humans of normal intelligence, even when external environmental conditions or sensorimotor capacities are atypical. Congenitally-deaf children deprived of training in lip reading or formal sign languages, for instance, invent their own rudimentary sign languages (Goldin-Meadow & Mylander, 1991). (See chap. 8 by Slobin, this volume, for discussion of this phenomenon.) This suggests that some species-typical cognitive capacities may be so critical for survival that evolution has provided mechanisms to assure their development even in unusual circumstances. Species-typical, self-generated behaviors, for example, can help assure exposure to essential environmental input.

Much of the information reaching the developing human brain does derive from self-generated behaviors, including species-typical behaviors such as babbling, the social smile, facial imitation, vocal turn-taking behaviors, and repetitive actions on objects. By the second half of the first year, infants also spontaneously group objects together, experiment with physical relationships between objects, engage in rhythmic pounding activities, and attend simultaneously to parental vocalizations and objects (Case, 1985; Langer, 1986; Piaget, 1952). These behaviors generate input that assists the development of language and of cognitive skills such as logic, mathematics, rhythm, classification, and understandings of physical and spatial causality. Human-reared apes fail to demonstrate some of these behaviors, such as babbling or vocal turning activities, while other behaviors, including some object manipulation behaviors, appear to be less frequent or to develop at later ages in the apes. Thus, human infants possess behavioral propensities not possessed by apes that channel or canalize human brain development in human-specific directions (Gibson, 1990).

Other species no doubt also have infantile behaviors that help channel development in species-normative patterns. Indeed, Gould and Marler (1987) described a similar phenomenon, which they termed learning by instinct, that is, genetic propensities to engage in behaviors that facilitate particular learning experiences. The main distinction between Gould and Marler's concept of learning by instinct and the infantile channeling, epigenetic processes described here is scope. What is developmentally channeled in humans and many other mammals is not merely learning, but also the functional organization of the brain and, hence, sensori-motor and cognitive capacities.

The neural control of human infantile "channeling" behaviors is un-known. Myelination and other developmental data, however, suggest that at least some of them are mediated by subcortical structures or by the motor and premotor cortex, rather than by the neocortical association ar-eas, which have often been considered the primary seats of higher intelli-gence. In both rhesus monkeys and humans, the brain stem, thalamus, and basal ganglia mature in advance of the neocortex, and within the neo-cortex, the motor and sensory areas mature in advance of the association areas (Gibson, 1991). The human brain stem is already reasonably well myelinated at birth and motor cortical layers are beginning to myelinate at that time. In contrast, myelin is first seen within the cortical substance of the association areas only during the latter half of the second year of life (Conel, 1939–1967), and these areas do not become completely myelinated until adolescence (Yakovlev & Lecours, 1967). Most of these subcortical areas are greatly enlarged in humans as compared to great apes (Stephan, Frahm, & Baron, 1981), and, thus, have the increased in-formation-processing capacities that would be needed to mediate in-creases in human infantile behavioral capacities. Given that the general sequence of brain myelination is the same throughout the vertebrate or-der (Gibson, 1991) and that the brains of altricial mammals and birds are far less mature at birth or hatching than the human brain (Portman, 1967), subcortical structures are also the most likely mediators of species-typical infantile behaviors in many higher vertebrates.

EXPANDED HUMAN COGNITIVE CAPACITIES

Even when apes are reared in human homes, their greatest accomplish-ments in spheres such as language, tool use, music, and theory of mind fall far short of the average human accomplishment. As a general guide-line, adult apes perform at about the level of a 3- to 4-year-old human child on humanlike cognitive tasks, including Piagetian tasks (Parker & Gibson, 1979; Parker & McKinney, 1999). An examination of the basic processes that result in expansion of human cognitive skills during later childhood years can, thus, potentially provide insight into the basic neurological mechanisms differentiating adult human and ape intelligence.

In humans, early infantile behaviors are holistic and stereotyped in form. For example, under appropriate conditions, a human newborn will simultaneously reach and grasp for an object using one stereotyped movement involving simultaneous closure of all fingers (Bower, 1974). The infant cannot differentiate the reach and grasp into component units such as reaching, grasping, and moving individual fingers in isolation. At a later age, when infants develop motor differentiation capacities, they can also construct new, varied, action patterns that combine individual arm, hand, and finger movements, such as first grasping and then reaching or pointing and then grasping. These motor differentiation and construction capacities when applied to many body parts eventually provides older children and adults with the motor flexibility and creativity needed for tool use, dance, gymnastics, speech, and music. Much human motor construction is also hierarchical in the sense that newly constructed speech, dance, or gymnastic movements can then serve as embedded subcomponents of still more complex and also highly varied motor routines.

Similar analytical and combinatorial processes characterize cognitive domains. With maturation, cognitive processes also eventually manifest a hierarchical structure in which newly constructed behaviors or thoughts serve as subcomponents of more complex mental constructs. Neo-Piagetian analyses by Robbie Case indicate that the human child's ability to construct complex cognitive schemes from diverse component parts increases in a quantitative fashion through adolescence and serves as the foundation for maturational increases in linguistic, mathematical, scientific, social, and other cognitive capacities (Case, 1985).

Increasing motor and mental constructional abilities also characterize great-ape maturation (Gibson, 1990, 1996b) but apes fail to reach human levels of mental constructional capacity. This factor along with the absence of humanlike channeling behaviors apparently accounts for the failure of adult apes to progress beyond a 3- to 4-year-old human child in performance on Piagetian tasks. In the linguistic domain, the performance of the most accomplished apes lags even further behind that of a human child. The most accomplished great-ape linguists manifest English comprehension approximately equivalent to that of a human 2½-year-old child (Savage-Rumbaugh et al., 1998), and their abilities to construct gestural or pictorial "sentences" never progresses much beyond the "two word" stage typically acquired by human infants in the second year of life (Greenfield & Savage-Rumbaugh, 1990).

That differing degrees of mental constructional capacity form the basis of ape–human cognitive differences is evident from adult behaviors. Apes manufacture simple tools, such as termiting sticks, from individual components but they do not construct tools of diverse components (Gibson, 1983). Nor do they construct objects that subsequently serve as subcomponents of more complex tools or architectural structures. In contrast, humans construct many tools from previously manufactured component parts. Even very simple tools, such as hafted spears, for example, are constructed from stone points, wooden shafts, cordage, and resin. Some component parts of a constructed spear, such as cordage or wooden shafts, can also be used in the construction of tents, rafts, hammers, and other, often quite complex tools or architectural creations. In other words, tool making can also be hierarchical in that manufactured units can then serve as embedded subcomponents of still more complex constructions. Similarly although some trained apes use rudimentary gestural and visual symbols and can combine these symbols in regular syntacticlike patterns, they do not construct phrases embedded into a hierarchical sentence structure (Greenfield & Savage-Rumbaugh, 1990). Their vocabularies remain sparse and their sentences short.

The realization that critical ape–human differences involve degrees of mental constructional capacity accords well with current understandings of ape–human neural differences, especially if one accepts Case's hypothesis that expanded neural information-processing capacities form the basis of expanded mental constructional capacity during the maturation of the human child (Case, 1985).

Adult human brains are approximately three to four times as large as those of adult great apes, and consequently provide for greatly increased information-processing capacities. Although most major brain structures are larger in humans than in apes, the neocortex (Passingham, 1975), especially the neocortical association areas and the cerebellar hemispheres are proportionately the most enlarged (MacLeod, 2000; MacLeod, Zilles, Schleicher, Rilling, & Gibson, 2003). They also have functional properties that contribute to human mental constructional capacities. The cerebellar hemispheres mediate the ability to shift attention from one item to another, a necessary prerequisite to combining several items into new constructs (Allen, Buxton, Wong, & Courchesne, 1997). The frontal association areas provide for working memory, that is, for the ability to keep several items of information in mind simultaneously (Goldman-Rakic, 1987). They also subserve abilities to inhibit irrelevant stimuli, ideas, and actions and to plan sequences of actions in pursuit of predetermined goals (Fuster, 1989). The parietal association areas have been suggested to have synthetic or associative functions, as, for instance, in the association of sounds and visual images in order to conceptualize word meanings (Geschwind, 1965); or in the association of proprioceptive and visual information to create body images (Luria, 1966). In other words, in the terms used in this chapter, the parietal association areas may be involved in the construction of concepts from component parts.

Other areas that have expanded in human evolution, including the hippocampus and basal ganglia, may also contribute to our enhanced mental- and motor-constructional skills by providing expansions in our declarative and procedural memory systems. Judging by behavioral evidence, the human motor and premotor cortices also have expanded information-processing capacities in comparison to those of great apes in cortical areas controlling movements of the oral cavity. These motor areas provide the ability to make discrete movements of the tongue, lips, and uvula essential for speech. Thus, our increased mental constructional skills are probably best viewed as a product of coordinated interactions of many expanded neural regions.

The comparative neural and behavioral evidence, thus, suggests that the neural foundations of expanded human cognition involve increased information-processing capacities in subcortical structures that mediate human infantile channeling behaviors and in varied cortical and subcortical structures that provide for enhanced human mental-constructional capacities. These views contrast with classic views that human and great apes brains differ qualitatively in varied respects. In particular, humans were long thought to be the only animals with functionally lateralized brains and with Broca's area. These assertions have not borne the test of time. Cytoarchitectural techniques long ago demonstrated the presence of Broca's area in monkeys (Bailey, Von Bonin, & McCulloch, 1950), and the great apes are now known to possess lateral asymmetries of Broca's area and of the planum temporale similar to those of humans (Cantalupo & Hopkins, 2001; Gannon, Holloway, Broadfield, & Braun, 1998; Gilissen, 2001). The best documented differences between apes and humans continue to remain those related to overall brain size, to the sizes of most neural structures, and to quantitative parameters that correlate with brain size such as increased neural connectivity.

HUMAN COGNITION AND THE FOSSIL RECORD

The recognition that differences between ape and human mental constructional capacities reflect, in part, quantitative differences in brain size has significant implications for our interpretations of the human fossil record. It suggests that human ancestors with ape-size brains would have had apelike mental constructional capacities, that those with intermediate brain sizes would have had mental constructional capacities intermediate between those of humans and apes, and that those with modern brain size would have had modern intellectual capacities. The fossil record accords with this interpretation.

Approximately 2.5 million years ago, early hominids in East Africa used stones as hammerstones to chip sharp-edged flakes from other stones (see Asfaw et al., 1999), a tool-making technique that has been mastered by the bonobo, Kanzi (Schick et al., 1999). Although the precise manufacturer of the stone tools is uncertain, the region was inhabited by several hominid species of the genus, *Australopithecus*, all of whom had brain sizes of approximately 400 to 500 cc (Wolpoff, 1996–1997), slightly larger than average great-ape brain size, but well within the range of great-ape variation. By approximately 1.8 million years ago, *Homo ergaster*, an early African representative of own genus had acquired a distinctly larger brain of about 800 cc (Walker & Leakey, 1993). At about this time, hominids also began manufacturing new, more sophisticated, stone tools called Acheulian hand axes. Unlike predecessor stone-flake tools, hand axes, especially the later ones, were made to a predetermined form including bilateral symmetry (Wynn, 1979), a sharp anterior point, and an enlarged posterior hand-hold area. They also routinely conformed to specific geometric proportions (Gowlett, 1996). In keeping, however, with brain sizes still considerably smaller than those of modern humans, they were not yet constructing tools of diverse components.

By at least 150,000 years ago (see White et al., 2003), brain size had reached its modern external form and size of about 1300 cc in both Neanderthals and Anatomically-Modern Humans (AMHs). Indeed, Neanderthal brain sizes were somewhat larger than our own. By this time, both Neanderthals and AMHs were also constructing tools and shelters. Specifically, both manufactured small pointed stone tools, called Mousterian points, that were sometimes attached to wooden shafts (Churchill, 1993; Shea, 1989, 1993). They also constructed tents and other shelters and may have been making clothing (Tattersall, 1999). The combined evidence of modern brains and constructed tools suggests that both

Neanderthals and early AMHs possessed essentially modern human intellectual abilities (Gibson, 1996a).

Nonetheless, some authors continue to assert that fully-modern intelligence was reached only about 50,000 years ago or later, long after the achievement of modern brain size and form (Mellars, 1996; Mithen, 1996; Noble & Davidson, 1996). One archaeologist goes so far as to say the human brain functionally experienced a functional reorganization 50,000 years ago despite the lack of evidence of any changes in brain shape or size (Klein, 1999). Opinions such as these, which relegate not only Neanderthals but also early modern humans to the status of our intellectual inferiors, derive from evidence of an artistic and technological explosion in Europe that began about 40,000 years ago during Upper Paleolithic times as evidenced by an apparent explosion of cave art, mobile art of bone and stone, and finely-crafted stone tool points (Mellars, 1989). However, although rare, some evidence exists for art and finely-crafted stone tools prior to 50,000 years ago, both among European Neanderthals and among African AMHs (Brooks et al., 1995; Marshack, 1989, 1991; McBrearty & Brooks, 2000). This suggests that the sudden appearance about 40,000 years ago of art and improved technology reflected processes other than sudden changes in basic neurological capacities. That artistic and technological explosions can reflect alternate processes is evidenced, of course, by the numerous such explosions that have occurred in the last millennia, none of which can be attributed to evolutionary changes in the brain.

Creative thought derives from the ability to combine and recombine concepts in a seemingly infinite variety of ways, that is, from mental constructional processes such as those described here (Boden, 1998). Once invented, new creations such as tools, artistic, and scientific traditions serve as part of the fundamental core of cultural knowledge encountered by new generations, often during their formative years when these traditions can impact brain developmental processes. New tools and traditions can also function as building blocks for later inventions, thus providing a ready mechanism for progressive change. The neurological processes essential for creative change would surely have been in place by the time the human brain reached its modern size and form at least by 150,000 years ago and possibly long before that.

In and of itself, the presence of modern mental constructional capacities would not necessarily have immediately spurred extraordinarily rapid changes, such as those that occurred at the onset of the Upper Paleolithic or in more recent times. Even in modern human populations, creativity is associated with high-population density, social stratification, and a sedentary lifestyle (Mithen, 1998). Hence, the sudden creative explosion in the Upper Paleolithic may have been spurred by changed social or environmental circumstances (also see Lewis-Williams, 2002).

By Upper Paleolithic times, people engaged, at least some of the time, in specialized foraging endeavors including the seasonal harvesting of spawning salmon and the seasonal felling of large numbers of migrating caribou (Mellars, 1973, 1989); and it has been suggested that seasonal pursuits of foraging bonanzas may have encouraged much of the cultural change that occurred at that time. Such activities benefit from specialized tools such as fishing hooks and long-distance projectiles and these can be easily carried by people on specialized foraging expeditions. In contrast, earlier peoples who foraged more opportunistically would have been better served carrying a few general purpose tools, such as hand axes or spears. The seasonal acquisition of large foraging bonanzas would have permitted the increased population density known to have occurred in Upper Paleolithic times and provided seasonal excesses of food that could have been stored for future potentially leisure periods. In turn, increased population density and leisure time would have provided an opportunity for the emergence of specialists, such as skilled artisans and toolmakers, and for the social stratification that encourages the production of status symbols, such as jewelry.

If this scenario is correct and modern intelligence and mental constructional skills were reached long prior to the flowering of the Upper Paleolithic period, one must ask what conditions served as the initial selective agents for the evolution of human creative abilities. Paleoclimatic evidence may provide the answer. Humans evolved in habitats characterized by rapid and repeated environmental change, and, hence, were subject to what Richard Potts has termed *variability selection*—the ability to survive in conditions of rapidly varying ecological change (Potts, 1996, 1998). Although these conditions would not have directly selected for artistic talent or for the ability to created finely-crafted tools, they would have selected for highly plastic brains and for the capacity to devise novel solutions to novel problems, rather than for the numerous special-purpose neural processes dedicated to solving highly-specific problems. The phenomena of neural epigenesis and mental construction provide us with precisely these capacities, which we need to survive in ever changing worlds.

SUMMARY

Abundant evidence suggests that in vertebrate, including mammalian, brain maturation is largely an epigenetic process, that is, it reflects interactions between genes and environment, as Piaget, a follower of Waddington, aptly recognized. Although neural plasticity and epigenesis strongly contribute to human mental flexibility, the widespread presence of these processes in other vertebrates, including fish, birds, and mammals, indicates that they are not sufficient to explain the greater intelligence of humans in comparison to most other animals, nor, for that matter, can they explain differences in intelligence among animals.

Modern primatological evidence suggests that great apes possess the rudiments of many behaviors once thought to be unique to the human species, but humans are able to apply much greater amounts of neural information-processing capacity to multiple behavioral domains. As a result, humans have greater mental constructional ability, that is, greater ability to differentiate concepts and actions into component parts and to combine and recombine differentiated units into new, more complex, highly-varied constructions. These capacities reflect the increased size of numerous neural structures in the human brain, including the neocortex, basal ganglia, cerebellum. The realization that increased mental-constructional capacities rather than the presence or absence of entire behavioral domains differentiates humans from apes supports Darwin's concepts that mental differences between apes and humans are matters of degree, rather than kind, and that they may have emerged gradually in human evolution—as suggested by archaeological record. Mental construction, of course, was first emphasized by Piaget and later elaborated by neo-Piagetians such as Case. Thus, the recognition of the importance of increased mental construction in human evolution provides an important intellectual link between Piagetian and Darwinian frameworks. Differences in mental constructional capacity may also underlie apparent differences in intelligence between great apes and monkeys (Byrne, 1995) and among vertebrates in general (Gibson, 1990). If so, this would also accord with Darwinian views of differences in intelligence among animals are also matters of degree.

Increased mental constructional skills, however, are the not the only factor differentiating ape and human minds. Human infants possess a repertoire of species typical infantile behaviors such as babbling, vocal turn-taking, and repetitive actions on objects. These appear to be mediated by early developing subcortical and motor cortical areas, and they generate environmental input that channels the development of the plastic human brain in species-typical directions, thereby assuring that all humans develop language, tool-using, and other species-typical cognitive capacities necessary for survival. All species, of course, have typical infantile behaviors. Consequently, the channeling of brain growth in species-typical directions by infantile behaviors could well be a generalized phenomenon (Gibson, 1990).

REFERENCES

Allen, G., Buxton, R. B., Wong, E. C., & Courchesne, E. (1997). Attentional activation of the cerebellum independent of motor involvement. *Science, 275*, 1940–1943.

Andrew, P. (2002). The earliest origins and subsequent evolution of lateralization. In L. Rogers & P. Andrew (Eds.), *Comparative vertebrate lateralization* (pp. 70–93). Cambridge, England: Cambridge University Press.

Asfaw, B., White, T., Lovejoy, C. O., Latimer, B., Simpson, S., & Suwa, G. (1999). *Australopithecus garhi*: A new species of early hominid from Ethiopia. *Science, 284*, 629–635.

Bailey, P., Von Bonin, G., & McCulloch, W. S. (1950). *The isocortex of chimpanzee.* Urbana: University of Illinois Press.

Bates, C., & Killackey, H. (1985). The organization of the neonatal rat's trigeminal complex and its role in the formation of central trigeminal patterns. *Journal of Comparative Neurology, 240*, 265–287.

Boden, M. A. (1998). What is creativity? In S. Mithen (Ed.), *Creativity in human evolution and prehistory* (pp. 22–60). London: Routledge.

Bower, T. G. R. (1974). *Development in infancy.* New York: W. H. Freeman.

Brooks, A. S., Helgren, D. M., Cramer, J. S., Franklin, A., Hornyak, W., Keating, J. M., Klein, R. G., Rink, W. J., Schwarcz, H., Smith, J. N. L., Stewart, K., Todd, N. E., Verniers, J., & Yellen, J. E. (1995). Dating and context of three middle stone-age sites with bone points in the Upper Semliki Valley, Zaire. *Science, 268*, 548–553.

Byrne, R. W. (1995). *The thinking ape: Evolutionary origins of intelligence.* Oxford: Oxford University Press.

Byrne, R., & Russon, A. (1998). Learning by imitation: A hierarchical approach. *Behavior and Brain Sciences, 21*, 667–672.

Cantalupo, C., & Hopkins, W. D. (2001). Asymmetric Broca's area in great apes. *Nature, 414*, 505.

Case, R. (1985). *Intellectual development: Birth to adulthood.* New York: Academic Press.

Changeaux, J. P. (1985). *Neuronal man.* New York: Pantheon Books.

Churchill, S. E. (1993). Weapon technology, prey size selection, and hunting methods in modern hunter-gatherers: Implications for hunting in the Palaeolithic and Mesolithic. In G. L. Peterkin, H. M. Bricker, & P. Mellars (Eds.), *Hunting and animal exploitation in the later Palaeolithic and Mesolithic of Eurasia* (pp. 11–24). Washington, DC: American Anthropological Association.

Conel, J. L. (1939–1967). *The postnatal development of the human cerebral cortex* (Vols. 1–8). Cambridge, MA: Harvard University Press.

Cosmides, L., & Tooby, J. (1992). Cognitive adaptations for social exchange. In J. Barkow, L. Cosmides, & J. Tooby (Eds.), *The adapted mind* (pp. 163–228). New York: Oxford University Press.

Cowell, P. E., Waters, N. S., & Denenberg, V. H. (1997). Effects of early environment on the development of functional laterality in Morris maze performance. *Laterality, 2*, 221–232.

Damerose, E., & Vauclair, J. (2002). Posture and laterality in human and nonhuman primates: Asymmetries in maternal handling and the infant's early motor asymmetries. In L. J. Rogers & R. J. Andrew (Eds.), *Comparative vertebrate lateralization* (pp. 306–362). Cambridge: Cambridge University Press.

Darwin, C. (1871). *The descent of man and selection in relation to sex.* London: John Murray.

Darwin, C. (1872). *The expression of the emotions in man and animals.* London: John Murray.

Deng, C., & Rogers, L. (2002). Factors affecting the development of lateralization in chicks. In L. J. Rogers & R. J. Andrew (Eds.), *Comparative vertebrate lateralization* (pp. 206–246). Cambridge: Cambridge University Press.

Diamond, M. C. (1998). *Enriching heredity.* New York: The Free Press.

Diamond, M. C., Krech, D., & Rosensweig, M. R. (1964). The effects of an enriched environment on the histology of the rat cerebral cortex. *Journal of Comparative Neurology, 123*, 111–120.

Duscheck, J. (2002, October). The interpretation of genes. *Natural History, 111*, 52–59.

Fuster, J. M. (1989). *The prefrontal cortex* (2nd ed.). New York: Raven Press.

Gannon, P. J., Holloway, R. L., Broadfield, D. C., & Braun, A. R. (1998). Asymmetry of chimpanzee planum temporale: Humanlike pattern of Wernicke's brain language area homolog. *Science, 279*, 220–222.

Gardner, R. A., & Gardner, B. T. (1969). Teaching sign language to a chimpanzee. *Science, 165*, 664–672.

Geschwind, N. (1965). Disconnection syndromes in animals and man. *Brain, 88*, 237–294.

Gibson, K. R. (1983). Comparative neurobehavioral ontogeny and the constructionist approach to the evolution of the brain, object manipulation, and language. In E. DeGrolier (Ed.), *Glossogenetics: The origin and evolution of language* (pp. 37–62). London: Harwood Academic Publishers.

Gibson, K. R. (1990). New perspectives on instincts and intelligence: Brain size and the emergence of hierarchical mental constructional skills. In S. T. Parker & K. R. Gibson (Eds.), *"Language" and intelligence in monkeys and apes; Comparative developmental perspectives* (pp. 97–128). Cambridge: Cambridge University Press.

Gibson, K. R. (1991). Myelination and behavioral development: A comparative perspective on questions of neoteny, altriciality, and intelligence. In K. R. Gibson & A. Petersen (Eds.), *Brain maturation and cognitive development: Comparative and cross-cultural perspectives* (pp. 29–64). Hawthorne, NY: Aldine de Gruyter.

Gibson, K. R. (1996a). The biocultural human brain: Seasonal migrations, and the emergence of the European Upper Paleolithic. In P. Mellars & K. R. Gibson (Eds.), *Modelling the early human mind* (pp. 33–47). Cambridge, England: McDonald Institute for Archaeological Research.

Gibson, K. R. (1996b). The ontogeny and evolution of the brain, cognition, and language. In A. Lock & C. R. Peters (Eds.), *Handbook of symbolic evolution* (pp. 407–430). Oxford: Clarendon Press.

Gibson, K. R. (2002). Evolution of human intelligence: The roles of brain size and mental construction. *Brain, Behavior and Evolution, 59,* 10–20.

Gibson, K. R., & Jessee, S. (1999). Language evolution and the expansion of multiple neurological processing areas. In B. J. King (Ed.), *The origins of language: What non-human primates can tell us* (pp. 189–228). Santa Fe, NM: School of American Research Press.

Gilissen, E. (2001). Structural symmetries and asymmetries in human and chimpanzee brains. In D. Falk & K. R. Gibson (Eds.), *Evolutionary anatomy of the primate cerebral cortex* (pp. 187–225). Cambridge, England: Cambridge University Press.

Goldin-Meadow, S., & Mylander, C. (1991). Levels of structure in a communication system developed without a language model. In K. R. Gibson & A. C. Petersen (Eds.), *Brain maturation and cognitive development: Comparative and cross-cultural perspectives* (pp. 315–344). Hawthorne, NY: Aldine de Gruyter.

Goldman-Rakic, P. S. (1987). Circuitry of the prefrontal cortex and the regulation of behavior by representational knowledge. In F. Plum & V. Mountcastle (Eds.), *Handbook of physiology* (pp. 373–417). Bethesda, MD: American Psychological Society.

Goodall, J. (1964). Tool use and aimed throwing in a community of free-ranging chimpanzees. *Nature, 201,* 1264–1266.

Gould, J. L., & Marler, P. M. (1987). Learning by instinct. *Scientific American, 256*(1), 74–85.

Gowlett, J. A. J. (1996). Mental abilities of early *Homo*: Elements of constraint and choice in rule systems. In P. Mellars & K. Gibson (Eds.), *Modelling the early human mind* (pp. 191–216). Cambridge, England: McDonald Archaeological Institute.

Grabowski, T. J., & Damasio, A. (2000). Investigating language with neuro-imaging. In A. Toga & J. C. Mazziotta (Eds.), *Brain mapping: The systems* (pp. 425–461). New York: Academic Press.

Greenfield, P. M., & Savage-Rumbaugh, E. S. (1990). Grammatical combination in *Pan paniscus*: Process of learning and invention in the evolution and development of language. In S. T. Parker & K. R. Gibson (Eds.), *"Language" and intelligence in monkeys and apes: Comparative developmental perspectives* (pp. 540–578). Cambridge: Cambridge University Press.

Greenough, W. T., Black, J. E., & Wallace, C. (1987). Experience and brain development. *Child Development, 58*, 539–559.

Güntürkün, O. (2002). Ontogeny of visual asymmetry in chickens. In L. J. Rogers & R. J. Andrew (Eds.), *Comparative vertebrate lateralization* (pp. 247–273). Cambridge: Cambridge University Press.

Hallett, M. (2000). Plasticity. In J. C. Mazziotta, A. W. Toga, & R. J. Frackowiak (Eds.), *Brain mapping: The disorders* (pp. 569–586). New York: Academic Press.

Hubel, D., Wiesel, T. N., & LeVay, S. (1977). Plasticity of ocular dominance columns in monkey striate cortex. *Philosophical Transactions of the Royal Society of London, B278*, 377–409.

Kawai, M. (1965). Newly-acquired pre-cultural behavior of the natural troop of Japanese monkeys on Koshima Islet. *Primates, 6*, 1–30.

Killackey, H. P. (1979). Peripheral influences on connectivity in the developing rat trigeminal system. In R. Freeman & W. Singer (Eds.), *The developmental neurobiology of vision* (pp. 381–390). New York: Plenum Press.

Klein, R. (1999). *The human career: Human biological and cultural origins* (2nd ed.). Chicago: University of Chicago Press.

Langer, J. (1986). *The origins of logic: One to two years*. New York: Academic Press.

Lewis-Williams, D. (2002). *The mind in the cave*. London: Thames & Hudson.

Luria, A. (1966). *Higher cortical functions in man*. New York: Basic Books.

MacLeod, C. E. (2000). *The cerebellum and its part in the evolution of the human brain*. Unpublished doctoral dissertation, Simon Fraser University, Burnaby, Canada.

MacLeod, C. E., Zilles, K., Schleicher, A., Rilling, J. K., & Gibson, K. R. (2003). Expansion of the neocerebellum in *Hominoidea*. *Journal of Human Evolution, 44*, 401–429.

Marshack, A. (1989). Evolution of human capacity: The symbolic evidence. *Yearbook of Physical Anthropology, 32*, 1–34.

Marshack, A. (1991). The Taï plaque and calendrical notation in the Upper Paleolithic. *Cambridge Archaeological Journal, 1*, 25–61.

McBrearty, S., & Brooks, A. (2000). The revolution that wasn't: A new interpretation of the origin of modern human behavior. *Journal of Human Evolution, 39*, 453–563.

Mellars, P. (1973). The character of the middle-Upper Palaeolithic transition in southwest France. In C. Renfrew (Ed.), *The explanation of culture change* (pp. 255–276). London: Duckworth.

Mellars, P. (1989). Technological changes across the middle-Upper Paleolithic transition: Economic, social, and cognitive perspectives. In P. Mellars & C. Stringer (Eds.), *The human revolution: Behavioural and biological perspectives on the origin of modern humans* (pp. 338–365). Edinburgh: Edinburgh University Press.

Mellars, P. (1996). Symbolism, language and the Neanderthal mind. In P. Mellars & K. Gibson (Eds.), *Modelling the early human mind* (pp. 15–32). Cambridge, England: McDonald Archaeological Institute.

Merzenich, M., & Kaas, J. H. (1982). Reorganization of mammalian somatosensory cortex following peripheral nerve injury. *Trends in Neurosciences, 5,* 434–436.

Mithen, S. (1996). *The prehistory of the mind: A search for the origins of art, science, and religion.* London: Thames & Hudson.

Mithen, S. (Ed.). (1998). *Creativity in human evolution and prehistory.* London: Routledge.

Morgan, C. L. (1894). *An introduction to comparative psychology.* London: Walter Scott.

Neville, H. J. (1991). Neurobiology of cognitive and language processing: Effects of early experience. In K. R. Gibson & A. C. Petersen (Eds.), *Brain maturation and cognitive development* (pp. 355–380). Hawthorne, NY: Aldine de Gruyter.

Noble, W., & Davidson, I. (1996). *Human evolution, language, and mind: A psychological and archaeological inquiry.* Cambridge: Cambridge University Press.

Nudo, R. J., Wise, B. M., Sifuentes, F., & Milliken, G. W. (1996). Neural substrate for the effects of rehabilitation training on motor recovery after ischemic infarct. *Science, 272,* 1791–1794.

Parker, S. T., & Gibson, K. R. (1979). A model of the evolution of language and intelligence in early hominids. *The Behavioral and Brain Sciences, 2,* 367–407.

Parker, S. T., & McKinney, M. L. (1999). *Origins of intelligence: The evolution of cognitive development in monkeys, apes, and humans.* Baltimore: Johns Hopkins University Press.

Parker, S. T., Mitchell, R. W., & Boccia, M. (Eds.). (1994). *Self-awareness in animals and humans: Developmental perspectives.* Cambridge: Cambridge University Press.

Passingham, R. E. (1975). Changes in the size and organization of the brain in man and his ancestors. *Brain, Behavior, and Evolution, 11,* 73–90.

Piaget, J. (1952). *The origins of intelligence in children.* New York: International Universities Press.

Piaget, J. (1974). *Adaptation and intelligence: Organic selection and phenocopy.* Chicago: University of Chicago Press.

Pinker, S. (1994). *The language instinct.* New York: W. J. Morrow & Co.

Pons, T. P., Garraghty, P. E., & Mishkin, M. (1988). Lesion induced plasticity in the second somatosensory cortex of adult macaques. *Proceedings, National Academy of Science, 83,* 5279–5281.

Pons, T. P., Garraghty, P. E., Ommaya, A. K., Kaas, J. H., Taub, E., & Mishkin, M. (1991). Massive cortical reorganization after sensory deafferentiation in adult macaques. *Science, 252,* 1857–1860.

Portman, A. (1967). *Zoologie aus vier Jahrzehnten* [Zoology over four decades]. Munich: R. Piper & Verlag.

Potts, R. (1996). *Humanity's descent: The consequences of ecological instability.* New York: William & Morrow Co.

Potts, R. (1998). Variability selection in hominid evolution. *Evolutionary Anthropology, 7,* 81–96.

Rakic, P., & Kornack, D. R. (2001). Neocortical expansion and elaboration during primate evolution: A view from neuroembryology. In D. Falk & K. R. Gibson (Eds.), *Evolutionary anatomy of the primate cerebral cortex* (pp. 30–56). Cambridge: Cambridge University Press.

Rogers, L. J., & Andrew, R. J. (Eds.). (2002). *Comparative vertebrate lateralization.* Cambridge: Cambridge University Press.

Rumbaugh, D. M. (1977). *Language learning by a chimpanzee: The LANA Project.* New York: Academic Press.

Sadato, N., Pascual-Leone, A., Grafman, J., Ibanez, V., Derber, M. P., Dold, G., & Hallett, M. (1996). Activation of the primary visual cortex by Braille reading in blind subjects. *Nature, 380,* 526–528.

Savage-Rumbaugh, E. S., Shanker, S. G., & Taylor, T. J. (1998). *Apes, language, and the human mind.* Oxford: Oxford University Press.

Schick, K. D., Toth, N., Garufi, G., Savage-Rumbaugh, E. S., Rumbaugh, D., & Sevcik, R. (1999). Continuing investigations into the stone tool-making and tool-using capabilities of a bonobo (Pan paniscus). *Journal of Archaeological Science, 26,* 821–832.

Shea, J. (1989). A functional study of the lithic industries associated with hominid fossils in the Kebara and Qafzeh Caves, Israel. In P. Mellars & C. Stringer (Eds.), *The human revolution: Behavioural and biological perspectives on the origins of modern humans* (pp. 610–625). Edinburgh: Edinburgh University Press.

Shea, J. (1993). Lithic use wear evidence for hunting by Neanderthals and early modern humans from the Levantine Mousterian. In G. L. Peterkin, H. M. Bricker, & P. Mellars (Eds.), *Hunting and animal exploitation in the later Paleolithic and Mesolithic of Eurasia* (pp. 189–197). Washington, DC: American Anthropological Association.

Simons, D. J., & Land, P. W. (1987). Early experience of tactile stimulation influences organization of somatic sensory cortex. *Nature, 326,* 694–697.

Stephan, H., Frahm, H., & Baron, G. (1981). New and revised data on volumes of brain structures in insectivores and primates. *Folia Primatologica, 35*, 1–39.

Tattersall, I. (1999). *The last Neanderthal: The rise, success, and mysterious extinction of our closest human relatives* (Rev. ed.). Oxford: Westview Press.

Tooby, J., & Cosmides, L. (1992). The psychological foundations of culture. In J. Barkow, L. Cosmides, & J. Tooby (Eds.), *The adapted mind* (pp. 19–135). New York: Oxford University Press.

Von Noorden, G. K., & Crawford, M. L. (1992). The lateral geniculate nucleus in human strabismic amblyopia. *Investigative Ophthamology, 33*, 2729–2732.

Waddington, C. H. (1957). *The strategy of genes.* London: Allen & Unwin.

Walker, A., & Leakey, R. (Eds.). (1993). *The Nariokotome* Homo erectus *skeleton.* Cambridge, MA: Harvard University Press.

White, T., Asfaw, B., DeGusta, D., Tilbert, H., Richards, G. D., Suwa, G., & Howell, F. C. (2003). Pleistocene *Homo sapiens* from Middle Awash, Ethiopia. *Nature, 423*, 742–747.

Whiten, A., & Byrne, R. (1988). The manipulation of attention in primate tactical deception. In R. Byrne & A. Whiten (Eds.), *Machiavellian intelligence: Social expertise and the evolution of intellect in monkeys, apes, and humans* (pp. 211–223). Oxford: Clarendon Press.

Wolpoff, M. H. (1996–1997). *Human evolution.* New York: McGraw-Hill.

Wynn, T. (1979). The intelligence of later Acheulean hominids. *Man, 14,* 379–391.

Yakovlev, P. I., & Lecours, A. R. (1967). The myelinogenetic cycles of regional maturation of the brain. In A. Minkowski (Ed.), *Regional development of the brain in early life* (pp. 3–70). Oxford: Blackwell.

5

Cerebellar Anatomy and Function: From the Corporeal to the Cognitive

Carol Elizabeth MacLeod
Langara College, Vancouver, British Columbia, Canada

In thought, ideas and concepts are manipulated just as limbs are in movement. There would be no distinction between movement and thought once encoded in the neuronal circuitry of the brain; therefore, both movement and thought can be controlled with the same neural mechanisms.
—Ito (1993, p. 449)

When I first became engaged in the serious study of classical ballet as a young woman in Paris, I felt as if a large part of me had been awakened from sleep. I began to see the geometrical patterns in the Greek sculptures within the Louvre, mass and weight in the Maillol of the Jardin des Tuileries, transformations in the mimes of Marceau. I felt related to the world through my physical presence in a grand geometry of space, as if I could dance some complex and never-ending mathematical theorem. The sense of separation between mind and body was replaced by a feeling of completeness, in which intellect and emotions were acted out by a wiser source: the unconscious working through the gesture and the grand *jeté*.

Dance gave me a direct pathway to "embodiment," in which sensory and motor experiences grounded my conceptualization of the world. Although classical dance is a highly specialized and culturally specific expression of the unity of movement and thought, it is still a testament to the simple and profound relation between concrete and abstract experience, illustrated in the opening quotation by the great neuroscientist, Masao Ito, and developed in the corpus of Jean Piaget as he traced the development of logical relations and principles from the child's encounter with the physical world.

Piaget, through careful chronicling and testing of his observations of infants and children, constructed a grand theory of cognitive development in which schemata developed in one stage would form the basis for the more elaborate and far-reaching schemata of the next (Piaget, 1952, 1954, 1962). Inherent to his model are the processes of assimilation and accommodation, in which the individual absorbs elements of the concrete world into subjective experience, then integrates these new relations with existing concepts. In Piaget's view, cognitive growth is not an innately determined unfolding or flowering, but is rather a dialectical process of give and take between the individual and the world. As Masao Ito sees an ultimate resolution between movement and thought, shown in his opening quotation which refers specifically to the cerebellum, so too might the cerebellum act as a critical element in the modulation of the physical with the abstract in development. It is beyond the scope of this chapter to connect the ontogeny of the cerebellum and its circuitry with Piagetian stages, but an argument is made for the inclusion of the cerebellum when considering the fundamental relation between brain and sensorimotor development.

The connections between neuroanatomy and behavior are elusive because they are mediated by so many levels of processing, both within the highly specialized structures of the brain and between the brain and external experience. In this chapter, the most concrete aspects of the cerebellum, its gross anatomy, cytoarchitecture, and connectivity with the rest of the brain, will be related to its function to argue that the cerebellum acts as a mediator between peripheral and central nervous systems and hence ultimately affects the child's abstraction of logical principles from concrete experience. Cerebellar neuroanatomy is very generally described in the text, and the discussion focuses on cerebellar functions revealed by some germinal experiments. A more detailed explanation of the neuroanatomy is presented in the appendix in support of the central thesis that neuroanatomy is the key to the understanding of cerebellar function. In the second part of the chapter, the influence of the cerebellum in the evolution of ape and human cognition and the implications for niche construction will be explored in the context of an extensive volumetric study of the brains of monkeys, apes and humans. It is suggested that certain neocerebellar functions enabled the early hominoids to better exploit their early frugivorous niche, and that a wide complement of cerebellar processes facilitates the learning of skills used by extant great apes in a cognitively constructed environment.

THE CEREBELLUM

There is not movement without cognition, and there is not cognition without movement.
—Bloedel & Bracha (1997, p. 620)

Anatomy and Function

Early studies of cerebellar lesions by Gordon Holmes (1939) led to the understanding that the cerebellum is primarily concerned with movement, coordination, and muscle tonicity. Today, no one would deny the primacy of the cerebellum to motor functions, but neuroscientists such as Bloedel and Bracha have come to recognize the error in dissociating movement from cognition, because we do not move through the world without planning and awareness. Breakthroughs in computerized tomography in the last few years have also challenged the dichotomy between movement and thought. Functional magnetic imaging (fMRI) and positron emission tomography (PET) scans have revealed the participation of the cerebellum in a number of cognitive domains, including the planning of motor sequences, visuospatial problem solving, procedural learning and working memory, attention shifting, and even language (see Schmahmann, 1997, for a representative volume). This is no surprise to those who have studied the neuroanatomy of the cerebellum because of its extensive connections with the neocortex (H. C. Leiner, A. L. Leiner, & Dow, 1986, 1989, 1991; see Fig. 5.1).

The function of the cerebellum is implicit in its anatomy, specifically its longitudinal organization from medial to lateral. It is divided into three zones: archicerebellum (vermis), paleocerebellum (paravermis), and neocerebellum (hemispheres). Ontogenetic and phylogenetic growth occur from medial to lateral (Voogd, Feirabend, & Schoen, 1990), that is, the oldest part of the cerebellum is the most medial. These zones are functionally localized in some ways, with the more atavistic functions of equilibrium and balance associated with the medial cerebellum (archicerebellum), and the cognitive functions in higher primates associated with the lateral cerebellum. This functional localization has less to do

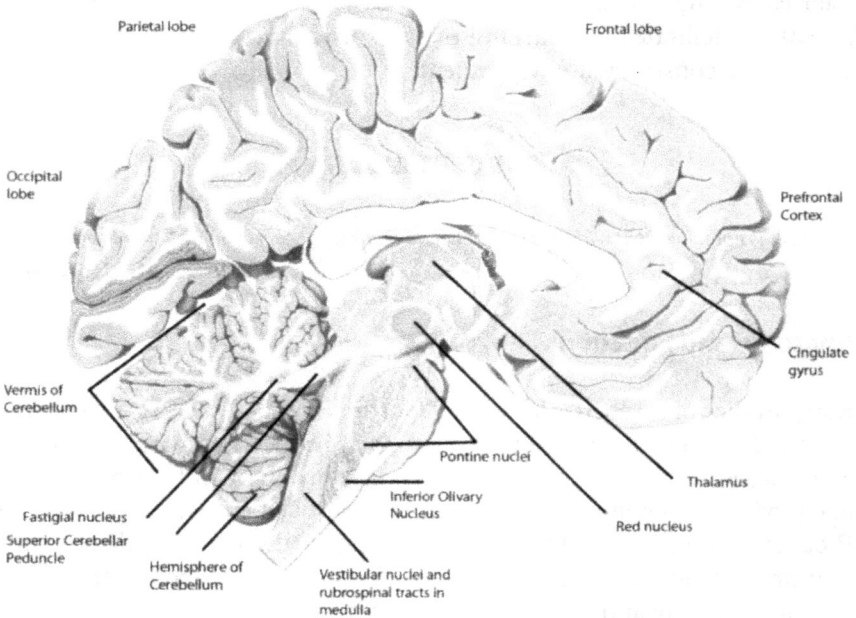

FIG. 5.1.   Sagittal section of brain taken slightly lateral of midline. Some of the major structures in cerebellar circuitry are illustrated. The vestibular nuclei are in the medulla, through which the rubrospinal tract also passes. (Adapted from Nieuwenhuys, Voogd, & Van Huijzen, 1988).

with information coming into the cerebellum than with the information it sends out via the nuclei that are specific to each zone (fastigial, interposed, and dentate nuclei). The neocerebellum sends outgoing information only through the dentate nucleus, which in turn projects to higher brain centers, including widespread areas of the neocortex that are clearly implicated in cognition. The dentate nucleus also sends information in a feedback loop down to the principal inferior olivary nucleus in the medulla. The principal inferior olivary nucleus projects specifically to the neocerebellum and to the dentate nucleus (Altman & Bayer, 1997), and is an essential element of cerebellar circuitry. These structures are important to the understanding of the experimental results detailing cerebellar function, and are illustrated in Fig. 5.2. The reader will find a more extensive description of cerebellar anatomy in the appendix.

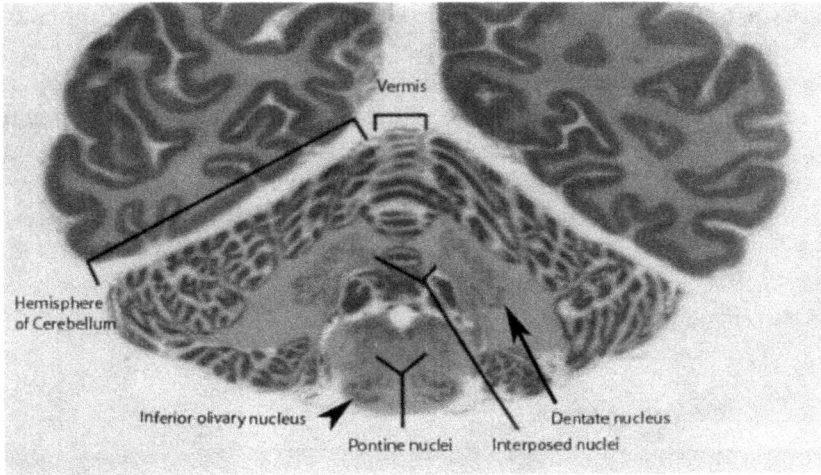

FIG. 5.2.   Major divisions of the cerebellum shown in this coronal section of a chimpanzee cerebellum from the Hirnforschung collection.

NEUROPSYCHOLOGICAL FUNCTIONS

Sensory and Motor Interface

There is mounting evidence that the cerebellum has a sensory component as much as a motor one. This is to be expected, given the direct inflow of information on touch, proprioception, and stereognosis to the cerebellum from peripheral receptors, and the cerebellum's extensive connections with cranial and brain-stem nuclei, and parietal and occipital lobes. The cerebellum is a multimodal processor, and it may very well make connections between sensory modalities that are very basic. In an experiment by Gao and colleagues (Gao et al., 1996), fMRI, which measures blood oxygenation change as a correlate of neural activity, was used to determine the participation of the dentate nucleus in finger-movement tasks and similar tasks with a substantial sensory-discrimination component. The neocerebellum and its dentate nucleus are traditionally associated with fine movements of the distal extremities (Thach, 1978), with the greatest involvement of the dentate nucleus predicted when the hand reaches for, grasps, raises, then drops an object. Gao et al.'s experimental results, however, found dentate activity to be the lowest during this task. It increased as a function of sensory discrimination, and was greatest when subjects grasped an object in each hand, then covertly determined whether the shapes of the two objects matched (i.e., subjects could not look at objects while in the scanner). Dentate activity also increased when subjects had to determine relative degrees of roughness of sandpaper. In both instances of sensory discrimination, the right dentate (with contralateral connections to the left hemisphere) was more active than the left. Based on this and other experimental work, Bower (1997a) proposed that the lateral hemispheres are involved with the more active and reactive exploratory data acquisition involving the somatosensory and auditory systems.

Visuospatial Integration

> The cerebellum has been shown to be specifically involved in rapid sequencing and planning in visual tasks, shifting visual attention, and solving visuospatial tasks with a cognitive component. Fully one quarter of the axons entering the cerebellum via the pons originate in the visuospatial areas of the posterior neocortex in primates. (Stein, Miall, & Weir, 1987)

The cerebellum is important in motor responses to visual stimuli. When Stein et al. (1987) injected local anesthetics into the cerebellar hemispheres of monkeys or cooled the cerebellum of monkeys trained to use joysticks to match a moving target, the monkeys' performance deteriorated rapidly. The authors postulated that the cerebellum uses "feed forward" to make use of information about the current speed of a target to predict where it will be by the end of the next movement, and helps program the amplitude of the next movement accordingly (Stein et al., 1987). They postulated that the monkeys could still use the occipito–frontal pathway to regulate parameters of their movements, but that this was "slower and less efficient than the cerebellar route" (Stein et al., 1987, p. 186). Slowed reaction time in generating sequences has also been found in human patients with cerebellar dysfunction (Inhoff, Diener, Rafal, & Ivry, 1989).

As with sensory discrimination, the visual guidance of movement seems to be more localized in the lateral cerebellum. Stein and colleagues (1987) postulated that the paravermal zone of the cerebellum (see Appendix) is probably concerned with controlling the actual execution of limb movements because there is a high correlation between its averaged discharges and the velocity of arm movements in a particular direction.

In fMRI images, the lateral part of the cerebellum (neocerebellum) is active in tasks requiring visual attention, even without any motor component whatsoever (Allen, Buxton, Wong, & Courchesne, 1997). The ability to shift attention rapidly is critically missing in autism (Courchesne et al., 1994). When given tasks requiring rapid shifts of attention between auditory and visual stimuli, autistic patients, and patients with acquired cerebellar lesions which block neocerebellar functions, were equally impaired compared to normal controls (Courchesne et al., 1994). Autistic patients have hypoplasia of vermal lobules VI and VII, which in turn affects *neocerebellar* ontogeny and function (Courchesne, Yeung-Courchesne, Press, Hesselink, & Jernigan, 1988).

Allen and colleagues (1997) suggested that the cerebellum anticipates events by recognizing learned sequences, then triggers changes in the neural responsiveness of systems expected to be needed in upcoming moments. "The cerebellum accomplishes this anticipatory function by encoding ('learning') sequences of multidimensional information about external and internal events" (Allen et al., 1997, p. 1943; cf. climbing fibre/mossy fibre interplay mentioned in appendix).

The dentate nucleus appears to participate in visually guided tasks that require problem solving more than simple visual tasks alone. Kim, Ugurbil, and Strick (1994) studied fMRI activation in the dentate nucleus during these two task conditions. The simple visually-guided task required the subject to move four pegs from one end of a pegboard to the other. The cognitive version of this task, called the "insanity task," required the subject to exchange the position of four blue pegs with four red pegs on the board by following a set of rules. The solution required continual decision making during the task and was far from evident. All seven subjects showed a substantial bilateral activation of the dentate nucleus during the insanity task that was considerably larger than the activation during the purely visually guided task, three to four times larger as measured by mean number of pixels activated in the dentate. This activation of the dentate nucleus signifies neocerebellar participation in visuospatial cognition. Although the authors of this study (Middleton & Strick, 1997a, 1997b) declined to speculate on the cortical targets of the dentate regions involved in the experiment, dentate projections to frontal and prefrontal areas of the neocortex, areas that would be active in deciding the best strategy to complete the insanity task, would be a likely anatomical substrate.

Linguistic Processing

It is well established that the cerebellum is involved in the motor aspects of speech production. Disturbance of the muscular control of speech, which results in slurred speech of abnormal rhythm, is termed *ataxic dysarthria*. It is commonly found when there is damage to the superior anterior vermal and paravermal regions, but not in the lateral and posterior regions (Fiez & Raichle, 1997). Disturbance in linguistic processing, however, is associated with the lateral cerebellum. That the cerebellum has any role to play in language other than the mechanics of speech production has been an unexpected discovery facilitated by neuroimaging studies.

In a PET study of language processing, subjects were asked to think of and say aloud appropriate verbs for presented nouns (Petersen, Fox, Posner, Mintun, & Raichle, 1989). An area within the right lateral cerebellum was activated when subjects generated verbs from nouns, but not when they read aloud or repeated auditorily-presented nouns, where only the medial cerebellum was activated. These findings have been corroborated by further studies, including the case of a male patient with a localized lesion in the right lateral cerebellar hemisphere (Fiez, Peterson, Cheney, & Raichle, 1992). The patient, RC1, scored poorly against 35 controls in generating appropriate verbs from nouns. For example, when given the noun "blanket," control subjects responded with "wrap," and RC1 with "warm." His scores were average to above average on standard tests used to evaluate cognitive functions, and his language was fluent and grammatically correct. However, when asked to generate such things as category labels, attributes, synonyms and words that begin with the same phoneme, the patient did poorly (Fiez et al., 1992). Consistent with Kim's findings on cerebellar involvement in the "Insanity task," the patient was impaired in the Tower of Toronto puzzle, in which subjects move a set of discs by shifting them among three pegs according to a set of restrictive rules. A score of less than 9 is considered impaired; RC1's score was 1 (Fiez et al., 1992). The extensive connections between the posterior parietal lobe and the cerebellum might prove to be the link with both deeper linguistic functions and visuospatial reasoning.

Fiez and colleagues (1992) also noted that RC1 was unable to detect errors in his own language performance, and unlike the controls, did not show any improvement with practice; that is, he showed a deficit in learning the task. In a PET-activation study of generating verbs from nouns, Raichle and colleagues recorded activity in the right lateral cerebellum as well as the left frontal lobe and anterior cingulate gyrus in normal subjects (Raichle et al., 1994). After 10 minutes of practice, the responses to the task became quicker, more accurate, and more stereotyped. The PET scan showed that activity decreased in the cerebellum and prefrontal cortex, but increased in the Sylvian-insular area bilaterally. Other experimental evidence showed greater cerebellar activation in visuospatial tasks and working memory tasks when the tasks were first being learned (Fiez & Raichle, 1997).

Procedural Learning and Working Memory

These results and other experimental data have led many researchers to hypothesize that the cerebellum is important in working memory (Fiez et al., 1996) and procedural learning (Doyon, 1997; Molinari, Petrosini, & Grammaldo, 1997). Rats with cerebellar lesions do not have difficulty solving the Morris water maze if they have learned the correct path preoperatively. However, they are unable to acquire the correct "map" postoperatively, and swim aimlessly around and around (Molinari et al., 1997). Certainly the aspect of task learning should be taken into account when evaluating experimental evidence that shows cerebellar participation in cognitive and linguistic tasks. The importance of the cerebellum may be more closely related to learning tasks, whether motor, sensory, visuospatial, or linguistic, than to the more automatic, habitual expression of cognition in the different modalities.

Motor Activity

The work of Thach (1978) refined the clinical model of cerebellar function to differentiate between the lateral and medial cerebellum in planning versus execution of movements. Activity in the dentate nucleus and the lateral cerebellum is important in the neurological processes of planning movements, whereas the medial cerebellum is concerned with the process of executing the movement. Thach suggests that the cerebellum would contribute to cognitive activity in a similar way through context linkage and planning of mental response sequences. "The prefrontal and premotor areas could still plan without the help of the cerebellum, but not so automatically, rapidly, stereotypically, so precisely linked to context, or so free of error. Nor would their activities improve optimally with mental practice" (Thach, 1996, p. 411).

Both the execution and the planning of movements entails timing. Cerebellar lesions result in a slowing down of processing time, but not an interruption of processing. This has been explored by Keele and Ivry (1990) in their studies of neurological patients, including those with bilateral cerebellar atrophy or olivo-ponto-cerebello atrophy. The cerebellar patients were timed for the accuracy of their tapping intervals, and for their perception of interval variability in sets of tones. They showed perturbation on both tasks, suggesting that the cerebellum is critical for accurate timing computations.

Precisely how movement and thought are related through the cerebellum, and how the cerebellum itself might be fundamental to the process of cognitive development, remain elusive goals at this point in our understanding. Nonetheless, the evidence for cerebellar participation in cognitive activities is overwhelming, with the frontal–cerebellar and occipito–parietal connections especially intriguing. Adele Diamond (2000) noted that motor development is as protracted as cognitive development, and that it is not until adolescence that the child achieves fine motor control, bimanual coordination, and visuomotor skills; and shows finesse in the ability to accurately represent transformations, flexibly manipulate information held in mind, and incorporate multiple aspects to problem solving. She proposes that these two sides to maturation are related through the mechanism of the cerebellum, since fMRI and PET studies of the last 10 years consistently find a link between the lateral cerebellum and the frontal lobes, particularly dorsolateral prefrontal cortex, in the completion of cognitive tasks. Further, cognitive deficits in developmental disorders such as autism and Attention Deficit Hyperactivity Disorder (ADHD) are accompanied by serious motor dysfunctions. Given that the cerebellum and the basal ganglia are connected to neocortical functioning, the study of development should incorporate both motor and cognitive stages and their interactions (Diamond, 2000). That is, if movement cannot be separated from cognition, then the study of the development of abstract principles should not be separated from the study of the coordinated expression of thought in the motor realm.

With improvements in technology, particularly in neuroimaging, it will be possible to trace connectivity and developing structures in the maturing human brain without intrusion, something that has been impossible until now, when research on cortico–cerebellar connections has been confined to animal models. Similarly, the growing awareness of cerebellar participation in cognitive tasks will alert developmental psychologists to a wider range of neural substrates of observed developmental stages. The study of the connection between ontogenetic development of the brain and stages of cognition is the most promising means by which the very concrete realities of axons and neurotransmitters can be related to the models of developmental stages (Gibson, 1977, 1991).

Gibson (1977) discriminated certain neurological capacities that underlie the development of sensorimotor stages. These include the ability to differentiate perceptual data into its fine components in the sensory realm, and the capacity to engage in increasingly finer motor activities through the more precise use of motor neurons. The growing infant shows an increasing ability to perceive simultaneous stimuli within and across modalities, and can coordinate more complex actions in either a simultaneous and sequential manner. These sensorimotor aspects of cognition are related by Gibson (1977) to the maturation of cerebral structures, but can also be related to cerebellar function as shown through the discussion just presented and in the appendix of circuitry, functional neuroanatomy, and neuropsychology of the cerebellum.

THE EVOLUTION OF THE HOMINOID BRAIN

Cerebellum Study

Anthropologists strive to understand the human brain in a wider comparative context, particularly that of our closest primate relatives, the apes. Major developments in human brain anatomy can be traced to the great apes and even the lesser apes, implying shared anatomical structures in the common hominoid ancestor. One measure of comparative anatomy is through relative volumes of structures in monkey, ape and human brains, with the assumption that larger volumes imply greater participation of that structure in neural processing (Jerison, 1973).

I undertook such a comparative study targeting the cerebellum. This study expanded the existing database for cerebellar structures by tenfold, enabling a statistically-viable contrast between the monkey and hominoid sample within multiple regression analysis (MacLeod, 2000; MacLeod, Zilles, Schleicher, & Gibson, 2001a; MacLeod, Zilles, Schleicher, Rilling, & Gibson, 2003). It determined that there was a major shift in brain proportions within the primate order with the evolution of the hominoids, which includes the lesser apes, the great apes, and humans.

The study used two very different sources of data. One was a set of 47 *in vivo* magnetic resonance scans of monkeys, apes and humans from the Yerkes Regional Primate Research Center in Atlanta, Georgia. Another was from 50 postmortem fixed brains with a comparable distribution of

TABLE 5.1
Sample Distribution

Genus	Sample Number Hirnforsch	Sample Number Yerkes
Homo	8	6
Gorilla	3	2
Pan paniscus	2	4
Pan troglodytes	7	7
Pongo	4	4
Hylobates	5	4
Erythrocebus	1	0
Macaca	2	5
Cercopithecus	2	0
Cercocebus	1	4
Papio	2	2
Ateles	3	0
Alouatta	2	0
Cebus	2	4
Saimiri	2	4
Aotus	4	0

primate species from the Institute for Brain Research in Düsseldorf, Germany (Table 5.1). The fixed brains displayed much finer detail than the MR scans, enabling measurement of certain nuclei, but because of shrinkage from the fixation and embedding process, the volumes had to be corrected to absolute values. Postmortem brain volumes were divided by fixed brain volumes to give a correction factor, which was then applied to all of the structures measured. Comparison of the final volumes from both the MRI and histological data showed that they were not significantly different in spite of the divergent techniques employed to extract the volumes.

The structures measured represented a rough model of a functionally-integrated cerebellar circuitry. These volumes included the whole brain, the cerebellum, the vermis and hemispheres, the dentate nucleus, and its partner, the principal inferior olivary nucleus. This enabled not only the testing of differential expansion of component structures, but also the hypothesis that a functionally integrated set of neural structures would expand as a unit (Barton & Harvey, 2000; Whiting & Barton, 2003). Finlay and Darlington (1995), in their model of developmental constraint, demonstrated in a large sample size of primates, insectivores, and bats that brain structures expand quite predictably with the increase of the brain as a whole, differing only in their rate of expansion shown by their slopes when regressed against total brain volume. On such a large scale, it is almost impossible to see expansion of structures beyond their predicted allometry, especially when the regression points for apes are overwhelmed by the rest of the sample and thus cannot have an significant effect on the determination of the slope. Whereas the Stephan sample from which the Finlay and Darlington data were drawn has only three ape volumes (gibbons, gorillas, and chimpanzees), my data set used volumes from 42 individual apes, enough to form a regression line that could be compared with the regression line derived from monkeys alone.

The results demonstrated that the cerebellum, particularly the cerebellar hemispheres, underwent a dramatic increase in apes and humans. The cerebellar hemispheres regressed against the vermis are 2.7 times larger than expected in a hominoid brain over that of a monkey (Fig. 5.3), and maintain a large differential increase when regressed against the rest of

HEMISPHERE TO VERMIS

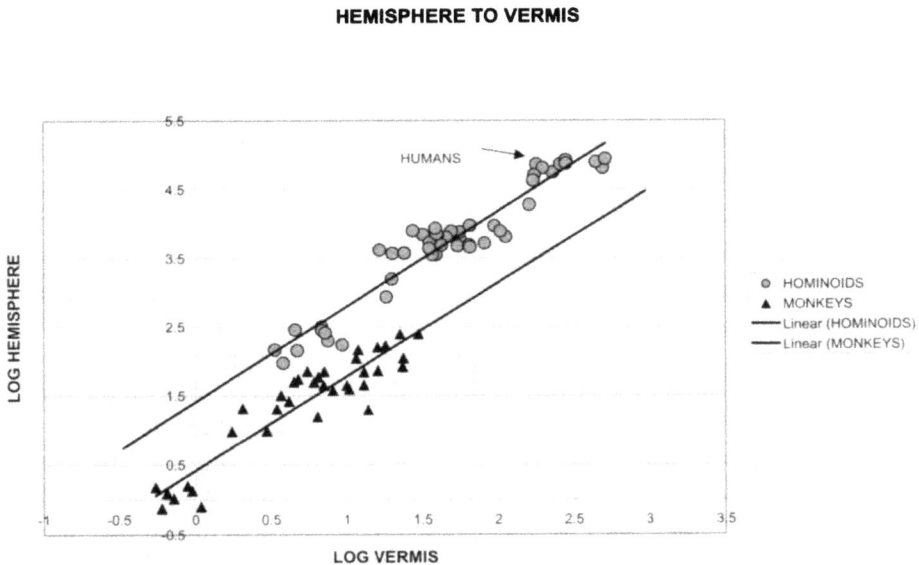

FIG. 5.3.   Logged cerebellar hemisphere to vermis volumes for the combined Yerkes and Hirnforschung samples. *SE* is .268, with an r^2 value of .968. Regression formula for monkeys is $y' = 0.367 + 1.4588x$, and for hominoids is $y' = 1.465 + 1.365x$. The hominoid regression line, which included humans in the regression, is significantly different from the monkey regression line ($p < .001$). Regressions were calculated with Systat 7, and represented graphically with Excel for Mac, v.X.

the brain (MacLeod, 2000; MacLeod et al., 2001a, 2003). The principal inferior olive (PIO) is 40% larger than expected in hominoids over monkeys when regressed against cerebellar volume, but the dentate nucleus does not take part in the differential increase of the lateral cerebellar complex (Fig. 5.4).

This is an unexpected finding, but until this study, our knowledge of the dentate expansion in primate evolution was based on subjective impressions of gross anatomy (Larsell & Jansen, 1970; Tilney & Riley, 1928), or studies that were constrained by a very small sample of ape specimens (Matano, Baron, Stephan, & Frahm, 1985). Thus, although my study demonstrated a definite grade shift in the lateral cerebellum between apes and monkeys, it also showed that functionally integrated structures do not always increase as a unit, but components of the system may expand in a mosaic fashion in evolution (MacLeod, Schleicher, & Zilles, 2001b). Brain structures still exhibit an extraordinary "discipline" in their allometric scaling in showing high correlation co-efficients when x predicts y, but they show significant breaks from expected values under more detailed analysis, with differential expansion of some structures in one grade over another.

Possible Significance of Lateral Cerebellar Expansion for Niche Selection

At the most pervasive level, the increase in lateral cerebellar cortex in the ancestor to apes and humans enabled the processing and integration of much more information from the auditory, proprioceptive, tactile, and visual modalities. Presumably this precognitive processing freed the neocortex to make more elaborate cortical links both directly and via the cerebellum (e.g., parietal to cerebellum to frontal). The ability to perceive and process information from diverse sources and to organize an integrated response to this complexity is the essence of intelligent behavior.

As the medial and anterior cerebellum are the locus of the execution of motor patterns, and the lateral cerebellum is the locus for the planning of these movements (Thach, 1997), the increase in the ratio of lateral to medial cerebellum would imply that the early hominoids had an augmented capacity for complexity of movement, for the cognitive aspects of movement in structuring their niche. If cerebellar expansion were linked to motor coordination alone, there would have been a comparable increase in the vermis, associated with balance and coordination.

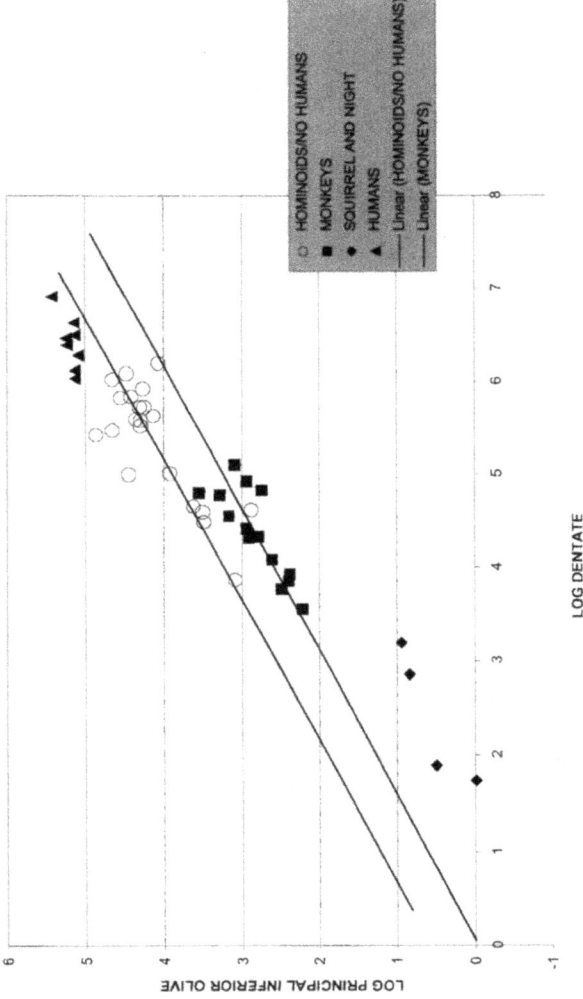

FIG. 5.4. Regression showing the differential expansion of the principal inferior olivary nucleus compared to the dentate nucleus. Squirrel and owl monkeys do not fit with the rest of the monkey sample, and humans have a larger principal inferior olive than expected when regressed against the dentate nucleus. *SE* is .294, and the r^2 value is .91. The formula for the ape regression line is: .550 + .664x, and for the monkey regression line is: −.048 + .654x. The principal inferior olive is almost twice the size expected in an ape over a monkey with a dentate of equal size.

159

The grade shift of the differential expansion of the neocerebellum is uniform in all hominoids, including the lesser apes, and thus appears to have taken place in the stem hominoid ancestor (MacLeod et al., 2003), although this particular ancestor has not been specifically identified from the fossil record. Current fossil evidence indicates that the possible ancestors to extant hominoids were not suspensory primates but quadrupedal branch walkers (Benefit, 1999). A variety of locomotory styles appeared in the early Miocene, leading Fleagle (1999) to describe the skeletal anatomy as versatile, comparable to the living spider monkeys or chimpanzees. All hominoids have upper body adaptations to suspending from branches, but it is possible that lesser apes and the rest of the hominoids evolved these adaptations after the two lines split, with substantial parallel evolution in the hominoid trunk and forelimb (Larson, 1998). Thus, the expansion of the lateral cerebellum may not have been coincident with suspensory locomotion, but could have contributed to its later expression because of its capacity to plan movements (see Rilling & Insel, 1998) and its visuospatial capacity to strategize movement paths. Suspensory locomotion may have led to greater efficiency in gathering fruit when hanging under branches because it offered more movement patterns in a three-dimensional space, an idea that invites more experimental investigation.

Povinelli and Cant (1995) argued that macaque locomotion is stereotyped compared to orangutan clambering, which requires the planning and execution of unusually flexible locomotor patterns, in part to avoid fatal accidents when heavy-bodied animals fall out of trees! They link the clambering of the great apes (and not the lesser apes) with a theory of self-awareness, because it would require the ape to conceive of itself as a causal agent in traveling through a three-dimensional space. This "self-conceptual ability" is evidenced by mirror self-recognition, a trait shown by the great ape/human clade, but not other primates. The evolution of this ability would have come much later in the Miocene with the appearance of the first large-bodied hominoids recognized as the likely ancestors to great apes and humans. Bodies of their size would have been too ponderous for agile, quick, and safe arboreal locomotion; movement would have required a more conscious decision-making process of arranging new motor schemata while moving in an arboreal environment (Povinelli & Cant, 1995). It follows from this analysis that the putative relation between movement and thought could be expressed in phylogenetic development by more complex and interlinked motor schemata facilitated by an expanded lateral cerebellum, although the bringing of movement plans and decisions to conscious awareness would be a property of the neocortex, and not the cerebellum. Povinelli and Cant (1995) placed the cognitive breakthrough of self-conception with the evolution of larger bodied apes, to the exclusion of the lesser apes. Anyone who has observed the graceful acrobatics of the lesser apes would not question their agile superiority to other primates, but whether they structure their space with greater cognitive complexity because of the expansion of the lateral cerebellum needs to be examined more thoroughly than was done with observations of siamang in the study of Povinelli and Cant.

Early Miocene hominoids were largely frugivorous, as inferred from habitat and dentition (Andrews & Martin, 1992; Fleagle, 1999; Kay & Ungar, 1997). It is possible that the lateral cerebellum bestowed an advantage in maintaining hominoid frugivory throughout the drying trends of the mid-Miocene that gave way to greater seasonality (see Potts, in press). Frugivory is associated with relatively larger home ranges than those of folivorous primates (Clutton-Brock & Harvey, 1979; Milton, 1981), even in the case of the gibbon, where body and group size are reduced. Larger ranges require enhanced memory capacity to return to fruiting trees that are patchily dispersed in space and time (Milton, 1981). Mapping the fruiting trees would recruit areas of the right inferior-parietal lobes, areas that are associated with visuospatial cognition and map making (Petrides & Iversen, 1979), and that have extensive afferents to the cerebellum (see Appendix). Experimental evidence of cerebellar participation in visuospatial problem solving (Kim et al., 1994) would also infer cerebellar participation in mapping fruit trees.

The cerebellum is relevant to the great ape feeding strategy of extractive foraging, or removing the nutrients from matrices in which they are embedded or encased. Whereas many species use specialized anatomical manipulators for this purpose and a few species use tools, great apes engage in intelligent tool use and other forms of object manipulation for this purpose (in captivity if not in the wild; Gibson, 1986; Parker & Gibson, 1977; Russon, 1998; Van Schaik, Fox, & Sitompul, 1996). As practiced by wild orangutans, chimpanzees, and humans, extractive foraging often requires a complex sequence of manipulations in which some steps are embedded in others in a hierarchical mental construction (Gibson, 1990; Greenfield, 1991). Procedural learning is necessary to successful extractive foraging, tool use, and other complex routines involving planning (Gibson, 1999; Parker & Milbrath, 1993). The neocortex appears to play a major role in the mediation of hierarchical mental construction (Gibson, 1990), with the inferior frontal gyrus specifically implicated in the hierarchical organization of object and phonological combination in children (Greenfield, 1991). However, frontal-cerebellar connections, as well as cerebellar involvement in the sensorimotor interface, visuospatial processing, and procedural learning suggest that the neocortex may do so in cooperation with the neocerebellum. Increased lateral cerebellar cortex and the olivary learning circuit could have helped hominoids to exploit foods that required a learning component of motor sequences necessary to nutrient extraction, enabling chimpanzees to crack open nuts or extract termites from mounds. Extractive foraging would have been even more complex in the environment of suspensory locomotion where foliage must be negotiated in three-dimensional space to obtain access to foods; arboreal positioning then becomes part of the process of hierarchical problem solving in feeding (Russon, 1998).

Although not all great apes use tools for extractive foraging in the wild, their feeding patterns may require similar cognitive operations. Gorillas, for example, exhibit elaborate patterns of complex manual food processing with a number of techniques specific to certain foods (R. Byrne & J. M. E. Byrne, 1993). Common foods eaten by the mountain gorilla are "defended" by such things as stings, prickles, hard casing, or surfaces covered with tiny hooks; gorilla feeding strategy relies on dextrous movements in a series of logical steps to process the food for eating (R. Byrne, 1995). The traditional role of the lateral cerebellar hemispheres in finger movements would support the motor aspects of the hominoid manipulation of food, but its sensory components would also contribute to complex foraging in the hominoids. If the cerebellum acts as a sensorimotor interface as shown by the experimental work of Gao and colleagues (1996), then the sensitive probings of the food in processing and the instant feedback that would be necessary to avoid getting pricked from the spiny foods so prominent in gorilla feeding would be well served by the cerebellum.

Great apes are the only nonhuman primates to exhibit true imitation and teaching by demonstration (Parker, 1996a), an extension of their extractive foraging strategies (Gibson, 1986; Parker & Gibson, 1977). The ability to teach and learn in natural settings has been well documented in chimpanzees, the most salient examples coming from the Tai chimpanzees (Boesch, 1991), whereas orangutans have been the subject of very meticulous studies of imitation (Russon, 1996). All great apes have demonstrated competence in learning visually-based language systems under laboratory conditions (R. A. Gardner & B. T. Gardner, 1969; Miles, 1983; Patterson, 1978; Premack, 1972; Rumbaugh, Gill, & von Glasersfeld, 1973). Apprenticeship requires focus of visual attention and recognition of important contextual changes, attributes of the lateral cerebellum shown by Allen and colleagues (1997). The inability to shift attention rapidly, particularly between visual and auditory modalities, is an impairment common to both cerebellar and autistic patients (Courchesne et al., 1994). Attention shifting is probably crucial to the joint attention between mother and infant that has been noted as an essential element of human language learning (Tomasello & Todd, 1983). Indeed, how can teaching and learning take place without joint attention, to which the lateral cerebellum would contribute?

Ultimately, the abilities of the great apes in natural and laboratory conditions to transmit learned behavior from one generation to the next (Boesch, 1996; Van Schaik et al., 2003; Whiten et al., 1999) attests to a niche that is more cognitively complex than that of monkeys, with augmented neocerebellar capacity a factor in the great ape adaptive dimension.

CONCLUSION

A comparison of volumes of the cerebellar structures in monkeys, apes, and humans shows a grade shift between monkeys and hominoids in the size, and it is presumed, importance of these structures, giving rise to a slightly differently organized brain when considering the relative proportion of its components. If natural selection favored expanded neocerebellar function, then advantages in procuring food in the arboreal environment of the early hominoids would have accrued, with adjunct improvements in locomotor efficiency. The association of the lateral cerebellum with the planning of complex movement, visuospatial and sensorimotor integration, and procedural learning, would imply that the increase in the lateral cerebellum provided more than a postural advantage, but one that enabled hominoids to integrate themselves corporeally and cognitively with their frugivorous niche. What may have begun as a particular feeding adaptation in the early Miocene could then have served as the springboard for the superior cognitive skills that we recognize in great apes in the laboratory and in the wild.

We know these skills enable great apes to pass to higher sensorimotor stages than monkeys into the preoperations period (Parker, 1996b; Russon, Bard, & Parker, 1996), but the gap between behavior and what we know of the cerebellum precludes a precise model of how cerebellar functions might facilitate these transformations. A most fruitful approach to reconciling behavior and neuroanatomy is a developmental one. Fairbanks (2000), building on the work of Byers and Walker (1995), compared three types of play across primate species, including our own, and correlates play with sensorimotor development. Each type of play (activity, object, and social), peaks at about the same time relative to weaning and the eruption of permanent molars in primate species for which there are adequate data. (Also see Parker, 2002, linking molar eruption to cognitive development in monkeys, apes and humans.) These periods can be correlated with myelination (Gibson, 1991), selective retention of synapses in the cerebellum (Pysh & Weiss, 1979) and the motor cortex (Rakic, Bourgeois, Eckenhoff, Zecevic, & Goldman-Rakic, 1986), peak levels of cerebral glucose metabolism (Jacobs et al., 1995; Moore, Cherry, Pollack, Hovda, & Phelps, 1999), muscle fiber differentiation of slow and fast fire types, and establishment of permanent pathways for muscle innervation (Purves & Lichtman, 1985). Thus, the physical and cognitive aspects of play can be connected with neurological development as the act of playing influences the formation of the growing brain (Fairbanks, 2000). Adding further research on cerebellar growth and development to the studies connecting brain development and myelination with Piagetian stages in both human and nonhuman primates (Gibson, 1977, 1991; Parker & Gibson, 1977; Parker & McKinney, 1999) will ultimately lead to a greater understanding of the neurological basis of cognitive development. Yoking motor and cognitive development with the cerebellum in mind will lead us even closer to that elusive ideal of the reconciliation of the abstract with the concrete.

ACKNOWLEDGMENTS

Grateful acknowledgment to Dr. Kathleen Gibson, my thesis advisor on the cerebellum project, for her inspiration and guidance. I also thank the organizers of the meetings of the Jean Piaget Society, Berkeley, 2001, for their generosity and encouragement. Thanks to Jonas Langer, Constance Milbrath, Sue Parker, and Anne Russon for their incisive suggestions in clarifying the meaning of the text.

APPENDIX
CEREBELLAR ANATOMY

Although the cerebellum is organized into three major zones, its zones are not localized, independent units, but function according to their connections with other parts of the brain. The *vermis*, the oldest part of the cerebellum, has extensive connections with the vestibular nuclei responsible for balance and equilibrium. Additional afferents from cranial nerves concerning eye movements also project to the vermis, and reticular nuclei send afferents to the vermal and paravermal zones of the cerebellum; together they enable an integration of visual data and proprioception (internal body awareness of position) to maintain equilibrium. However, the vermis and paravermis are implicated in much more than balance and equilibrium. The cerebellum has connections with the cingulate gyrus, septal nucleus, and the hippocampus, limbic structures important in social and emotional life. These connections explain the interruption in normal affective behavior that is often found with cerebellar lesions involving the vermis (Schmahmann & Sherman, 1998). Decreased cerebellar posterior vermis size is associated with fragile-X syndrome (Mostofsky et al., 1998), and with autism, specifically with hypoplasia of vermal lobules VI and VII (Courchesne et al., 1988).

It is in the lateral part of the cerebellum, the neocerebellum, that cognitive activity is most apparent in PET and fMRI scans. This follows from the neurocircuitry. The dentate nucleus sends information from the neocerebellum through the superior cerebellar peduncles (Fig. 5.5). This information projects via the thalamus to primary sensorimotor and premotor areas of the neocortex, including the frontal eye field (areas 4 and 6; Altman & Bayer, 1997). Recent work has shown that there are important thalamic projections from the cerebellum to the prefrontal cortex, to the oculomotor cortex (Middleton & Strick, 1994, 1997a, 1997b), temporal and posterior parietal lobes, as well as the paralimbic cerebral cortex (Schmahmann & Pandya, 1995). Conversely, the neocerebellum receives a massive input from the neocortex via the pontine nuclei through the middle cerebellar peduncle, including projections from the prefrontal and frontal lobes, the superior temporal gyrus, primary and association regions of the parietal and occipital lobes, and from the posterior parahippocampal gyrus and the cingulate gyrus (Schmahmann, 1996). The prefrontal cortex is associated with planning and strategy, and conscious control of behavior. The superior temporal gyrus is crucial to the decoding of speech, especially on the left side; association areas of the parietal and occipital lobes are the loci for fundamental visuospatial understanding, and many of these areas have been recruited for linguistic functions in the evolution of the modern human brain (Kolb & Whishaw, 1990). The connections between the cerebellum and neocortex are not proportionally reciprocal, however, because there is a significant afferent component traveling to the cerebellum from the parietal and occipital

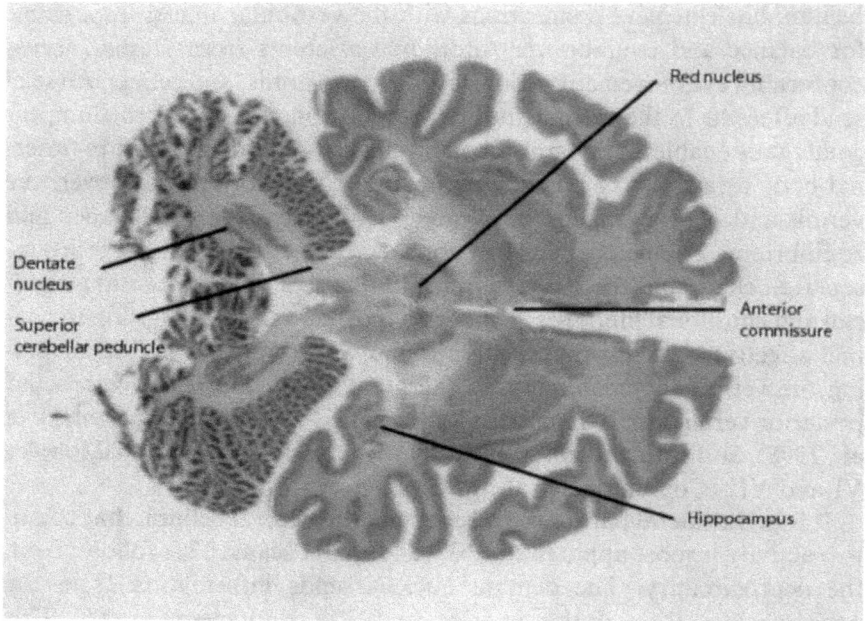

FIG. 5.5. Horizontal view of chimpanzee brain showing axons originating in
the dentate nucleus and passing through the superior cerebellar peduncle to the
red nucleus and thalamus. The septal nuclei cannot be shown in this section,
but are slightly superior to the anterior commissure.

lobes (Glickstein, 1997; Stein et al., 1987), whereas the most significant
efferent component of the neocerebellum projects to the frontal and prefrontal lobes (Chan-Palay, 1977; Middleton & Strick, 1997b). This suggests that the cerebellum can provide important connections between
parts of the neocortex (Glickstein, 1997), and may even participate in the
integration of cognitive activity at a high level. Compare the estimated 40
million axonal projections to the cerebellum from the neocortex with the
mere 2 million fibers in the descending corticospinal tracts, and it becomes clear that neither the neocortex nor the cerebellum can be conceived as operating in isolation (H. C. Leiner et al., 1986).

   All parts of the cerebellum are in direct contact with the body through
the dorsal columns of the spinal cord. On their way to the neocortex, the
dorsal columns send information from peripheral receptors to the cerebellum about fine touch, vibration, stereognosis (imaging an object
through touch), and proprioception (Altman & Bayer, 1997). The response of the cerebellum to the periphery is not direct, but is through de-

scending intermediate nuclei of the vestibular and reticular tracts or through the rubrospinal tract of the red nucleus (Altman & Bayer, 1997). Most importantly, the neocerebellar response to the periphery is primarily through the mediation of the neocortex. Thus, the cerebellum functions as a partner with the neocortex, processing information but never sending direct commands to the body except through intermediaries.

The Nature of Afferents and Efferents

What is the nature of the information sent by the cerebellum? Research cannot yet adequately answer this question, but the actual mechanism of the cerebellum is one of inhibition. The only cells to send information out of the cerebellum are the large Purkinje cells. The greater mass of the cerebellum is composed of mossy fibers (white matter), and their targets, the granule cells (Fig. 5.6). The axons of the granule cells reach upward past the Purkinje cells to the surface of the cerebellar cortex, where they bifur-

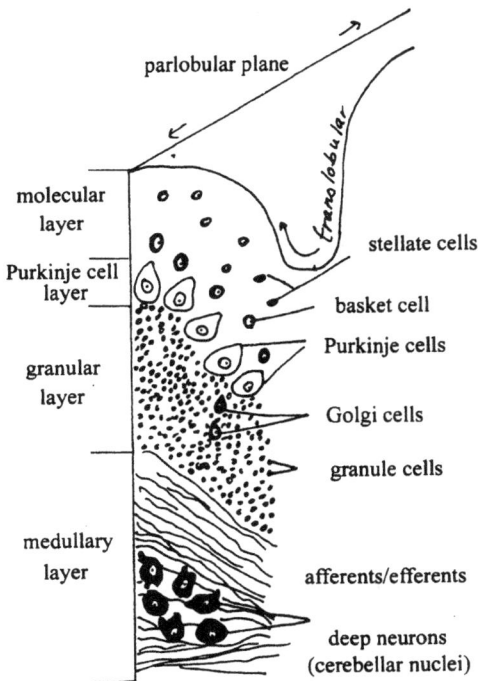

FIG. 5.6.   The cytoarchitecture of the cerebellar cortex. Medullary layer is white matter (adapted from Altman & Bayer, 1997).

cate into unmyelinated parallel fibers. These stimulate the Purkinje cells to fire only after an enormous number of synaptic contacts with the granule cells. Basket and stellate cells regulate the Purkinje output into a beam of excitation that travels to the appropriate zonal nucleus, and these signals are inhibitory (Eccles, Ito, & Szentágothai, 1967). The dentate nucleus, for example, is excited by collaterals of afferent fibers, these same fibers that have excited the Purkinje cells, but the firing of the Purkinje cell axons in the nucleus inhibits the excitatory target pattern. Thus, the information is presented in binary form by the nucleus to the neocortex as a reversed figure–ground pattern in which the "important" information about the figure is ignored. Of course, if it were a simple question of reversing the "image," the elaborate processing in the cerebellum would be unnecessary.

The very simplicity of the cerebellar cortex enables the cerebellum to integrate, process, and recontextualize information from diverse parts of the nervous system in all modalities. In contrast to the cytoarchitecture of the neocortex, the structure of cerebellar cortex is very uniform, with no functionally localized areas distinguished by cell shape and distribution (Eccles et al., 1967). Although there is a somatotopy (projection of particular body regions to a regional area of the cortex) to the medial cerebellum (Schmahmann, 1996), the pattern of projection onto the cerebellar cortex is not an integrated one, but one of "fractured somatotopy." The incoming mossy fibers break tactile information up into a mosaic, placing the receptors for such things as the whiskers of a rat in disparate parts of the cerebellar cortex, and providing them with a new context (Welker, 1987). The pontine nuclei also break up and recontextualize information to the cerebellum from the cerebral cortex. Axons from neurons in many parts of the pontine nuclei converge in a small volume of cerebellar cortex (convergence) but also sent pontocerebellar collaterals to different lobules and sides of the cerebellum (divergence; Brodal & Bjaalie, 1997). Whether fractured somatotopy is characteristic of all information, especially cognitive, remains to be explored, but there exists the enticing possibility that the cerebellum recontextualizes not only tactile but all types of information to provide a richer environment for neocortical associations.

Mossy fibers are not the only afferents to the cerebellum. The inferior olivary nucleus sends climbing fibers to the cerebellar cortex and collaterals to specific deep cerebellar nuclei. The inferior olivary nucleus may be important to learning, and a special role for this nucleus can be deduced from the behavior of the climbing fibers. Purkinje cells of the cerebellar cortex react quite differently to the climbing fibers than they do to the mossy fibers. The climbing fibers innervate the Purkinje cell with up to 300 synaptic connections; under such conditions, the Purkinje cell cannot help but fire. The relation between mossy fibers, which represent the lion's share of afferents to the cerebellum, and climbing fibers, originating only from the olive, is still not completely understood. Through electroencephalograph (EEG) recordings, it can be seen that the complex spike firing pattern of climbing fibers, distinguished from the simple spike pattern of mossy fibers, appears most frequently when a new motor sequence is being learned (Gilbert & Thach, 1977). The climbing fibers might condition the Purkinje cells to recognize an important pattern from mossy fiber input (Bloedel & Ebner, 1985; Marr, 1969), or erase the usual response of the Purkinje cells to a mossy fiber pattern to enable error correction (Albus, 1971; Ito & Kano, 1982). As there is an accumulating body of evidence from neuropsychology that the cerebellum is important in procedural learning, the particular function of these climbing fibers would likely figure significantly in this process. The difficulty is in connecting the nature of cell activity with the clinical and experimental evidence of cerebellar function.

The number of fibers in the cerebellar peduncles, and the number of neurons in the cerebellar cortex (which are more than the cerebral cortex; Glickstein, 1993), attest to the amount of information that the cerebellum processes from both central and peripheral nervous systems (H. C. Leiner et al., 1986). Cerebellar connections with the rest of the brain are ubiquitous, making it impossible to attribute any one function solely to the cerebellum. The corollary is that the cerebellum appears to participate in a number of cognitive functions once thought to be isolated to the neocortex. From the brief synopsis of cerebellar circuitry already presented, it is evident that the cerebellum is able to process both ascending information from the dorsal columns and at the same time, information from primary, associative, and limbic cortices. Particular areas of sensory activity are recontextualized in several new areas in the cerebellar cortex (fractured somatotopy). Yet if the incoming information is reorganized, the outgoing information is more strictly defined by parasagittal strips and beams of excitation. The principal inferior olivary nucleus projects specifically to these parasagittal strips (Voogd, Gerrits, & Hess, 1987), thus playing a role in both the output of the Purkinje cells to the nuclei, and the input of relevant information to the Purkinje cells through the climbing fibers and to the cerebellar nuclei through collaterals.

It has been said that the cerebellum cannot generate new patterns, but merely modulate the patterns provided by the neocortex in the motor, cognitive, and affective dimensions (Schmahmann, 1996). Yet this modulation is itself a creative process, one which could be important to the processes of assimilation and accommodation that take place at much higher levels.

REFERENCES

Albus, J. S. (1971). A theory of cerebellar function. *Mathematical Biosciences, 10*, 25–61.

Allen, G., Buxton, R. B., Wong, E. C., & Courchesne, E. (1997). Attentional activation of the cerebellum independent of motor involvement. *Science, 275*, 1940–1943.

Altman, J., & Bayer, S. A. (1997). *Development of the cerebellar system in relation to its evolution, structure, and functions*. New York: CRC Press.

Andrews, P., & Martin, L. (1992). Hominoid dietary evolution. In A. Whiten & E. M. Widdowson (Eds.), *Foraging strategies and natural diet of monkeys, apes and humans* (pp. 39–49). Oxford: Clarendon Press.

Barton, R. A., & Harvey, P. H. (2000). Mosaic evolution of brain structure in mammals. *Nature, 405*, 1055–1058.

Benefit, B. R. (1999). Victoriapithecus: The key to old world monkey and catarrhine origins. *Evolutionary Anthropology, 7*(5), 155–174.

Bloedel, J. R., & Bracha, V. (1997). Duality of cerebellar motor and cognitive functions. In J. D. Schmahmann (Ed.), *The cerebellum and cognition (International review of neurobiology*, Vol. 41, pp. 613–636). San Diego: Academic Press.

Bloedel, J. R., & Ebner, T. J. (1985). Climbing fiber function: Regulation of Purkinje cell responsiveness. In J. R. Bloedel, J. Dichgans, & W. Precht (Eds.), *Cerebellar functions* (pp. 247–259). Berlin: Springer-Verlag.

Boesch, C. (1991). Teaching in wild chimpanzees. *Animal Behaviour, 41*, 530–532.

Boesch, C. (1996). Three approaches for assessing chimpanzee culture. In A. E. Russon, K. A. Bard, & S. T. Parker (Eds.), *Reaching into thought; The minds of the great apes* (pp. 404–429). Cambridge, England: Cambridge University Press.

Boesch, C., & Tomasello, M. (1998). Chimpanzee and human cultures. *Current Anthropology, 39*(5), 591–614.

Bower, J. M. (1997a). Is the cerebellum sensory for motors sake, or motor for sensorys sake: The view from the whiskers of a rat. In C. E. de Zeeuw, P. Strata, & J. Voogd (Eds.), *The cerebellum: From structure to control* (pp. 463–498). Amsterdam: Elsevier.

Bower, J. M. (1997b). Control of sensory data acquisition. In J. D. Schmahmann (Ed.), *The cerebellum and cognition (International review of neurobiology*, Vol. 41, pp. 489–513). San Diego: Academic Press.

Brodal, P., & Bjaalie, J. G. (1997). Salient anatomic features of the cortico-ponto-cerebellar pathway. In C. E. de Zeeuw, P. Strata, & J. Voogd (Eds.), *The cerebellum: From structure to control* (pp. 227–250). Amsterdam: Elsevier.

Byers, J. A., & Walker, C. (1995). Refining the motor training hypothesis for the evolution of play. *American Naturalist, 146*, 25–40.

Byrne, R. (1995). *The thinking ape: Evolutionary origins of intelligence.* Oxford: Oxford University Press.

Byrne, R., & Byrne, J. M. E. (1993). Complex leaf-gathering skills of mountain gorillas (*gorilla g. beringei*) variability and standardization. *American Journal of Primatology, 31*, 241–261.

Chan-Palay, V. (1977). *Cerebellar dentate nucleus: Organization, cytology and transmitters.* Berlin: Springer-Verlag.

Clutton-Brock, T. H., & Harvey, P. H. (1979). Home range size, population density and phylogeny in primates. In I. S. Berstein & E. O. Smith (Eds.), *Primate ecology and human origins* (pp. 201–214). New York: Garland Press.

Courchesne, E., Yeung-Courchesne, R., Press, G. A., Hesselink, J. R., & Jernigan, T. L. (1988). Hypoplasia of cerebellar vermal lobules VI and VII in autism. *New England Journal of Medicine, 318*(21), 1349–1354.

Courchesne, E., Townsend, J., Akshoomoff, N. A., Saitoh, O., Yeung-Courchesne, R., Lincoln, A. J., James, H. E., Haas, R. H., Schreibman, L., & Lau, L. (1994). Impairment in shifting attention in autistic and cerebellar patients. *Behavioral Neuroscience, 108*(5), 848–865.

Diamond, A. (2000). Close interrelation of motor development and cognitive development and of the cerebellum and prefrontal cortex. *Child Development, 71*(1), 44–56.

Doyon, J. (1997). Skill learning. In J. D. Schmahmann (Ed.), *The cerebellum and cognition (International review of neurobiology*, Vol. 41, pp. 273–294). San Diego: Academic Press.

Eccles, J. C., Ito, M., & Szentágothai, J. (1967). *The cerebellum as a neuronal machine.* New York: Springer-Verlag.

Fairbanks, L. A. (2000). The developmental timing of primate play: A neural selection model. In S. T. Parker, J. Langer, & M. L. McKinney (Eds.), *Biology, brains, and behavior: The evolution of human development* (pp. 131–158). Santa Fe, NM: School of American Research Press.

Fiez, J. A., Petersen, S. E., Cheney, M. K., & Raichle, M. E. (1992). Impaired non-motor learning and error detection associated with cerebellar damage. *Brain, 115*, 155–178.

Fiez, J. A., Raife, E. A., Balota, D. A., Schwwarz, J. P., Raichle, M. E., & Petersen, S. E. (1996). A positron emission tomography study of the short-term maintenance of verbal information. *Journal of Neuroscience, 16*, 808–822.

Fiez, J. A., & Raichle, M. E. (1997). Linguistic processing. In J. D. Schmahmann (Ed.), *The cerebellum and cognition (International review of neurobiology*, Vol. 41, pp. 233–254). San Diego: Academic Press.

Finlay, B. L., & Darlington, R. B. (1995). Linked regularities in the development and evolution of mammalian brains. *Science, 268*, 1578–1584.

Fleagle, J. C. (1999). *Primate adaptation and evolution* (2nd ed.). Toronto: Academic Press.

Gao, J.-H., Parsons, L. M., Bower, J. M., Xiong, J., Li, J., & Fox, P. T. (1996). Cerebellum implicated in sensory acquisition and discrimination rather than motor control. *Science, 272*, 545–547.

Gardner, R. A., & Gardner, B. T. (1969). Teaching sign language to a chimpanzee. *Science, 165*, 664–672.

Gibson, K. R. (1977). Brain structure and intelligence in macaques and human infants from a Piagetian perspective. In S. Chevalier-Skolnikoff & F. E. Poirier (Eds.), *Primate bio-social development: Biological, social, and ecological determinants* (pp. 113–157). New York: Garland.

Gibson, K. R. (1986). Cognition, brain size and the extraction of embedded food resources. In J. C. Else & P. C. Lee (Eds.), *Primate ontogeny, cognition and social behaviour* (pp. 93–105). Cambridge: Cambridge University Press.

Gibson, K. R. (1990). New perspectives on instincts and intelligence: Brain size and the emergence of hierarchical mental constructional skills. In S. T. Parker & K. R. Gibson (Eds.), *"Language" and intelligence in monkeys and apes: Comparative developmental perspectives* (pp. 97–128). Cambridge: Cambridge University Press.

Gibson, K. R. (1991). Myelination and behavioral development: A comparative perspective on questions of neoteny, altriciality and intelligence. In K. R. Gibson & A. C. Petersen (Eds.), *Brain maturation and cognitive development: Comparative and cross-cultural perspectives* (pp. 29–63). Hawthorne, NY: Aldine de Gruyter.

Gibson, K. R. (1999). Social transmission of facts and skills in the human species: Neural mechanisms. In H. O. Box & K. R. Gibson (Eds.), *Mammalian social learning: Comparative and ecological perspectives* (pp. 351–366). Cambridge: Cambridge University Press.

Gilbert, P. F. C., & Thach, W. T. (1977). Purkinje cell activity during motor learning. *Brain Research, 128*, 309–328.

Glickstein, M. (1993). Motor skills but not cognitive tasks. *Trends in Neurosciences, 16*(11), 450–451.

Glickstein, M. (1997). Mossy-fibre sensory input to the cerebellum. In C. I. Zeeuw, P. Strata, & J. Voogd (Eds.), *The cerebellum: From structure to control* (pp. 251–262). Amsterdam: Elsevier.

Greenfield, P. (1991). Language, tools and brain: The ontogeny and phylogeny of hierarchically organized sequential systems. *Behavioral and Brain Sciences, 14*, 531–595.

Holmes, G. (1939). The cerebellum of man. *Brain, 62*, 1–30.

Inhoff, A. W., Diener, H. C., Rafal, R. D., & Ivry, R. B. (1989). The role of cerebellar structures in the execution of serial movements. *Brain, 112*, 565–581.

Ito, M. (1993). Movement and thought: Identical control mechanisms by the cerebellum. *Trends in Neurosciences, 16*(11), 448–450.

Ito, M., & Kano, M. (1982). Long lasting depression of parallel fiber-Purkinje cell transmission induced by conjunctive stimulation of parallel fibers and climbing fibers in the cerebellar cortex. *Neuroscience Letters, 33*, 253–258.

Jacobs, B., Chugani, H. T., Allada, V., Chen, S., Phelps, M. E., Pollack, D. B., & Raleigh, M. J. (1995). Developmental changes in brain metabolism in sedated rhesus macaques and vervet monkeys revealed by positron emission tomography. *Cerebral Cortex, 3*, 222–233.

Jerison, H. J. (1973). *Evolution of the brain and intelligence.* New York: Academic Press.

Kay, R. F., & Ungar, P. S. (1997). Dental evidence for diet in some Miocene Catarrhines with comments on the effects of phylogeny on the interpretation of adaptation. In D. R. Begun, C. V. Ward, & M. D. Rose (Eds.), *Function, phylogeny, and fossils: Miocene hominoid evolution and adaptations* (pp. 131–151). New York: Plenum Press.

Keele, S. W., & Ivry, R. (1990). Does the cerebellum provide a common computation for diverse tasks? A timing hypothesis. *Annals of the New York Academy of Science, 608*, 179–211.

Kim, S.-G., Ugurbil, K., & Strick, P. L. (1994). Activation of a cerebellar output nucleus during cognitive processing. *Science, 265*, 949–951.

Kolb, B., & Whishaw, I. Q. (1990). *Fundamentals of human neuropsychology.* New York: W. H. Freeman.

Larsell, O., & Jansen, J. (1970). *The comparative anatomy and histology of the cerebellum from monotremes through apes* (Vol. 2). Minneapolis: University of Minnesota Press.

Larson, S. G. (1998). Parallel evolution in the hominoid trunk and forelimb. *Evolutionary Anthropology, 6*(3), 87–99.

Leiner, H. C., Leiner, A. L., & Dow, R. S. (1986). Does the cerebellum contribute to mental skills? *Behavioral Neuroscience, 100*(4), 443–454.

Leiner, H. C., Leiner, A. L., & Dow, R. S. (1989). Reappraising the cerebellum: What does the hindbrain contribute to the forebrain? *Behavioral Neuroscience, 103*(5), 998–1008.

Leiner, H. C., Leiner, A. L., & Dow, R. S. (1991). The human cerebro-cerebellar system: Its computing, cognitive, and language skills. *Behavioural Brain Research, 44*, 113–128.

MacLeod, C. E. (2000). *The cerebellum and its part in the evolution of the hominoid brain.* Unpublished doctoral dissertation, Simon Fraser University, Burnaby, BC, Canada.

MacLeod, C. E., Zilles, K., Schleicher, A., & Gibson, K. R. (2001a). The cerebellum: An asset to hominoid cognition. In B. M. F. Galdikas, N. E. Briggs, L. K. Sheeran, G. L. Shapiro, & J. Goodall (Eds.), *All apes great and small. Volume I: African apes* (pp. 35–53). New York: Kluwer Academic.

MacLeod, C. E., Schleicher, A., & Zilles, K. (2001b). Natural selection and fine neurological tuning. *American Journal of Physical Anthropology, 32*(Suppl.), 100.

MacLeod, C. E., Zilles, K., Schleicher, A., Rilling, J. K., & Gibson, K. R. (2003). Expansion of the neocerebellum in *Hominoidea. Journal of Human Evolution, 44*, 401–429.

Marr, D. (1969). A theory of cerebellar cortex. *Journal of Physiology, 202*, 437–470.

Matano, S., Baron, G., Stephan, H., & Frahm, H. D. (1985). Volume comparisons in the cerebellar complex of primates II. Cerebellar nuclei. *Folia Primatologica, 44*, 182–203.

Middleton, F. A., & Strick, P. L. (1994). Anatomical evidence for cerebellar and basal ganglia involvement in higher cognitive function. *Science, 266*, 458–461.

Middleton, F. A., & Strick, P. L. (1997a). Cerebellar output channels. In J. D. Schmahmann (Ed.), *The cerebellum and cognition (International review of neurobiology*, Vol. 41, pp. 61–82). San Diego: Academic Press.

Middleton, F. A., & Strick, P. L. (1997b). Dentate output channels: Motor and cognitive components. In C. I. DeZeeuw, P. Strata, & J. Voogd (Eds.), *The cerebellum: From structure to control, progress in brain research* (Vol. 114, pp. 553–568). Amsterdam: Elsevier.

Miles, H. L. (1983). Apes and language: The search for communicative competence. In J. de Luce & H. T. Wilder (Eds.), *Language in primates* (pp. 43–61). New York: Springer-Verlag.

Milton, K. (1981). Distribution patterns of tropical plant foods as an evolutionary stimulus to primate mental development. *American Anthropologist, 83*, 534–548.

Molinari, M., Petrosini, L., & Grammaldo, L. G. (1997). Spatial event processing. In J. D. Schmahmann (Ed.), *The cerebellum and cognition (International review of neurobiology*, Vol. 41, pp. 217–230). San Diego: Academic Press.

Moore, A. H., Cherry, S. R., Pollack, D. B., Hovda, D. A., & Phelps, M. E. (1999). Application of positron emission tomography to determine glucose utilization in conscious infant monkeys. *Journal of Neuroscience Methods, 88*, 123–133.

Mostofsky, S. H., Mazzacco, M. M., Aakalu, G., Warsofsky, I. S., Denckla, M. B., & Reiss, A. L. (1988). Decreased cerebellar posterior vermis size in fragile X syndrome: Correlation with neurocognitive performance. *Neurology, 50*(1), 121–130.

Nieuwenhuys, R., Voogd, J., & Van Huijzen, C. (1988). *The human central nervous system; A synopsis and atlas* (3rd rev. ed.). Berlin: Springer-Verlag.

Parker, S. T. (1996a). Apprenticeship in tool-mediated extractive foraging: The origins of imitation, teaching, and self-awareness in great apes. In A. E. Russon, K. A. Bard, & S. T. Parker (Eds.), *Reaching into thought: The minds of the great apes* (pp. 348–370). Cambridge, England: Cambridge University Press.

Parker, S. T. (1996b). Using cladistic analysis of comparative data to reconstruct the evolution of cognitive development in Hominids. In E. P. Martins (Ed.), *Phylogenies and the comparative method in animal behavior* (pp. 361–398). Oxford: Oxford University Press.

Parker, S. T. (2002). Evolutionary relationships between molar eruption and cognitive development in anthropoid primates. In N. Minugh-Purvis & K. MacNamara (Eds.), *Human evolution through developmental change* (pp. 305–316). Baltimore, MD: The Johns Hopkins University Press.

Parker, S. T., & Gibson, K. R. (1977). Object manipulation, tool use and sensorimotor intelligence as feeding adaptations in Cebus monkeys and great apes. *Journal of Human Evolution, 6,* 623–641.

Parker, S. T., & McKinney, M. (1999). *Origins of intelligence.* Baltimore, MD: The Johns Hopkins University Press.

Parker, S. T., & Milbrath, C. (1993). Higher intelligence, propositional language, and culture as adaptations for planning. In K. R. Gibson & T. Ingold (Eds.), *Tools, language and cognition in human evolution* (pp. 314–333). Cambridge: Cambridge University Press.

Patterson, F. G. (1978). The gestures of a gorilla: Language acquisition in another pongid. *Brain and Language, 5,* 72–97.

Petersen, S. E., Fox, P. T., Posner, M. I., Mintun, M., & Raichle, M. E. (1989). Positron emission tomographic studies of the processing of single words. *Journal of Cognitive Neuroscience, 1*(2), 153–170.

Petrides, M., & Iversen, S. K. (1979). Restricted posterior parietal lesions in the rhesus monkey and performance on visuospatial tasks. *Brain Research, 161,* 63–77.

Piaget, J. (1952). *The origins of intelligence in children.* New York: International Universities Press.

Piaget, J. (1954). *The construction of reality in the child.* New York: Basic Books.

Piaget, J. (1962). *Play, dreams, and imitation in childhood.* London: Routlege & Kegan Paul.

Potts, R. (in press). Paleoenvironments and the evolution of adaptability in apes. In A. Russon & D. Begun (Eds.), *Evolution of great ape intelligence.* Cambridge: Cambridge University Press.

Povinelli, D., & Cant, J. G. H. (1995). Arboreal clambering and the evolution of self-conception. *The Quarterly Review of Biology, 70*(4), 393–421.

Premack, D. (1972). Language in chimpanzees. *Science, 172*, 808–822.

Purves, A., & Lichtman, J. W. (1985). Elimination of synapses in the developing nervous system. *Science, 210*, 153–157.

Pysh, J. J., & Weiss, G. M. (1979). Exercise during development induces an increase in Purkinje cell dendritic tree size. *Science, 206*, 230–231.

Raichle, M. E., Fiez, J. A., Videen, T. O., MacLeod, A.-M. K., Pardo, J. V., Fox, P. T., & Petersen, S. E. (1994). Practice-related changes in human brain functional anatomy during nonmotor learning. *Cerebral Cortex, 4*, 8–26.

Rakic, P., Bourgeois, J.-P, Eckenhoff, M. F., Zecevic, N., & Goldman-Rakic, P. S. (1986). Concurrent overproduction of synapses in diverse regions of the primate cerebral cortex. *Science, 232*, 232–235.

Rilling, J. K., & Insel, T. R. (1998). Evolution of the cerebellum in primates: Differences in relative volume among monkeys, apes and humans. *Brain, Behavior and Evolution, 52*, 308–314.

Rumbaugh, D. M., Gill, T. V., & von Glasersfeld, E. C. (1973). Reading and sentence completion by a chimpanzee (*Pan*). *Science, 9*, 343–347.

Russon, A. E. (1996). Imitation in everyday use: Matching and rehearsal in the spontaneous imitation of rehabilitant orangutans (*Pongo pygmaeus*). In A. E. Russon, K. A. Bard, & S. T. Parker (Eds.), *Reaching into thought: The minds of the great apes* (pp. 152–176). Cambridge: Cambridge University Press.

Russon, A. E. (1998). The nature and evolution of intelligence in orangutans (*Pongo pygmaeus*). *Primates, 39*(4), 485–503.

Russon, A. E., Bard, K. A., & Parker, S. T. (Eds.). (1996). *Reaching into thought: The minds of the great apes.* Cambridge: Cambridge University Press.

Schmahmann, J. D. (1996). From movement to thought: Anatomic substrates of the cerebellar contribution to cognitive processing. *Human Brain Mapping, 4*, 174–198.

Schmahmann, J. D. (Ed.). (1997). *The cerebellum and cognition (International review of neurobiology*, Vol. 41). San Diego: Academic Press.

Schmahmann, J. D., & Pandya, D. N. (1995). Prefrontal cortex projections to the basilar pons in rhesus monkey: Implications for the cerebellar contribution to higher function. *Neuroscience Letters, 199*, 175–178.

Schmahmann, J. D., & Sherman, J. C. (1998). The cerebellar cognitive affective syndrome. *Brain, 121*, 561–579.

Stein, J. F., Miall, R. C., & Weir, D. J. (1987). The role of the cerebellum in the visual guidance of movement. In M. Glickstein, C. Yeo, & J. Stein (Eds.), *Cerebellum and neuronal plasticity* (pp. 175–192). New York: Plenum.

Thach, W. T. (1978). Correlation of neural discharge with pattern and force of muscular activity, joint position, and direction of intended next movement in motor cortex and cerebellum. *Journal of Neurophysiology, 41*, 654–686.

Thach, W. T. (1996). On the specific role of the cerebellum in motor learning and cognition: Clues from PET activation and lesion studies in man. *Behavioral and Brain Sciences, 19*, 411–431.

Thach, W. T. (1997). Context-response linkage. In J. D. Schmahmann (Ed.), *The cerebellum and cognition (International review of neurobiology*, Vol. 41, pp. 599–611). San Diego: Academic Press.

Tilney, F., & Riley, H. A. (1928). *The brain from ape to man: A contribution to the study of the evolution and development of the human brain* (Vols. 1 & 2). New York: Paul B. Hoeber.

Tomasello, M., & Todd, J. (1983). Joint attention and lexical acquisition style. *First Language, 4,* 197–212.

Van Schaik, C. P., Ancrenaz, M., Borgen, G., Galdikas, B., Knott, C., Singleton, I., Suzuki, A., Utami, S., & Merrill, M. (2003). Orangutan cultures and the evolution of material culture. *Science, 299,* 102–105.

Van Schaik, C. P., Fox, E. A., & Sitompul, A. F. (1996). Manufacture and use of tools in wild Sumatran orangutans. *Naturwissenschaften, 4,* 186–188.

Voogd, J., Feirabend, H. K. P., & Schoen, J. H. R. (1990). Cerebellum and precerebellar nuclei. In G. Paxinos (Ed.), *The human nervous system* (pp. 321–386). San Diego: Academic Press.

Voogd, J., Gerrits, N. M., & Hess, D. T. (1987). Parasagittal zonation of the cerebellum in macaques: An analysis based on acetylcholinesterase histochemistry. In M. Glickstein, C. Yeo, & J. Stein (Eds.), *Cerebellum and neuronal plasticity* (pp. 15–39). New York: Plenum.

Welker, W. (1987). Spatial organization of somatosensory projections to granule cell cerebellar cortex: Functional and connectional implications of fractured somatotopy (summary of Wisconsin studies). In J. S. King (Ed.), *New concepts in cerebellar neurobiology* (pp. 239–280). New York: Alan R. Liss.

Whiten, A., Goodall, J., McGrew, W. C., Nishida, T., Reynolds, V., Sugiyama, Y., Tutin, C. E. G., Wrangham, R. W., & Boesch, C. (1999). Cultures in chimpanzees. *Nature, 399,* 682–685.

Whiting, B. A., & Barton, R. A. (2003). The evolution of the cortico-cerebellar complex in primates: Anatomical connections predict patterns of correlated evolution. *Journal of Human Evolution, 44*(1), 3–10.

6

From Mirror Neurons to
the Shared Manifold Hypothesis:
A Neurophysiological Account
of Intersubjectivity

Vittorio Gallese
University of Parma

The extensive and pervasive social habits of primates are the result of a very long evolutionary path. We readily ascribe intelligence to other animals, while being simultaneously inclined to think that—cognitively speaking—humans "do it better." We are and we feel that we are different from other animals, even from our closest relatives—apes. There are many differences indeed. The crucial one is the capacity to "read" the mind of others, which most people ascribe just to humans. The context is therefore between species confined to behavior reading and our species, which makes use of a different level of explanation—mind reading.

But mind reading, whatever it might be, certainly does not include all there is in our cognitive life. There are many possible ways to characterize our cognitive life; there are many possible ways to "live" our lives and to look at them. We can put ourselves on a scale and check our body weight, complaining about how much we had for supper. Or we can think about what someone else should not have thought about us. In both instances, we do not experience any identity shift. We do not feel different when we are checking our body weight with respect to when we entertain counterfactual, third-person metarepresentations. Quite rightly so, what does change is not the individual organism. What changes is the type of *relation* in which each individual organism (as part of a biological system) engages itself during the various possible kinds of interaction with the world outside. There are almost infinite levels at which we may decide to act in the world. And there are almost infinite levels at which others may do the same. We can take a swim, plant a tree, get a doctoral degree, or dream of thinking about Ulysses. And we know that others look the same, do the same, think the same, or that they do not. These levels of interaction pertain to different identities, different persons whom, nevertheless, we recognize and represent as similar to us.

One could adopt a solipsistic view and claim that it is just because all individuals are the same that we should not indulge in speculations on the relevance of others' minds to define cognition. *Solipsism* implies that focusing on a single individual's mind is all that is required to define what a mind is and how it works.

But I will not follow the solipsistic view. I maintain that our cognitive way to deal with life is but one expression of the global and richer experience we have of the world by way of interacting with it. Within the network of interactions shaping our take of the world, the social dimension seems to play a very powerful role. Social behavior is not peculiar to primates; it is found across species as far apart in evolution as humans and ants. Social interactions certainly play different roles with different modalities across different species; nevertheless, common to all social species and to all social cultures, of whatever complexity, is the notion of identity of the individuals within those species and cultures. It follows that all levels of interaction that can be employed to characterize cognition in single individuals must intersect or overlap one to another to enable the development of mutual recognition and intelligibility.

It has been cleverly demonstrated that most of the words and sentences that we employ in everyday language to describe our transactions with the world are deeply rooted in our embodied experience of the world (see Lakoff, 1987; Lakoff & Johnson, 1984, 1999; Lakoff & Nuñez, 2000). Still, the dominant view in cognitive science is to reiteratively put most efforts in clarifying what are the formal rules structuring a solipsistic mind. A much less effortful and deep inquiry has been devoted to investigate what triggers the sense of identity that we experience among the many sides of our self and between us and the thousands of other selves populating the world we share.

There are indeed at least two types of identity waiting for an explanation: (a) The identity we experience as individual organisms, the sense marking our uniqueness among other beings (*i-Identity*); (b) The identity we experience in other social individuals, the sense of "being like you" (*s-*

Identity). Our possibility to see or think that we are different from other living and nonliving objects is determined by our capacity to entertain i-Identity. But it is s-Identity that enables us to have the possibility of entertaining a meaningful dialogue with others.

Is there any primacy effect that regulates the development of these different types of identity? How do individual and social identities relate to each other? In this chapter, I concentrate on the functional aspects and the neural underpinnings of s-Identity. The point I want to make is that beside—and likely before—the ascription of any intentional content to others, we entertain a series of "implicit certainties" about the content-bearing individuals whom we confront. These implicit certainties are constitutive of the intersubjective relation, and deal with the sense of oneness, of identity with the other, that enables the possibility to ascribe any content, whatever it might be, to the individual with whom we are interacting. I analyze, from a neuroscientific perspective, what are the functional mechanisms at the basis of the implicit certainties enabling intersubjective relations, and what might be the neural mechanisms underpinning them.

CONTROL STRATEGIES AND REPRESENTATIONS

Several developmental psychology studies have shown that the capacity of infants to establish relations with others is accompanied by the registration of behavioral invariance. As pointed out by Stern (1985), this invariance encompasses unity of locus, coherence of motion, and coherence of temporal structure. This experience-driven process of constant remodeling of the biological system is one of the building blocks of cognitive development, and it capitalizes on coherence, regularity, and predictability. Identity guarantees all these features, henceforth, its high social adaptive value.

Anytime we meet someone we do not just "perceive" that someone to be, broadly speaking, similar to us. We are implicitly aware of this similarity, because we literally embody it. Meltzoff and Brooks (2001) convincingly suggested that the analogy between infant and caregiver is the starting point for the development of (social) cognition.

The seminal study of Meltzoff and Moore (1977) and the subsequent research field it opened (see Meltzoff, 2002; Meltzoff & Moore, 1997), showed that newborns as young as 18 hours are perfectly capable of reproducing mouth movements (e.g., tongue protrusion) displayed by the adult they are facing. That particular part of their body replies, though not in a reflexive way (see Meltzoff & Moore, 1977, 1994), to movements displayed by the equivalent body part of someone else. More precisely, this means that newborns set into motion, and in the "correct" way, a part of their body they have no visual access to, but which nevertheless acts by matching an observed behavior; visual information is then transformed into motor information.

This apparently innate mechanism has been defined as *active intermodal mapping* (AIM; see Meltzoff & Moore, 1997). Intermodal mapping provides representational frames not constrained by any particular mode of interaction, be it visual, auditory, or motor (Meltzoff, 2002). Modes of interaction as diverse as seeing, hearing, or doing something must therefore share some peculiar feature making the process of equivalence carried out by AIM possible.

The issue then consists in clarifying the nature of this peculiar feature and the possible underlying mechanisms. The relational character intrinsic to any biological system–environment interaction appears to be a good candidate. Our environment is composed of a variety of lifeless, more or less compliant, forms of matter; and of a variety of "alive stuff," whose peculiar character is more and more focused by the infant's immature eye. Individuals confront themselves with all possible kinds of external objects, by virtue of their peculiar status as biological systems, thus by definition are constrained in their peculiar "modes of interaction" (see Gallese, 2003a, 2003b, 2003c).

Any organism–world interaction requires a control system implementing a control strategy. The control system needs to map the peculiar states of the organism in relation with external objects or states of affairs. Interestingly enough, biological control strategies share with modes of interaction the relational character. As modes of interaction, control strategies are intrinsically relational in that they model the interaction between organism and environment to better control it. But a model is indeed a form of representation. If the functional strategy of biological control systems is that of modeling organism–world interactions, this step allows a relation of interdependence, if not superposition, between behavior control and representation to be established (see Gallese, 2000a, 2000b, 2000c, 2003). This relation holds for both organism–object and organism–organism modes of interaction, and it seems to be established at the very onset of our life, when a self-conscious, subjective perspective is not yet fully established. Yet, the absence of a fully self-conscious subject does not preclude the presence of a primitive "self–other shared space"; this space is shared by the infant with the living others. The shared space enables the social bootstrapping of cognitive and affective development because it provides an incredibly powerful tool to detect and incorporate coherence, regularity, and predictability in the course of the interactions of the individual with the social environment.

Once the crucial bonds with the world of others are established, this space carries over to the adult conceptual faculty of socially mapping sameness and difference ("I am a different subject"). Within intersubjective relations, the other is a living oxymoron, being just a different self. My proposal is that s-Identity, the "selfness" we readily attribute to others, the inner feeling of "being like you" triggered by our encounter with others, is the result of the preserved shared intersubjective space. Self–other physical and epistemic interactions are shaped and conditioned by the same bodily and environmental constraints. This common relational character is underpinned at the level of the brain, by neural networks that compress the redundant "who done it," "who is it" specifications, and realize a narrower content state (see also Gallese, 2003c). A content specifies what kind of interaction or state is at stake; this narrower content is shared just because, as we have learned from developmental psychology, the shareable character of experience and action is the earliest constituent of our life.

Before presenting empirical evidence in support of my hypothesis, it is necessary to clarify what are the conditions under which the neuroscientific level of description would appear reasonably apt to support it. The following conditions should do the job: (a) Evidence of a neural representational format capable of achieving sameness of content in spite of the multiple and different ways content might be originated by its referents; (b) Indifference of the representational format to the peculiar perspectival spaces from which referents project their content. In other words, indifference to self–other distinctions; (c) Persistence of the same representational format also in adulthood.

The posited important role of identity relations in constraining the cognitive development of our social mind provides a strong motivation to investigate from a neuroscientific perspective, the functional mechanisms (and their neural underpinnings) at the basis of the self–other identity. In the next sections, I review neuroscientific evidence for a mechanism that appears to be in a good position for satisfying all three conditions. Before doing so, however, I think that some clarification about the relationship between biological control systems and "neural representations" is required.

The traditional view holds that mental representations and their semantic content are *extrinsic* properties, that is, they are relational because they depend on objects and facts of the world. In contrast, according to the same view, causative properties—those related to the control and production of overt behavior—are considered to be *intrinsic*, and nonrelational (for an extensive discussion of this point, see Kim, 1998). My objection to this dichotomous account is that the causative properties in a sensory-motor loop, those involved in behavior production, actually also belong to the target object. I want to question the whole conceptual distinction between extrinsic/semantic properties and intrinsic/causative properties. The philosophical intuition behind my thesis is that causative properties, conceived of as the firing of sensory–motor neural networks, are content properties, and that—at least for some forms of mental content—their meaning is literally constituted by the way they are "enacted" by a situated and functionally-grounded organism.

Throughout the chapter, I speak of "neural representations of . . ." Also, this expression requires some qualification. I do not refer to a merely symbolic, abstract equivalence between a real entity in the world and a computational code, which in principle can be multiply instantiated in whatever substrate. The expression, "the neural representation of . . ." has to be qualified according to what I take to be its original meaning, that of biological control.

To clarify, I use an example. Let us suppose that we are asked to reduce the rate of our heartbeat by about 10%. At first, it would seem almost impossible to achieve such a goal. However, if we can visualize our heartbeat on a monitor plugged to an electrocardiograph, with some practice, we will be able to reduce the frequency at which our heart beats to the desired level. There is no magic in it; we are simply using a well-known practice named *biofeedback*. What does this trivial example tell us? It tells us that a very efficient way to control a given variable (the heart rate in our example) is to produce a copy, a representation of that variable. According to this perspective, the notion of *representation* is freed from its abstract connotation—typical of the representational–computational account of the mind—and is respelled within a naturalistic perspective. This account of representation stresses its preconceptual and prelinguistic roots. This point will hopefully become clearer once I have specified the double executive/representational character of the motor system.

ACTION PERCEPTION

In our daily life, we are constantly exposed to the actions of the individuals inhabiting our social world. We can immediately tell whether a given observed act or behavior is the result of a purposeful attitude or rather the unpredicted consequence of some accidental event totally unrelated to the agent's will. In other words, we are able to understand the behavior of others in terms of their mental states. Yet, because our starting point to attribute intentions to others must necessarily rely on the observation of their overt behavior, I analyze how actions are represented and implicitly understood. The main aim of my arguments is to show that, far from being exclusively dependent on mentalistic/linguistic abilities, the capacity to understand others as intentional agents is deeply grounded in the relational nature of our interactions with others.

The perception of our social environment relies basically on vision. It is through vision that we are able to recognize different individuals, to locate them in space, and to record their behavior. About 50 years of neuroscientific investigation have clarified many aspects of vision, from the transduction processes carried out by receptors in the retina, to the different stages along which visual images are processed and analyzed by the brain. In this section, I confine my review to data obtained from monkeys. The problem of how the human brain processes actions is addressed later in the chapter.

The most accepted model of how the brain analyzes visual information maintains that visual processing is carried out in a piecemeal fashion, with specialized cortical regions "dedicated" to the analysis of shape, color, and motion (for a comprehensive neuroscientific account of vision, see Zeki, 1993).

Motion analysis is crucial to discriminate and recognize observed actions performed by other individuals. Area MT, or V5 (Zeki & Shipp, 1988), which in the monkey is located in the caudal part of the ventral bank of the Superior Temporal Sulcus (STS), is one of the most studied among the so-called extrastriate visual areas. Several electrophysiological studies have shown that area MT is specialized for the analysis of visual motion (Desimone & Ungerleider, 1986; Dubner & Zeki, 1971; Maunsell

& Van Essen, 1983; Van Essen, Maunsell, & Bixby, 1981; Zeki, 1974). Interestingly enough, however, very little efforts have been comparatively devoted to the investigation of where and how biological motion is analyzed and processed in the brain.

Since the mid-1980s, David Perrett and co-workers (Oram & Perrett, 1994; Perrett et al., 1989; Perrett, Mistlin, Harries, & Chitty, 1990), have filled this gap with a series of seminal works showing that in a cortical sector buried within the anterior part of the STSa of the monkey, there are neurons selectively activated by the observation of various types of body movements such as walking, turning the head, stretching the arm, bending the torso, and so on (for review, see Carey, Perrett, & Oram, 1997; Jellema & Perrett, 2001). Particularly interesting are neurons responsive to goal-related behaviors; these neurons do not respond to static presentations of hands or objects, but require the observation of a meaningful, goal-related, hand–object interaction in order to be triggered (Perrett, Mistlin, Harries, & Chitty, 1990). Comparable hand actions without target object or hand movements without physical contact with the object in view do not evoke any response. The responses of these neurons generalize across different viewing conditions including distance, speed, and orientation. Incidentally, it must be noted that no attempt has been made by these authors to test the responsiveness of these neurons during active movements of the monkey. The responses of some of these neurons have been shown not to be sensitive to form, so that even light dot displays moving with biologically plausible kinematics are as good as true limbs and hands in evoking the neurons' discharge (Jellema & Perrett, 2001; Oram & Perrett, 1994). Neurons responding to complex biological visual stimuli such as walking or climbing were reported also in the amygdala (see Brothers & Ring, 1992).

Altogether these results provide strong evidence supporting the notion that distinct specific sectors of the visual system are selectively involved in the representation of behaviors of others. Visual representation, however, is not understanding. A visual representation of a given stimulus does not necessarily convey all the information required to assign a meaning to it, and therefore to understand such a stimulus. A purely visual representation of a behavior does not allow for is to code/represent it as an intended, goal-related behavior.

I now briefly present some empirical results that may help in elucidating the neural mechanisms at the basis of a more comprehensive account of implicit action understanding.

The most rostral sector of the ventral premotor cortex of the macaque monkey controls hand and mouth movements (Hepp-Reymond, Hüsler, Maier, & Qi, 1994; Kurata & Tanji, 1986; Rizzolatti et al., 1988; Rizzolatti, Scandolara, Gentilucci, & Camarda, 1981). This sector, which has specific histochemical and cytoarchitectonic features, has been termed area F5 (Matelli, Luppino, & Rizzolatti, 1985). A fundamental functional property of area F5 is that most of its neurons do not discharge in association with elementary movements, but are active during actions such as grasping, tearing, holding or manipulating objects (Rizzolatti et al., 1988).

What is responded to is the relation, in motor terms, between the organism and the external object of the interaction. Furthermore, this relation is of a very special kind: a relation projected to an expected success. A hand reaches for an object, it grasps it, and does things with it. F5 neurons become active only if a particular type of interaction (e.g., hand–object, mouth–object, or both) is executed, and they cease firing when the relation leads to a different state of the organism (e.g., to take a piece of food, to throw an object away, to break it, to bring it to the mouth, to bite it, etc.). Particularly interesting in this respect are grasping-related neurons that fire any time the monkey successfully grasps an object, regardless of the effector employed, be it in either of his two hands, or the mouth, or both (Rizzolatti et al., 1988; see also Rizzolatti, Fogassi, & Gallese, 2000).

The independence between the nature of the effector involved and the end-state that the same effector attains constitutes an *abstract* kind of representation (Gallese, 2003c). The firing of these neurons instantiate the same content (the new end state the organism will attain), even if differently mediated. In accord with information theory, a narrower content state has been reached by compressing redundant information such as "which effector," or "which dynamic parameters" should be involved in the course of the interaction. This compression process is not cognitive per se. It is just an information compression process. Nevertheless, by employing a mentalist language we could describe this neural mechanism in terms of goal-representation (see Rizzolatti, 1988; Rizzolatti, Fogassi, & Gallese, 2000).

Beyond purely motor neurons, which constitute the overall majority of all F5 neurons, area F5 contains also visuomotor neurons. They have motor properties that are indistinguishable from those of the aforementioned purely motor neurons, while they have peculiar "visual" proper-

ties. A class of visuomotor neurons discharge when the monkey observes
an action made by another individual and when it executes the same or a
similar action. We defined these as *mirror neurons* (Gallese, Fadiga,
Fogassi, & Rizzolatti, 1996; Rizzolatti, Fadiga, Gallese, & Fogassi,
1996a; see also Rizzolatti et al., 2001).

MIRROR NEURONS AND ACTION REPRESENTATION

In order to be activated by visual stimuli, mirror neurons require an interaction between the action's agent (human being or a monkey) and its object. Control experiments showed that neither the sight of the agent alone
nor of the object alone were effective in evoking the neuron's response.
Similarly, much less effective were mimicking the action without a target
object or performing the action by using tools (Gallese et al., 1996).

Frequently, a strict congruence was observed between the observed action effective in triggering the neuron and the effective executed action. In
about 30% of the recorded neurons, the effective observed and executed
actions corresponded both in terms of the general action (e.g., grasping)
and in terms of the way in which that action was executed (e.g., precision
grip). In the remaining neurons, only a general congruence was found
(e.g., any kind of observed and executed grasping elicited the neuron's response). This latter class of mirror neurons is particularly interesting because they appear to generalize across different ways of achieving the
same goal, thus enabling a more abstract type of action coding.

As we have already seen, neurons responding to the observation of
complex actions such as grasping or manipulating objects, have been described by Perrett and co-workers in the cortex buried within the STS.
These neurons, whose visual properties are for many aspects similar to
those of mirror neurons, could constitute the mirror neurons' source of
visual information. The STS region, however, has no direct connection
with area F5, but has links with the anterior part of the inferior parietal
lobule (area PF or 7b), which, in turn, is reciprocally connected with area
F5 (Matelli et al., 1986; see also Rizzolatti et al., 1998). Area PF, or 7b, is
located on the convexity of the inferior parietal lobule. Area PF, through
its connections with STSa on one hand, and F5 on the other, could play
the role of an "intermediate step" within a putative cortical network for
implicit action understanding, by feeding to the ventral premotor cortex
visual information about action as received from STSa.

In a new series of experiments, we decided to better clarify the nature and the properties of such a cortical matching system in the monkey brain. The results of this study showed that about one third of the PF recorded neurons responded both during action execution and action observation (Gallese, Fogassi, Fadiga, & Rizzolatti, 2001). All PF mirror neurons responded to the observation of actions in which the experimenter's hand(s) interacted with objects. Similarly to what was observed in F5, PF mirror neurons neither responded to object presentation nor to observed actions performed using tools. Observed mimed actions evoked weaker, if any, responses. What these experiments show is that the "mirror" system, matching action observation on action execution, is not a prerogative of the premotor cortex, but extends also to the posterior parietal lobe.

On the basis of these findings, it appears that the sensorimotor integration process supported by the F5-PF fronto–parietal cortical network instantiates an "internal copy" of actions utilized not only to generate and control goal-related behaviors, but also to provide—at a preconceptual and prelinguistic level—a meaningful account of behaviors performed by other individuals.

Several studies using different methodologies have demonstrated the existence of a similar matching system also in humans (see Buccino et al., 2001; Cochin, Barthelemy, Lejeune, Roux, & Martineau, 1998; Decety et al., 1997; Fadiga, Fogassi, Pavesi, & Rizzolatti, 1995; Grafton, Arbib, Fadiga, & Rizzolatti, 1996; Hari et al., 1998; Iacoboni et al., 1999; Rizzolatti et al., 1996b). It is interesting to note that brain-imaging experiments in humans have shown that during hand-action observation a cortical network composed of sectors of Broca's region, STS region, and the posterior parietal cortex is activated (Buccino et al., 2001; Decety & Grèzes, 1999; Decety et al., 1997; Grafton et al., 1996; Iacoboni et al., 1999; Rizzolatti et al., 1996b). Given the homology between monkey's area F5 and Broca's region (see Gallese, 1999; Matelli & Luppino, 1997; Rizzolatti & Arbib, 1998) it appears that even a part of the human brain traditionally considered to be unique to our species, nevertheless shares with its nonhuman precursor area, a similar functional mechanism. In other words, Broca's region appears to be not only involved in speech control, but also, similarly to monkey's area F5, in a prelinguistic analysis of others' behavior.

A recent brain imaging study (Buccino et al., 2001) showed that when we observe goal-related behaviors executed with effectors as different as the mouth, the hand, or the foot, different specific sectors of our premotor cortex become active. These cortical sectors are those same sectors that are active when we perform the same actions. Whenever we look at someone performing an action, beside the activation of various visual areas, there is a concurrent activation of the motor circuits that are recruited when we ourselves perform that action. Although we do not overtly reproduce the observed action, nevertheless, our motor system becomes active as if we were executing the same action we are observing.

According to this perspective, to perceive an action is equivalent to internally simulating it. This implicit, automatic, and unconscious process of motor simulation enables the observer to use his or her own resources to penetrate the world of the other without the need of explicitly theorizing about it. A process of action simulation automatically establishes a direct implicit link between agent and observer. Action is therefore a suitable candidate principle enabling social bonds to be initially established (Gallese, 2003a).

SELF–OTHER IDENTITY AND SHARED MULTIMODAL CONTENT

So far I have presented neuroscientific evidence demonstrating that in adult individuals (both monkeys and humans), a mirror-matching neural mechanism enables the representation of content independently for the self–other distinction, thus satisfying the last two criteria I posited to be necessary to empirically ground my working hypothesis. The first criterion, however, (sameness of content regardless of the multiple and different ways it might be originated by its referents), has not yet been addressed.

In a recent study, we investigated whether in the monkey premotor cortex there are neurons that discharge when the monkey makes a specific hand action and also when it hears the corresponding action-related sounds. The results showed that the monkey premotor cortex contains neurons that discharge when the monkey executes an action, sees, or just hears the same action performed by another agent. We have defined these neurons *audiovisual mirror neurons* (Keysers et al., 2003; Kohler et al., 2001, 2002). They respond to the sound of actions and discriminate between the sounds of different hand or mouth transitive actions, compatible with the monkey's natural behavioral repertoire. Audiovisual mirror neurons, however, do not respond to other similarly interesting sounds such as arousing noises, or monkeys' and other animals' vocalizations. The actions whose sounds evoke the strongest responses when heard also trigger the strongest responses when observed or executed. It does not significantly differ at all for the activity of this neural network if matter of facts of the world such as noisy actions are specified at the motor, visual, or auditory level. Such a neural mechanism enables the monkey to represent the end-stage of the interaction independently from its different modes of presentation: sounds, images, or willed effortful acts of the body. All modes of presentation of the event are blended within a circumscribed, informationally lighter, level of semantic reference.

Sameness of content is shared with different organisms. This shared semantic content is the product of modeling the observed behavior as an action with the help of a matching equivalence between what is observed or heard and what is executed.

Mirror neurons instantiate a multimodal representation of organism–organism relations. They map this multimodal representation across different spaces inhabited by different actors. These spaces are blended within a unified common intersubjective space, which paradoxically does not segregate any subject. This space is *we-centric*.

The shared intentional space underpinned by the mirror-matching mechanism is not meant to identify an agent and an observer; as organisms we are equipped with plenty of systems, from proprioception to the expectancy created by the inception of any activity, capable to tell self from other. The shared space instantiated by mirror neurons simply blends the interactive individuals within a shared implicit semantic content. The self–other identity parallels the self–other dichotomy.

As shown by some results of developmental psychology (see Meltzoff, 2002), the you–me analogy is heavily relying on action and action imitation, but is not confined to the domain of action. It is a global dimension, which comprises all aspects defining a life form, from its peculiar body to its peculiar affect. This global dimension encompasses a broad range of implicit certitudes we entertain about other individuals. In the following sections, I discuss many different forms of interaction, all contributing to our global shared experiential dimension with others. I try to recompose within an integrated neuroscientific framework all these multidimensional articulations of the self–other relationships, by introducing a new conceptual tool: the shared manifold of intersubjectivity.

SELF–OTHER IDENTITY AND EMPATHY

The self–other identity goes beyond the domain of action. It incorporates the domain of sensations, affect, and emotions. The affective dimension of interindividual relations has very early attracted the interest of philosophers because it was recognized as a distinctive feature of human beings. In the 18th century, Scottish moral philosophers identified our capacity to interpret the feeling of others in terms of "sympathy" (see Smith, 1759/ 1976). But it was during the second half of the 19th century that these issues acquired a multidisciplinary character, being tackled in parallel by philosophers and the scholars of a new discipline—psychology.

"Empathy" is a later English translation (see Titchener, 1909) of the German word, *Einfühlung*. It is commonly held that *Einfühlung* was originally introduced by Theodore Lipps (1903a) into the vocabulary of the psychology of aesthetic experience, to denote the relationship between a work of art and the observer, who imaginatively projects himself or herself into the contemplated object.

But the origin of the term is actually older. As pointed out by Prigman (1995), Robert Vischer introduced the term in 1873 to account for our capacity to symbolize the inanimate objects of nature and art. Vischer was strongly influenced by the ideas of Lotze, who in 1858 already proposed a mechanism by means of which humans are capable of understanding inanimate objects and other species of animals by "placing ourselves into them" [*sich mitlebend . . . versetzen*]).

Lipps (1903b), who wrote extensively on empathy, extended the concept of *Einfühlung* to the domain of intersubjectivity that he characterized in terms of inner imitation of the perceived movements of others. When I am watching an acrobat walking on a suspended wire, Lipps (1903b) noted, "I feel myself so inside of him" [*Ich Fühle mich so in ihm*]). We can see here a first suggested relation between imitation (though, "inner" imitation, in Lipps' words) and the capacity of understanding others by ascribing them feelings, emotions, and thoughts.

Phenomenology has further developed the notion of *Einfühlung*. A crucial point of Husserl's thought is the relevance he attributes to intersubjectivity in the constitution of our cognitive world. Husserl's rejection of solipsism is clearly epitomized in his fifth *Cartesian Meditations* (1977), and even more in the posthumously published, *Ideen II* (1989), where he emphasizes the role of others in making our world objective. It is through a shared experience of the world, granted by the presence of other individuals, that objectivity can be constituted. Interestingly enough, according to Husserl, the bodies of self and others are the primary instruments of our capacity to share experiences with others. What makes the behavior of other agents intelligible is the fact that their body is experienced not as a material object (*Körper*), but as something alive (*Leib*), something analogous to our own experienced acting body.

From birth onwards the *Lebenswelt*, the world inhabited by living things, constitutes the playground of our interactions. Empathy is deeply grounded in the experience of our lived body, and it is this experience that enables us to directly recognize others, not as bodies endowed with a mind, but as persons like us. Persons are rational individuals. What we now discover is how a rationality assumption can be grounded in bodily experience. According to Husserl, there can be no perception without awareness of the acting body.

The relationship between action and intersubjective empathic relations becomes even more evident in the works of Edith Stein and Merleau-Ponty. In her book, *On the Problem of Empathy* (1912/1964), Edith Stein, a former pupil of Husserl, clarified that the concept of *empathy* is not confined to a simple grasp of the other's feelings or emotions. There is a more basic connotation of empathy; the other is experienced as another being as oneself through an appreciation of similarity. An important component of this similarity resides in the common experience of action. As Edith Stein points out, if the size of my hand were given at a fixed scale, as something predetermined, it would become very hard to empathize with any other types of hand not matching these predetermined physical specifications. However, we can perfectly recognize children's hands and monkeys' hands as such despite their different visual appearance. Furthermore, we can recognize hands as such even when all the visual details are not available, even despite shifts of our point of view, and even when no visual shape specifications are provided. Even if all we can see are just moving light-dot displays of people's behavior, we are not only able to recognize a walking person, but also to discriminate whether it is ourselves or someone else we are watching (see Cutting & Kozlowski, 1977). Because in normal conditions we never look at ourselves when walking, this recognition process can be much better accounted for by a mechanism in which the observed moving stimuli activate the observer's motor schema for walking, than solely by means of a purely visual process. This seems to suggest that our "grasping" of the meaning of the world doesn't exclusively rely on its visual representation, but is strongly influenced by action-related sensorimotor processes, that is we rely on our own embodied personal knowledge.

Merleau-Ponty in the *Phenomenology of Perception* (1945/1962) wrote: "The communication or comprehension of gestures come about through the reciprocity of my intentions and the gestures of others, of my gestures and intentions discernible in the conduct of other people. It is as if the other person's intention inhabited my body and mine his" (p. 185). Self and other relate to one another, because they both represent opposite extensions of the same correlative and reversible system, self–other.

The shared intersubjective space in which we live since birth continues to constitute a substantial part of our semantic space. When we observe other acting individuals, therefore facing their full range of expressive power (the way they act, the emotions and feelings they display), a meaningful embodied interindividual link is automatically established.

The discovery of mirror neurons in adult individuals shows that the very same neural substrates are activated when some of these expressive acts are both executed and perceived. Thus, we have a subpersonally instantiated common space. It relies on neural circuits involved in action control. The hypothesis I am putting forward here is that a similar mechanism could underpin our capacity to share feelings and emotions with others. Sensations and emotions displayed by others can be implicitly understood through a resonance mechanism similar to that instantiated by the mirror-matching system for actions.

THE SHARED MANIFOLD HYPOTHESIS

Throughout the chapter, I have emphasized that the establishment of self–other identity is a driving force for the cognitive development of more articulated and sophisticated forms of intersubjective relations. I have also proposed that the mirror-matching system could be involved in enabling the constitution of this identity. I think that the concept of *empathy* should be extended in order to accommodate and account for all different aspects of expressive behavior enabling us to establish a meaningful link between others and ourselves. This enlarged notion of empathy opens up the possibility to unify under the same account the multiple aspects and possible levels of description of intersubjective relations.

As we have seen, when we enter into relationship with others, there is a multiplicity of states that we share with them. We share emotions, our body schema, our being subject to somatic sensations such as pain, and so on. A comprehensive account of the richness of content we share with others should rest on a conceptual tool capable of being applied to all these different levels of description, while simultaneously providing their functional and subpersonal characterization.

I introduce this conceptual tool as the *shared manifold* of intersubjectivity (see Gallese, 2001, 2003a, 2003b). I posit that it is by means of this shared manifold that we recognize other human beings as similar to us. It is just because of this shared manifold that intersubjective communication, social imitation, and mind reading become possible (Gallese, 2003a). The shared manifold can be operationalized at three different levels: a phenomenological level, a functional level, and a subpersonal level.

The *phenomenological level* is the one responsible for the sense of similarity, of being individuals within a larger social community of persons like us, that we experience anytime we confront ourselves with other human beings. It could be defined also as the *empathic* level, provided that empathy is characterized in the enlarged way I am advocating here. Actions, emotions, and sensations experienced by others become meaningful to us because we can share them with others.

The *functional level* can be characterized in terms of "as if" modes of interaction enabling models of self or other to be created. The same functional logic is at work during both self-control and the understanding of others' behavior. Both are models of interaction, which map their referents on identical relational functional nodes. All modes of interaction share a relational character. At the functional level of description of the shared manifold, the relational logic of operation produces the self–other identity by enabling the system to detect coherence, regularity, and predictability, independently from their situated source.

The *subpersonal level* is instantiated as the level of activity of a series of mirror matching neural circuits. The activity of these neural circuits is, in turn, tightly coupled with multilevel changes within body states. We have seen that mirror neurons instantiate a multimodal intentional shared space. My hypothesis is that analogous neural networks might be at work to generate multimodal emotional and sensitive shared spaces (see Gallese, 2001, 2003a, 2003b; Goldman & Gallese, 2000). These are the shared spaces that allow us to implicitly appreciate, experience, and understand the emotions and the sensations we take others to experience. No systematic attempt has been produced so far to experimentally validate or falsify this hypothesis. Yet, there are clues that my hypothesis might be not so ill-founded.

Preliminary evidence suggests that in humans, a mirror-matching mechanism is at work in pain-related neurons. Hutchison et al. (1999) studied pain-related neurons in the human cingulate cortex. Cingulotomy procedures for the treatment of psychiatric disease provided an opportunity to examine prior to excision whether neurons in the anterior cingulate cortex of locally anesthetized but awake patients responded to painful stimuli. It was noticed that a neuron that responded to noxious mechanical stimulation applied to the patient's hand also responded when the patient watched pinpricks being applied to the examiner's fingers. Both applied and observed painful stimuli elicited the same response in the same neuron.

Calder, Keane, Manes, Antoun, and Young (2000) showed that a patient who suffered stroke damaging the insula and the putamen was selectively impaired in detecting disgust in many different modalities, such as facial signals, nonverbal emotional sounds, and emotional prosody. The same patient was also selectively impaired in subjectively experiencing disgust and therefore in reacting appropriately to it. Once the capacity to experience and express a given emotion is lost, the same emotion cannot be easily represented and detected in others.

Emotions constitute one of the earliest ways to acquire knowledge about the situation of the living organism, and therefore to reorganize it in the light of its relations with others. This points to a strong interaction between emotion and action. We dislike things that we seldom touch, look at, or smell. We do not "translate" these things into motor schemas suitable to interact with them (likely "tagged" with positive emotions), but rather into aversive motor schemas (likely tagged with negative emotional connotations). The coordinated activity of sensorimotor and affective neural systems results in the simplification and automatization of the behavioral responses that living organisms are supposed to produce in order to survive.

The strict coupling between affect and sensorimotor integration is demonstrated by a recent study by Adolphs, H. Damasio, Tranel, Cooper, and A. R. Damasio (2000) in which these authors review over 100 brain-damaged patients. Among other results, this study shows that patients who suffered damage to sensorimotor cortices are those scoring worse when asked to rate or name facial emotions displayed by viewed human faces.

Iacoboni and co-workers (Carr, Iacoboni, Dubeau, Mazziotta, & Lenzi, 2003) in a recent fMRI study on healthy participants, showed that both observation and imitation of facial emotions activate the same restricted group of brain structures including the premotor cortex, the insula, and the amygdala. It is possible to speculate that such a double activation pattern during observation and imitation of emotions could be due to the activity of a neural mirror-matching system.

Furthermore, in a recently published fMRI study (Wicker et al., 2003), we showed that the same region within the anterior insula becomes active both when healthy subjects experience disgust while inhaling disgusting odorants, and when they observe the disgusted facial expression of another individual.

My hypothesis also predicts the existence of "somatosensory mirror neurons" enabling the capacity when observing other bodies to map different body locations, and to refer them to equivalent locations of our body. New experiments both on monkeys and humans are on their way in our lab to empirically test this hypothesis.

CONCLUSIONS

Preliminary evidence suggests that the same neural structures that are active during sensations and emotions are active also when the same sensations and emotions are to be detected in others. It appears therefore that a whole range of different mirror matching systems may be present in our brain. This "resonance mechanism," originally discovered and described in the domain of actions, is likely a basic organizational feature of our brain.

These automatic, implicit and nonreflexive simulation mechanisms create a representation of emotion-driven, body-related changes. The activation of these "as if body loops" can likely be not only internally driven, but also triggered by the observation of other individuals (see Adolphs, 1999; Gallese, 2001; Goldman & Gallese, 2000).

The discovery of mirror neurons in the premotor cortex of monkeys and humans has unveiled a neural matching mechanism that, in the light of more recent findings, appears to be present also in a variety of nonmotor-related, human-brain structures. Much of what we ascribe to the mind of others when witnessing their behavior depends on the resonance mechanisms that their behavior triggers in us. The detection of intentions that we ascribe to observed agents and that we assume to underpin their behavior is constrained by the necessity for an intersubjective link to be established. Early imitation is but one early example of intersubjective link in action. The shared manifold I described in this chapter is a good candidate for determining and constraining this intersubjective link.

It should be added that the existence of the shared manifold of intersubjectivity does not entail that we experience others as we experience ourselves. The shared manifold simply enables and bootstraps mutual intelligibility. Thus, self–other identity is not all there as in intersubjectivity. As pointed out by Husserl (1977), if this were the case, others could not anymore be experienced as such (see also Zahavi, 2001). On the contrary, it is the alterity of the other that grounds the objective character of reality. The quality of our own self-experience of the "external world" and its content are constrained by the presence of other subjects that are intelligible while preserving their alterity character.

ACKNOWLEDGMENTS

Shorter and preliminary versions of this paper have been presented at the First International Conference of Social Cognitive Neuroscience, Los Angeles, April 2001; at the 31st Annual Meeting of the Jean Piaget Society, Berkeley, California, May 2001; at the Second Meeting of the Mc-Donnell Project in Philosophy and the Neurosciences, Tofino, Canada, June 2001; and at the International Conference on Imitation, Royaumont, France, May 2002. I wish to thank all audiences for the feedback I have received from them. This work was supported by the Eurocores Program of the European Science Foundation and by MIURST (Ministero Italiano per l'Università, la Ricerca Scientifica e Tecnologica).

REFERENCES

Adolphs, R. (1999). Social cognition and the human brain. *Trends in Cognitive Sciences, 3*, 469–479.

Adolphs, R., Damasio, H., Tranel, D., Cooper, G., & Damasio, A. R. (2000). A role for somatosensory cortices in the visual recognition of emotion as revealed by three-dimensional lesion mapping. *Journal of Neuroscience, 20*, 2683–2690.

Brothers, L., & Ring, B. (1992). A neuroethological framework for the representation of minds. *Journal of Cognitive Neuroscience, 4*, 107–118.

Buccino, G., Binkofski, F., Fink, G. R., Fadiga, L., Fogassi, L., Gallese, V., Seitz, R. J., Zilles, K., Rizzolatti, G., & Freund, H.-J. (2001). Action observation activates premotor and parietal areas in a somatotopic manner: An fMRI study. *European Journal of Neuroscience, 13*, 400–404.

Calder, A. J., Keane, J., Manes, F., Antoun, N., & Young, A. W. (2000). Impaired recognition and experience of disgust following brain injury. *Nature Neuroscience, 3*, 1077–1078.

Carey, D. P., Perrett, D. I., & Oram, M. W. (1997). Recognizing, understanding and reproducing actions. In M. Jeannerod & J. Grafman (Eds.), *Action and cognition* (*Handbook of Neuropsychology*, Vol. 11, sec. 16, pp. 111–130). Amsterdam: Elsevier Science.

Carr, L., Iacoboni, M., Dubeau, M. C., Mazziotta, J. C., & Lenzi, G. L. (2003). Neural mechanisms of empathy in humans: A relay from neural systems for imitation to limbic areas. *Proceedings of the National Academy of Science USA, 3*, 5497–5502.

Cochin, S., Barthelemy, C., Lejeune, B., Roux, S., & Martineau, J. (1998). Perception of motion and qEEG activity in human adults. *Electroencephalography and Clinical Neurophysiology, 107*, 287–295.

Cutting, J. E., & Kozlowski, L. T. (1977). Recognizing friends by their walk: Gait perception without familiarity cues. *Bulletin of the Psychonomic Society, 9*, 353–356.

Decety, J., & Grèzes, J. (1999). Neural mechanisms subserving the perception of human actions. *Trends in Cognitive Sciences, 3*, 172–178.

Decety, J., Grezes, J., Costes, N., Perani, D., Jeannerod, M., Procyk, E., Grassi, F., & Fazio, F. (1997). Brain activity during observation of actions. Influence of action content and subject's strategy. *Brain, 120*, 1763–1777.

Desimone, R., & Ungerleider, L. (1986). Multiple visual areas in the caudal superior temporal sulcus of the macaque. *Journal of Comparative Neurology, 248*, 164–189.

Dubner, R., & Zeki, S. (1971). Response properties and receptive fields of cells in an anatomically defined region of the superior temporal sulcus in the monkey. *Brain Research, 35*, 528–532.

Fadiga, L., Fogassi, L., Pavesi, G., & Rizzolatti, G. (1995). Motor facilitation during action observation: A magnetic stimulation study. *Journal of Neurophysiology, 73*, 2608–2611.

Gallese, V. (1999). From grasping to language: Mirror neurons and the origin of social communication. In S. Hameroff, A. Kazniak, & D. Chalmers (Eds.), *Towards a science of consciousness* (pp. 165–178). Boston, MA: MIT Press.

Gallese, V. (2000a). The acting subject: Towards the neural basis of social cognition. In T. Metzinger (Ed.), *Neural correlates of consciousness: Empirical and conceptual questions* (pp. 325–333). Cambridge, MA: MIT Press.

Gallese, V. (2000b). The inner sense of action: Agency and motor representations. *Journal of Consciousness Studies, 7*(10), 23–40.

Gallese, V. (2001). The "shared manifold" hypothesis: From mirror neurons to empathy. *Journal of Consciousness Studies, 8*(5–7), 33–50.

Gallese, V. (2003a). The manifold nature of interpersonal relations: The quest for a common mechanism. *Philosophical Transactions of the Royal Society London B, 358*, 517–528.

Gallese, V. (2003b). The roots of empathy: The shared manifold hypothesis and the neural basis of intersubjectivity. *Psychopathology, 36*(4), 171–180.

Gallese, V. (2003c). A neuroscientific grasp of concepts: From control to representation. *Philosophical Transactions of the Royal Society London B, 358,* 1231–1240.

Gallese, V., & Goldman, A. (1998). Mirror neurons and the simulation theory of mind-reading. *Trends in Cognitive Sciences, 12,* 493–501.

Gallese, V., Fadiga, L., Fogassi, L., & Rizzolatti, G. (1996). Action recognition in the premotor cortex. *Brain, 119,* 593–609.

Gallese, V., Fogassi, L., Fadiga, L., & Rizzolatti, G. (2001). Action representation and the inferior parietal lobule. In W. Prinz & B. Hommel (Eds.), *Attention and performance XIX* (pp. 334–355). Oxford: Oxford University Press.

Goldman, A., & Gallese, V. (2000). Reply to Schulkin. *Trends in Cognitive Sciences, 4,* 255–256.

Grafton, S. T., Arbib, M. A., Fadiga, L., & Rizzolatti, G. (1996). Localization of grasp representations in humans by PET: 2. Observation compared with imagination. *Experimental Brain Research, 112,* 103–111.

Hari, R., Forss, N., Avikainen, S., Kirveskari, S., Salenius, S., & Rizzolatti, G. (1998). Activation of human primary motor cortex during action observation: A neuromagnetic study. *Proceedings of the National Academy of Science USA, 95,* 15061–15065.

Hepp-Reymond, M.-C., Hüsler, E. J., Maier, M. A., & Qi, H.-X. (1994). Force-related neuronal activity in two regions of the primate ventral premotor cortex. *Canadian Journal of Physiology and Pharmacology, 72,* 571–579.

Husserl, E. (1977). *Cartesian meditations* (D. Cairns, Trans.). Dordrecht: Kluwer Academic.

Husserl, E. (1989). *Ideas pertaining to a pure phenomenology and to a phenomenological philosophy, second book: Studies in the phenomenology of constitution.* Dordrecht: Kluwer Academic.

Hutchison, W. D. (1999). Pain related neurons in the human cingulate cortex. *Nature Neuroscience, 2,* 403–405.

Iacoboni, M., Woods, R. P., Brass, M., Bekkering, H., Mazziotta, J. C., & Rizzolatti, G. (1999). Cortical mechanisms of human imitation. *Science, 286,* 2526–2528.

Jeannerod, M., Arbib, M. A., Rizzolatti, G., & Sakata, H. (1995). Grasping objects: The cortical mechanisms of visuomotor transformation. *Trends in Neuroscience, 18,* 314–320.

Jellema, T., & Perrett, D. I. (2001). Coding of visible and hidden actions. In W. Prinz & B. Hommel (Eds.), *Attention & performance XIX. Common mechanisms in perception and action* (pp. 356–379). Oxford: Oxford University Press.

Keysers, C., Kohler, E., Umiltà, M. A., Fogassi, L., Nanetti, L., & Gallese, V. (2003). Audio-visual mirror neurones and action recognition. *Experimental Brain Research, 153,* 628–636.

Kim, J. (1998). *Mind in a physical world.* Cambridge, MA: MIT Press.

Kohler, E., Keysers, C., Umiltà, M. A., Fogassi, L., Gallese, V., & Rizzolatti, G. (2002). Hearing, sounds, understanding actions: Action representation in mirror neurons. *Science, 297,* 846–848.

Kurata, K., & Tanji, J. (1986). Premotor cortex neurons in macaques: Activity before distal and proximal forelimb movements. *Journal of Neuroscience, 6,* 403–411.

Lakoff, G. (1987). *Women, fire, and dangerous things.* Chicago: University of Chicago Press.

Lakoff, G., & Johnson, M. (1984). *Metaphors we live by.* Chicago: University of Chicago Press.

Lakoff, G., & Johnson, M. (1999). *Philosophy in the flesh.* New York: Basic Books.

Lakoff, G., & Nuñez, R. (2000). *Where mathematics comes from: How the embodied mind brings mathematics into being.* New York: Basic Books.

Lipps, T. (1903a). *Grundlegung der Aesthetik.* Bamburg und Leipzig: W. Engelmann.

Lipps, T. (1903b). Einfulung, innere nachahmung und organenempfindung. *Archiv. F. die Ges. Psy.* (Vol. 1, part 2). Leipzig: W. Engelmann.

Martin, A., & Chao, L. L. (2001). Semantic memory and the brain: Structure and processes. *Current Opinion in Neurobiology, 11,* 194–201.

Matelli, M., Camarda, R., Glickstein, M., & Rizzolatti, G. (1986). Afferent and efferent projections of the inferior area 6 in the macaque monkey. *Journal of Comparative Neurology, 251,* 281–298.

Matelli, M., & Luppino, G. (1997). Functional anatomy of human motor cortical areas. In F. Boller & J. Grafman (Eds.), *Handbook of neuropsychology, Vol. 11.* Amsterdam: Elsevier Science.

Matelli, M., Luppino, G., & Rizzolatti, G. (1985). Patterns of cytochrome oxidase activity in the frontal agranular cortex of the macaque monkey. *Behavioral Brain Research, 18,* 125–137.

Maunsell, J., & Van Essen, D. (1983). Functional properties of neurons in middle temporal visual area of the macaque monkey, 1: Selectivity for stimulus direction, speed, and orientation. *Journal of Neurophysiology, 49,* 1127–1147.

Meltzoff, A. (2002). Elements of a developmental theory of imitation. In W. Prinz & A. Meltzoff (Eds.), *The imitative mind: Development, evolution and brain bases* (pp. 19–41). Cambridge: Cambridge University Press.

Meltzoff, A. N., & Brooks, R. (2001). "Like me" as a building block for understanding other minds: Bodily acts, attention, and intention. In B. F. Malle, L. J. Moses, & D. A. Baldwin (Eds.), *Intentions and intentionality: Foundations of social cognition* (pp. 171–191). Cambridge, MA: MIT Press.

Meltzoff, A. N., & Moore, M. K. (1977). Imitation of facial and manual gestures by human neonates. *Science, 198,* 75–78.

Meltzoff, A. N., & Moore, M. K. (1994). Imitation, memory, and the representation of persons. *Infant Behavior and Development, 17*, 83–99.

Meltzoff, A. N., & Moore, M. K. (1997). Explaining facial imitation: A theoretical model. *Early Development and Parenting, 6*, 179–192.

Merleau-Ponty, M. (1962). *Phenomenology of perception* (C. Smith, Trans.). London: Routledge. (Original work published 1945)

Perrett, D. I., Mistlin, A. J., Harries, M. H., & Chitty, A. J. (1990). Understanding the visual appearance and consequence of hand actions. In M. A. Goodale (Ed.), *Vision and action: The control of grasping* (pp. 163–342). Norwood, NJ: Ablex.

Prigman, G. W. (1995). Freud and the history of empathy. *International Journal of Psychoanalysis, 76*, 237–252.

Rizzolatti, G., Camarda, R., Fogassi, M., Gentilucci, M., Luppino, G., & Matelli, M. (1988). Functional organization of inferior area 6 in the macaque monkey: II. Area F5 and the control of distal movements. *Experimental Brain Research, 71*, 491–507.

Rizzolatti, G., Fadiga, L., Gallese, V., & Fogassi, L. (1996a). Premotor cortex and the recognition of motor actions. *Cognitive Brain Research, 3*, 131–141.

Rizzolatti, G., Fadiga, L., Matelli, M., Bettinardi, V., Paulesu, E., Perani, D., & Fazio, G. (1996b). Localization of grasp representations in humans by PET: 1. Observation versus execution. *Experimental Brain Research, 111*, 246–252.

Rizzolatti, G., Fogassi, L., & Gallese, V. (2000). Cortical mechanisms subserving object grasping and action recognition: A new view on the cortical motor functions. In M. S. Gazzaniga (Ed.), *The cognitive neurosciences, second edition* (pp. 539–552). Cambridge, MA: MIT Press.

Rizzolatti, G., Fogassi, L., & Gallese, V. (2001). Neurophysiological mechanisms underlying the understanding and imitation of action. *Nature Neuroscience Reviews, 2*, 661–670.

Rizzolatti, G., Luppino, G., & Matelli, M. (1998). The organization of the cortical motor system: New concepts. *Electroencephalography and Clinical Neurophysiology, 106*, 283–296.

Rizzolatti, G., Scandolara, C., Gentilucci, M., & Camarda, R. (1981). Response properties and behavioral modulation of 'mouth' neurons of the postarcuate cortex (area 6) in macaque monkeys. *Brain Research, 255*, 421–424.

Smith, A. (1976). *The theory of moral sentiments* (D. D. Raphael & A. L. Macfie, Eds.). Oxford: Oxford University Press. (Original work published 1759)

Stein, E. (1964). *On the problem of empathy*. The Hague: Martinus Nijhoff. (Original work published 1912)

Stern, D. N. (1985). *The interpersonal world of the infant*. London: Karnac Books.

Titchener, E. B. (1909). *Lectures on the experimental psychology of thought processes*. New York: Macmillan.

Van Essen, D., Maunsell, J., & Bixby, J. (1981). The middle temporal visual area in the macaque: Myeloarchitecture, connections, functional properties, and topographic organization. *Journal of Comparative Neurology, 199*, 293–326.

Wicker, B., Keysers, C., Plailly, J., Royet, J.-P., Gallese, V., & Rizzolatti, G. (2003). Both of us disgusted in my insula: The common neural basis of seeing and feeling disgust. *Neuron, 40*, 655–664.

Zahavi, D. (2001). Beyond empathy. Phenomenological approaches to intersubjectivity. *Journal of Consciousness Studies, 8*, 151–167.

Zeki, S. M. (1974). Functional organization of a visual area in the posterior bank of the superior temporal sulcus of the rhesus monkey. *Journal of Physiology, 236*, 549–573.

Zeki, S. (1993). *A vision of the brain*. Oxford: Blackwell.

Zeki, S., & Shipp, S. (1988). The functional logic of cortical connections. *Nature, 355*, 311–317.

Plasticity, Localization, and Language Development[1]

Elizabeth Bates
Formerly of University of California, San Diego

The term *aphasia* refers to acute or chronic impairment of language, an acquired condition that is most often associated with damage to the left side of the brain, usually due to trauma or stroke. We have known about the link between left hemisphere damage (LHD) and language loss for more than a century (Goodglass, 1993). For almost as long, we have also known that the lesion–symptom correlations observed in adults do not appear to hold for very young children (Basser, 1962; Lenneberg, 1967). In fact, in the absence of other complications, infants with congenital damage to one side of the brain (left or right) usually go on to acquire language abilities that are well within the normal range (Eisele and Aram, 1995; Feldman et al., 1992; Vargha-Khadem et al., 1994). To be sure, children with a history of early brain injury typically perform below neurologically intact age-matched controls on a host of language and nonlanguage measures, including an average full-scale IQ difference somewhere between 4 and 8 points from one study to another (especially in children with persistent seizures; Vargha-Khadem et al., 1994). Brain damage is not a good thing to have, and some price must be paid for

[1]Chapter reprinted from *The Changing Nervous System: Neurobehavioral Consequences of Early Brain Disorders* (pp. 214–253), by S. H. Broman and J. M. Fletcher (Eds.), 1999, New York: Oxford University Press. Copyright © 1999 by the Oxford University Press. Reprinted with permission.

wholesale reorganization of the brain to compensate for early injuries. But the critical point for present purposes is that these children are not aphasic, despite early damage of a sort that often leads to irreversible aphasia when it occurs in an adult.

In addition to the reviews by other authors cited above, my colleagues and I have also published several detailed reviews, from various points of view, of language, cognition, and communicative development in children with focal brain injury (e.g., Bates et al., 1997, 1998; Elman et al., 1996; Reilly et al., 1998; Stiles, 1995; Stiles et al., 1998; Stiles and Thal, 1993; Thal et al., 1991). As these reviews attest, a consensus has emerged that stands midway between the historical extremes of *equipotentiality* (Lenneberg, 1967) and *innate predetermination* of the adult pattern of brain organization for language (e.g., Curtiss, 1988; Stromswold, 1995). The two hemispheres are certainly not equipotential for language at birth; indeed, if they were it would be impossible to explain why left hemisphere dominance for language emerges 95%–98% of the time in neurologically intact individuals. However, the evidence for recovery from early LHD is now so strong that it is no longer possible to entertain the hypothesis that language *per se* is innately and irreversibly localized to perisylvian regions of the left hemisphere.

The compromise view is one in which brain organization for language emerges gradually over the course of development (Elman et al., 1996; Karmiloff-Smith, 1992) based on "soft constraints" that are only indirectly related to language itself. Hence the familiar pattern of language localization in adults is the product rather than the cause of development, an end product that emerges out of initial variations in the way that information is processed from one region to another. Crucially, these variations are not specific to language, although they do have important implications for how and where language is acquired and processed. In the absence of early brain injury, these soft constraints in the initial architecture and information-processing proclivities of the left hemisphere will ultimately lead to the familiar pattern of left hemisphere dominance. However, other "brain plans" for language are possible and will emerge when the default situation does not hold.

In the pages that follow, I do not intend to provide another detailed review of the outcomes associated with early brain injury; the reader is referred elsewhere for a more complete catalogue of such findings. What I do instead is to go beyond these findings to their implications for the nature and origins of language localization in the adult, providing an account of how this neural system might emerge across the course of development. With this goal in mind, the chapter is organized as follows: (*1*) a very brief review of findings from developmental neurobiology that serve as animal models for the kind of plasticity that we see in human children; (*2*) an equally brief illustration of results from retrospective studies of language development in the focal lesion population; (*3*) the distinction between prospective and retrospective studies, including a discussion of putative "critical periods" for language development; (*4*) an overview of prospective findings on language development in children with congenital lesions to one side of the brain; and (*5*) a new view of brain organization for language in the adult, an alternative to the static phrenological view that has dominated our thinking for two centuries, one that takes into account the role of experience in specifying the functional architecture of the brain.

DEVELOPMENTAL PLASTICITY: ANIMAL MODELS

Evidence for the plasticity of language in the human brain should not be surprising in light of all that has been learned in the last few decades about developmental plasticity of isocortex in other species (Bates et al., 1992; Deacon, 1997; Elman et al., 1996; Janowsky and Finlay, 1986; Johnson, 1997; Killackey, 1990; Mueller, 1996; Quartz and Sejnowski, 1998; Shatz, 1992). Without attempting an exhaustive or even a representative review, here are just a few of my favorite examples of research on developmental plasticity in other species, studies that provide animal models for the kind of plasticity that we have observed in the human case.

Isacson and Deacon (1996) have transplanted plugs of cortex from the fetal pig into the brain of the adult rat. These "foreigners" (called *xenotransplants*) develop appropriate connections, including functioning axonal links down the spinal column that stop in appropriate places. Although we know very little about the mental life of the resulting rat, no signs of pig-appropriate behaviors have been observed.

Stanfield and O'Leary (1985) have transplanted plugs of fetal cortex from one region to another (e.g., from visual to motor or somatosensory cortex). Although these cortical plugs are not entirely normal compared with "native" tissue, they set up functional connections with regions inside and outside the cortex. More importantly still, the transplants develop representations (i.e., cortical maps) that are appropriate for the region in which they now live and not for the region where they were born. ("When in Rome, do as the Romans do.")

Sur and his colleagues (see Pallas and Sur, 1993; Sur et al., 1990) have rerouted visual information from visual cortex to auditory cortex in the infant ferret. Although (again) the representations that develop in auditory cortex are not entirely normal, these experiments show that auditory tissue can develop retinotopic maps. It seems that auditory cortex becomes auditory cortex under normal conditions primarily because (in unoperated animals) it receives information from the ear; but if it has to, it can also process visual information in roughly appropriate ways.

Killackey and colleagues (1994) have modified the body surface of an infant rat by removing whiskers that serve as critical perceptual organs in this species. Under normal conditions, the somatosensory cortex of the rat develops representations ("barrel cells") that are isomorphic with input from the whisker region. In contrast, the altered animals develop somatosensory maps reflecting changes in the periphery, with expanded representations for the remaining whiskers; regions that would normally subserve the missing whiskers are reduced or absent (Killackey, 1990). In other words, the rat ends up with the brain that it needs rather than the brain that Nature intended.

Finally, in an example that may be closer to the experience of children with early focal brain injury, a recent study by Webster et al. (1995) shows that the "where is it" system (mediated in dorsal regions, especially parietal cortex, including area MT) can take over the functions of the "what is it" system (mediated in ventral regions, especially inferior temporal cortex, including area TE). When area TE is bilaterally removed in an adult monkey, that animal displays severe and irreversible amnesia for new objects, suggesting that this area plays a crucial role in mediating object memory and detection (i.e., the so-called what is it system). However, as Webster et al. (1995) have shown, bilateral removal of area TE in infant monkeys leads to performance only slightly below age-matched unoperated controls (at both 10 months and 4 years of age). If area TE is no longer available, where has the "what is it" system gone? By lesioning additional areas of visual cortex, Webster et al. (1995) showed that the object detection function in TE-lesioned infant monkeys is mediated by dorsal regions of extrastriate cortex that usually respond to motion rather than form (i.e., the "where is it" system). In other words, a major higher cognitive function can develop far away from its intended site, in areas that would ordinarily play little or no role in the mediation of that function.

These examples and many others like them have led most developmental neurobiologists to conclude that cortical differentiation and functional specialization are largely the product of input to the cortex, albeit within certain broad architectural and computational constraints (Johnson, 1997). Such findings provide a serious challenge to the old idea that the brain is organized into largely predetermined, domain-specific faculties (i.e., the phrenological approach). An alternative proposal that is more compatible with these findings is offered later.

LANGUAGE OUTCOMES IN CHILDREN WITH EARLY FOCAL BRAIN INJURY: RETROSPECTIVE FINDINGS

As noted earlier, retrospective studies of language outcomes in children with unilateral brain injury have repeatedly found that these children are not aphasic; they usually perform within the normal range, although they often do perform slightly below neurologically intact age-matched controls (cf. Webster et al., 1995). More importantly for our purposes here, there is no consistent evidence in these retrospective studies to suggest that language outcomes are worse in children with LHD than in children whose injuries are restricted to the right hemisphere. Without attempting an exhaustive review, three examples are given to illustrate these points.

Figure 7.1 presents idealized versus observed results for verbal versus nonverbal IQ scores in a cross-sectional sample of children with congenital injuries who were tested at various ages between 3 and 10 years. Figure 7.1A illustrates what we might expect if the left–right differences observed in adults were consistently observed in children: higher verbal than nonverbal IQ scores in children with right hemisphere damage (RHD), which means that these children should line up on the upper diagonal; higher nonverbal than verbal IQ scores in children with LHD, which means that these children ought to fall on the lower diagonal. These idealized scores were obtained by taking actual pairs of scores for individual children in our focal lesion sample and reversing any scores that were not in the predicted direction. In contrast with this idealized outcome, Figure 7.1B illustrates the actual verbal and performance IQ scores for 28 LHD and 15 RHD cases (note that there are no differences between these two groups in gender or chronological age and no mean differences in full-scale IQ). The actual data in Figure 7.1B illustrate several points. First, in line with other studies of this population, the mean full-scale IQ for the sample as a whole is 93.2, within the normal range but below the mean of 100 that we would expect if we were drawing randomly from the normal population. Second, the range of outcomes observed in the focal lesion population as

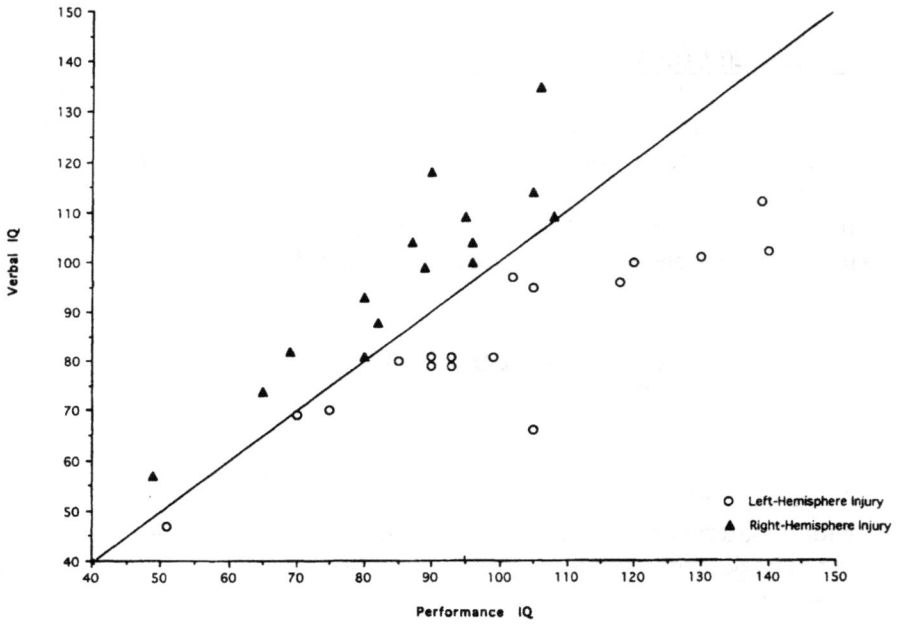

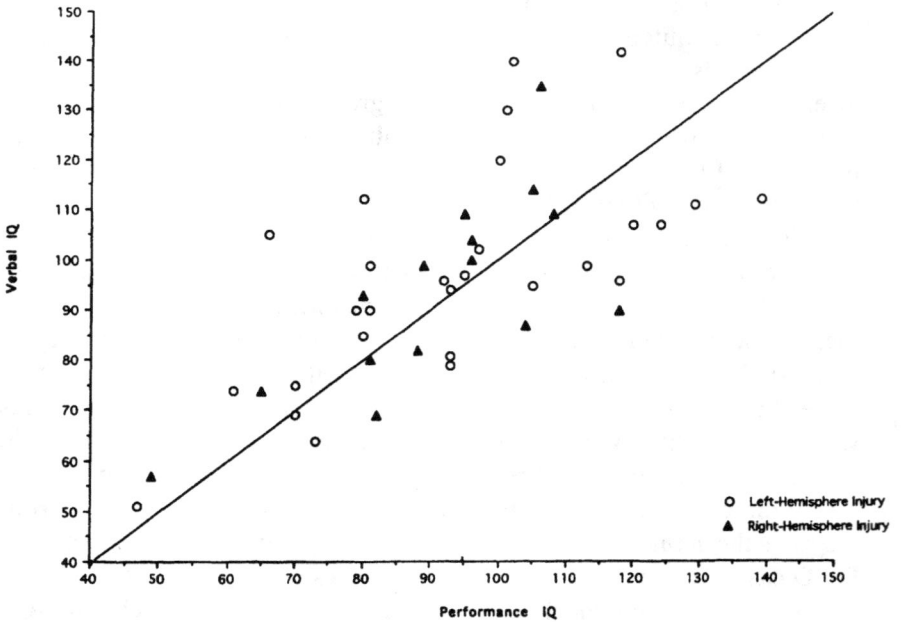

FIG. 7.1. *(Top)* Idealized relation between verbal and performance IQ in children with left versus right hemisphere injury. *(Bottom)* Observed relation between verbal and performance IQ in children with left versus right hemisphere injury. (Adapted from Bates et al., 1998).

a whole is extraordinarily broad, including some children who can be classified as mentally retarded (i.e., 16.3% of this sample have full-scale IQs at or below 80) and some with IQs over 120. Third, the correlation between the verbal and nonverbal subscales is relatively strong (+ 0.65, p < 0.0001), which means that verbal and nonverbal IQs do not dissociate markedly in this group. In fact, as we can clearly see from the difference between Figure 7.1A (predicted outcomes) and 7.1B (the outcomes actually observed in these children), there is absolutely no evidence in these data for a double dissociation between verbal and nonverbal IQ as a function of LHD versus RHD.

Figure 7.2 presents results from a more focused study of grammatical development, illustrating the number of different complex syntactic forms produced in a narrative discourse tasks by LHD, RHD, and neurologically intact controls who were tested between 6 and 12 years of age. Figure 7.2 demonstrates (once again) that children with focal brain injury perform within the normal range in production of complex syntax, even

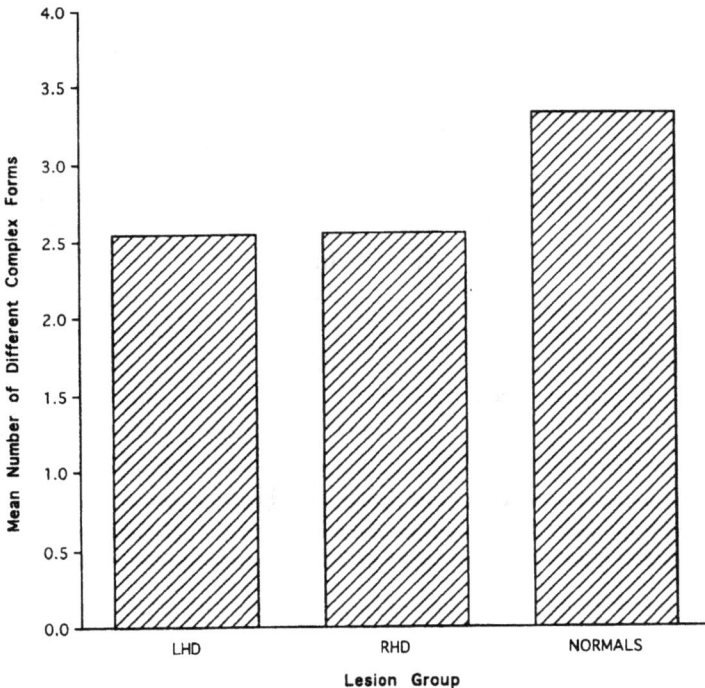

FIG. 7.2.   Number of different complex syntactic forms produced by children with left (LHD) versus right (RHD) hemisphere damage in a story-telling task (ages 6–12 years). (Adapted from Reilly et al., 1998).

though they do (as a group) score significantly below neurologically intact controls, In this respect, the Reilly et al. (1998) result for grammatical development in human children is remarkably similar to the findings reported by Webster et al. (1995) on the relative preservation of memory for novel objects in infant monkeys with bilateral TE lesions (i.e., performance roughly 10% below that of normal controls). In addition, Figure 7.2 shows that there is no evidence in this age range for a difference in syntactic production as a function of lesion side or site.

Finally, Figure 7.3 compares results for adults and 6–12-year-old children with LHD versus RHD on the same sentence comprehension task. All data are based on z-scores, with patients at each age level compared with the performance of age-matched normal controls (hence the difference in performance between normal adults and normal 6–12-year-old children is factored out of the results). In this particular procedure, subjects are asked to match each stimulus sentence to one of four pictured alternatives. Half the items are familiar phrases (well-known metaphors and figures of speech like "She took a turn for the worse"), and the other half are novel phrases matched to the familiar phrases in length and complexity. As Figure 7.3A shows, there is a powerful double dissociation between novel and familiar phrases in adult victims of unilateral brain injury: Adults with LHD score markedly better on the familiar phrases, while adults with RHD score better on the novel phrases. This is one example of a growing body of evidence challenging the old assumption that the left hemisphere is "the" language hemisphere, even in adults. The right hemisphere does make an important contribution to language processing, but its contribution is qualitatively different from that of the left hemisphere, involving a number of functions including emotionality, intonation contours, and (as this example illustrates) figurative, metaphorical, and/or formulaic speech (all forms of speech in which the meaning of the sentence as a whole goes beyond the meaning one would obtain by computing across the separate elements in the sentence). A comparison between Figure 7.3A and 7.3B helps to clarify three important points. First, children with focal injuries fare far better than adults with comparable damage when they are compared with age-matched controls. Second, the powerful double dissociation observed in adults is not observed in children. Third, novel sentences are more susceptible to the effects of brain injury than are familiar phrases in the child group, but RHD children actually perform below the LHD group in comprehension of novel sentences (significant by a one-tailed *t* test), the opposite of what we might expect if the adult pattern held for children with focal brain injury.

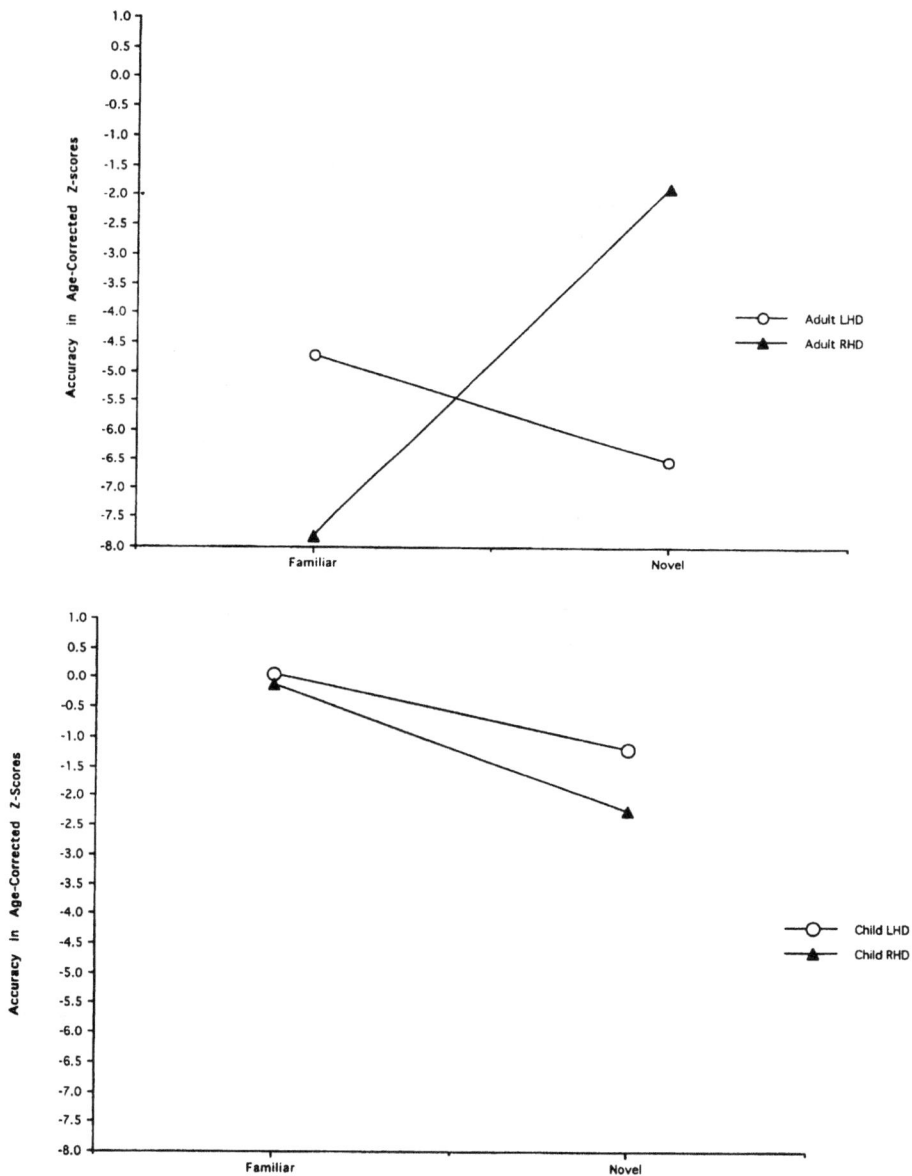

FIG. 7.3. *(Top)* Performance on familiar versus novel sentences in adults with left (LHD) versus right (RHD) hemisphere injury. *(Bottom)* Performance on familiar versus novel sentences in children with left versus right hemisphere damage. (Adapted from Kempler et al., 1999).

In short, whether we are talking about global measures like IQ or more subtle measures of sentence production and comprehension, children with LHD versus RHD do not display the profiles of impairment that we would expect based on the adult aphasia literature—at least not in these and other retrospective studies, with outcome measures at or above 6 years of age (i.e., beyond the point at which fundamental aspects of grammar and phonology are usually in place; Bates et al., 1995).

AGE OF LESION ONSET AND THE PROBLEM OF CRITICAL PERIODS

The distinction between retrospective and prospective studies is related to the controversial problem of "critical periods" for language, with special focus on the age at which a lesion is acquired. By definition, prospective studies focus on children whose lesions are acquired very early, preferably before the point at which language learning normally begins. In contrast, many retrospective studies collapse across children who acquired their lesions at different points across the course of language learning. Our own prospective studies are based exclusively on children with congenital injuries, defined to include pre- or perinatal injuries that are known to have occurred before 6 months of age, restricted to one side of the brain (left or right), confirmed through one or more forms of neural imaging (computed tomography or magnetic resonance imaging). Hence our results may differ from studies of children with injuries acquired at a later point in childhood.

What might those differences be? Unfortunately, there is very little empirical evidence regarding the effect of age of lesion onset on subsequent language outcomes. Only one fact is clear: that the outcomes associated with LHD are much better in infants than they are in adults. This means, of course, that plasticity for language must decrease markedly at some point between birth and adulthood (Lenneberg, 1967). But when does this occur, and how does it happen?

Many investigators have argued that this decrease in plasticity takes place at the end of a "critical period" for language, a window of opportunity that is also presumed to govern the child's ability to achieve native-speaker status in a second language (for discussions, see Bialystok and Hakuta, 1994; Curtiss, 1988; Elman et al., 1996; Johnson and Newport, 1989; Marchman, 1993; Oyama, 1993; Weber-Fox and Neville, 1996). So much has been said about this presumed critical period that a newcomer to the field (and many consumers within it) would be justified in assuming that we know a great deal about its borders (i.e., when it begins and when it comes to an end) and about the shape of the learning function in between these points. The very term *critical period* suggests that the ability to acquire a native language and/or the ability to recover from brain injury both come to a halt abruptly, perhaps at the same time, as the window of opportunity slams shut. The fact is, however, that we know almost nothing about the shape of this function. In fact, we are not even justified in assuming that the function is monotonic (i.e., that it gets progressively harder to learn a native language and progressively harder to recover from injuries to the left hemisphere).

With regard to the presumed critical period for recovery from brain injury, we are aware of only two large cross-sectional studies that have compared language and cognitive outcomes in children who acquired their lesions at different ages from congenital injuries (at or before birth) through early adolescence (Vargha-Khadem et al., unpublished results, cited with permission in Bates et al., 1999; Goodman and Yude, 1996). Figure 7.4 compares results from both these studies for verbal IQ. As Figure 7.4 indicates, the effect of age of injury is nonmonotonic in both stud-

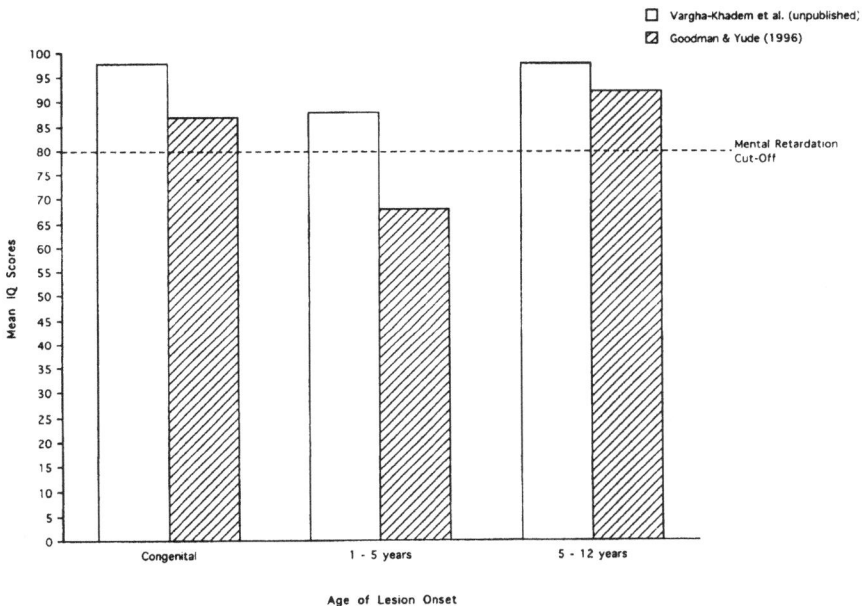

FIG. 7.4.   Relationship between age of lesion onset and IQ scores in two samples of children with focal brain injury.

ies: The worst outcomes are observed in children who suffered their injuries between approximately 1 and 4 years of age. In support of the critical period hypothesis, better outcomes are observed following congenital injuries. However, in direct contradiction to the critical period hypothesis, better outcomes are also observed in children whose injuries occurred between approximately 5 and 12 years of age, which means that there is no monotonic drop in plasticity. To some extent, these unpleasant wrinkles in the expected function could be due to uncontrolled differences in etiology (e.g., the factors leading to injury may differ at birth, 1–5 years, and later childhood). At the very least, however, these results ought to make us skeptical of claims about a straightforward critical period for recovery from brain injury.

Similar nonmonotonic findings have been reported in at least one study of second-language acquisition and first-language loss (Liu et al., 1992). To illustrate, compare the results in Figure 7.5 (adapted from a famous study of second-language acquisition) and Figure 7.6. Figure 7.5 illustrates results from a grammaticality judgment task administered to first- and second-language learners of English, comparing performances

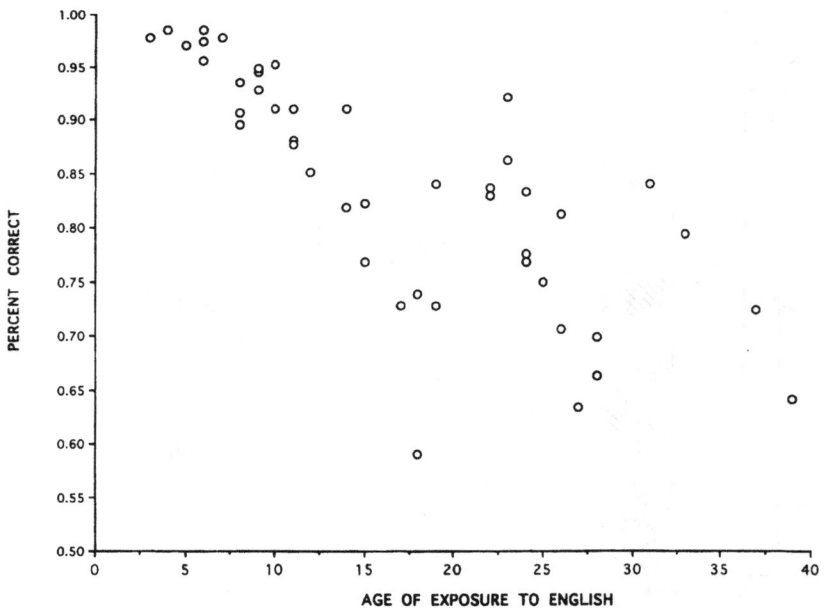

FIG. 7.5. Performance on a grammaticality judgment task in non-native speakers of English as a function of age of exposure to English. (Adapted from Johnson and Newport, 1989.)

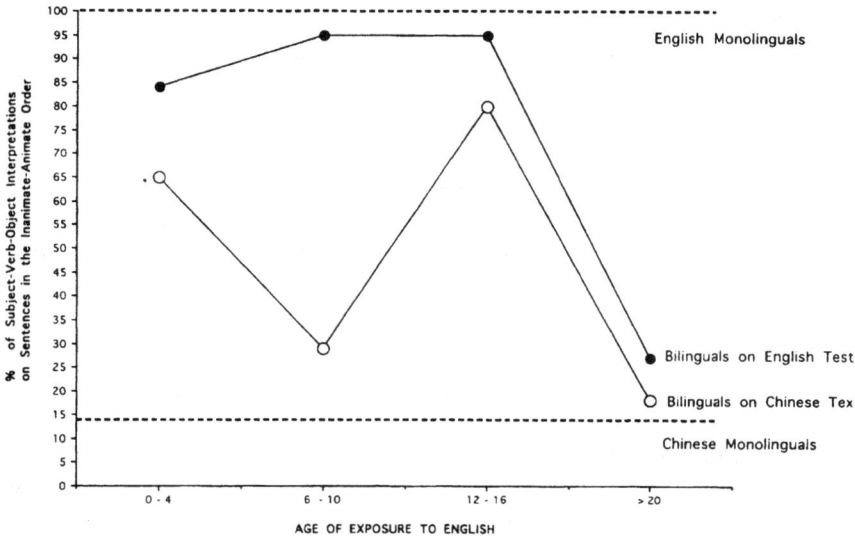

FIG. 7.6. "English-like" versus "Chinese-like" grammatical comprehension as a function of age of exposure to English. (Adapted from Liu et al., 1992.)

of individuals who arrived in the United States at different points spanning the period from birth to early adulthood. This well-known figure suggests that there is no single point at which the window of opportunity for second-language learning slams shut. However, it does provide evidence for a monotonic drop in language learning ability from birth to adolescence.

Consider, however, the results in Figure 7.6, based on a sentence interpretation task administered to Chinese–English bilinguals in both Chinese and English. In this task, subjects were able to use semantic or word order information to interpret "odd" sentences like "The rock chased the dog." Native speakers of English invariably choose the first noun, using word order to make their interpretation. Native speakers of Chinese invariably choose the second noun, ignoring word order in favor of semantic information. Both these strategies make perfect sense in terms of the information value of standard word order in these two languages (Chinese permits so much word order variation that a persistent word order strategy like the one used in English would not be very useful). Hence this little task serves as a useful litmus test for retention of the first language (L1) as well as acquisition of the second (L2). The interesting point for our purposes is that Chinese–English bilinguals often perform somewhere in between these two extremes, in one or both of their two languages, and these different "weightings" of word order and semantic information vary as a function of age of acquisition. Notice that results for English (L2) are generally in agreement with Johnson and Newport's results (1989) for a very different task: Although our results asymptote at an earlier point than those of Johnson and Newport (1989), they do provide evidence for a monotonic shift from "English-like" interpretations of English sentences in those who learned their English very early to "Chinese-like" interpretation of English sentences in those who learned their English relatively late. However, results for Chinese (L1) show a very different function, a nonmonotonic curve in which the best results (movement toward the second language without loss of the first language) are observed in those who are exposed to a second language sometime between 4 and 7 years of age.

Although this is a complex result, the point of this comparison for our purposes here is a simple one: There is no single "critical period" for language learning; results depend on many different factors, and the probability of a positive outcome can rise or fall at different points in development, in L2 learning, and in recovery from brain injury. This is when prospective studies can be particularly illuminating: By studying children during their first encounters with language and other forms of higher cognition, we can learn more about effects associated with the initial state of the brain, together with the processes of development and (re)organization that lead these children to a normal or near-normal outcome.

LANGUAGE OUTCOMES IN CHILDREN
WITH EARLY FOCAL BRAIN INJURY:
PROSPECTIVE FINDINGS

All theories that take some form of plasticity into account (including theories that assume a critical period) would lead us to expect relatively good outcomes in children with congenital injuries (i.e., the group that we have studied in our laboratory). Evidence for the developmental plasticity of language in this group has mounted in the last few years due in part to improved techniques for identifying children with early brain injury, including precise localization of the site and extent of damage through neuroradiology. In some cases, we have been able to identify such children in the first weeks of life, prior to the time when language acquisition would normally begin, permitting us to chart the course of language, cognition, and communicative development from the very beginning (Bates et al., 1997; Reilly et al., 1995; Stiles et al., 1998; Stiles and Thal, 1993), before the point at which alternative forms of brain organization have emerged.

In fact, the prospective studies that we have carried out so far provide compelling evidence for initial deficits and subsequent processes of recovery—phenomena that are not visible later on, when most retrospective studies take place. For example, prospective studies of nonverbal cognitive development by our colleague Joan Stiles have revealed subtle but consistent patterns of deficit in visuospatial cognition. For example, children with RHD appear to have difficulty perceiving and/or producing the global or configural aspects of a complex visual array; children with LHD are generally spared at the global level, but they have difficulty with the perception and/or production of local details (Note: I will return to this example later on, relating it to our findings for language.) These visuospatial deficits are qualitatively similar to those observed in LHD versus RHD adults, although they are usually more subtle in children, and they resolve over time as the children acquire compensatory strategies to solve the same problems (Stiles et al., 1998; Stiles and Thal, 1993).

If a similar result could be found in the domain of language, then we might expect (by analogy to the literature on adult aphasia) to find the following results in the first stages of language development:

Left hemisphere advantage for language: Children with LHD will perform below the levels observed in children with RHD on virtually all measures of phonological, lexical, and grammatical development, as well as measures of symbolic and communicative gesture.

The Broca pattern: By analogy to Broca's aphasia in adults, children with damage to the frontal regions of the left hemisphere will be particularly delayed in expressive but not receptive language and may (on some accounts) be particularly delayed in the development of grammar and phonology.

The Wernicke pattern: By analogy to Wernicke's aphasia in adults, children with damage to the posterior regions of the left temporal lobe will be particularly delayed in receptive language, perhaps (on some accounts) with sparing of grammar and phonology but selective delays in measures of semantic development.

Our group set out to test these three hypotheses in a series of prospective studies of early language development. In every case, we have uncovered evidence for early deficits, and these deficits do appear to be associated with specific lesion sites. However, in contrast with Stiles' findings for visuospatial cognition, results for language provide very little evidence for hypotheses based on the adult aphasia literature.

The first study (Marchman et al., 1991) focused on the emergence of babbling and first words in a small sample of five children with congenital brain injury, two with RHD and three with LHD, including one LHD case with injuries restricted to the left frontal region. All the children were markedly delayed in phonological development (babbling in consonant–vowel segments weeks or months behind a group of neurologically intact controls) and in the emergence of first words. However, three of the children moved up into the normal range across the course of the study. The two who remained behind had injuries to the posterior regions of the left hemisphere, results that fit with the first hypothesis (left hemisphere advantage for language) but stand in direct contradiction to both the Broca and the Wernicke hypotheses.

The second study (Thal et al., 1991) focused on comprehension and production of words from 12 to 35 months in a sample of 27 infants with focal brain injury based on a parental report instrument that was the predecessor of the MacArthur Communicative Development Inventories (MCDI) (Fenson et al., 1993, 1994). In complete contradiction to Hypothesis 1 (left hemisphere mediation of language) and Hypothesis 3 (the Wernicke hypothesis), delays in word comprehension were actually more likely in the RHD group. In line with Hypothesis 1, but against Hypothesis 2 (the Broca hypothesis), delays in word production were more likely in children with injuries involving the left posterior quadrant of the brain.

A more recent study built on the findings of Thal et al. (1991) with a larger sample of 53 children, 36 with LHD and 17 with RHD (Bates et al., 1997), and a combination of parent report (the MCDI) and analyses of free speech. This report is broken into three substudies, with partially overlapping samples. Study 1 used the MCDI to investigate aspects of word comprehension, word production, and gesture at the dawn of language development in 26 children between 10 and 17 months of age. Study 2 used the MCDI to look at production of both words and grammar in 29 children between 19 and 31 months. Study 3 used transcripts of spontaneous speech in 30 children from 20 to 44 months, focusing on mean length of utterance in morphemes (MLU). In all these studies, comparisons between the LHD and RHD groups were followed by comparisons looking at the effects associated with lesions involving the frontal lobe (comparing children with left frontal involvement to all RHD cases as well as LHD cases with left frontal sparing) and the temporal lobe (comparing children whose lesions include the left temporal lobe with all RHD cases and all LHD cases in which that region is spared). Results were compatible with those of Marchman et al. (1991) and Thal et al.

(1991), but were quite surprising from the point of view of lesion/symptom mappings in adult aphasia, as follows.

First, in a further disconfirmation of Hypotheses 1 and 3, Bates et al. (1997) report that delays in word comprehension and gesture were both more likely in children with unilateral damage to the right hemisphere at least likely in the 10–17-month window examined here. Further studies of gestural development in our laboratory have confirmed that the gestural disadvantage for RHD children is still present between 20 and 24 months (Stiles et al., 1998).

Second, in a partial confirmation of Hypothesis 2 (the Broca hypothesis), frontal involvement was associated with greater delays in word production and the emergence of expressive grammar between 19 and 31 months. However, in a surprising partial disconfirmation of Hypothesis 2, this frontal disadvantage was equally severe with either left frontal or right frontal involvement. In other words, the frontal lobes are important during this crucial period of development (which includes the famous "vocabulary burst" and the flowering of grammar), but there is no evidence for a left–right asymmetry in the frontal regions and hence no evidence in support of the idea that Broca's area has a privileged status from the very beginning of language development.

Third, in line with Hypothesis 1 (left hemisphere mediation of language) but in direct contradiction to Hypotheses 2 and 3 (analogies to Broca's and Wernicke's aphasia), delays in word production and the emergence of grammar were both more pronounced in children with injuries involving the left temporal lobe. In contrast with the above two findings (which only reached significance within a restricted period of development), this left temporal disadvantage was reliable across all three substudies by Bates et al. (1997) from the very first words (between 10 and 17 months of age) through crucial developments in grammar (between 20 and 44 months of age). Hence we do have evidence for the asymmetrical importance of Wernicke's area, but that evidence pertains equally to grammar and vocabulary (with no evidence of any kind for a dissociation between the two) and seems to be restricted to expressive language.

Reilly et al. (1998) conducted similar comparisons by lesion side and lesion site in a cross-sectional sample of 30 children with focal brain injury (15 LHD and 15 RHD) between 3 and 12 years of age; these results were also compared with performances by a group of 30 age-matched controls with no history of neurological impairment. Analyses were based on lexical, grammatical, and discourse measures from a well-known story-telling task. For children between 3 and 6 years of age, Reilly et al. (1998) replicated the specific disadvantage in expressive language for children with lesions involving the temporal region of the left hemisphere. However, this effect was not detectable in children between 6 and 12 years of age—even though all children in this study had the same congenital etiology. In fact, data for the older children provided no evidence of any kind for an effect of lesion side (left versus right) or lesion site (specific lobes within either hemisphere). The only effect that reached significance in older children was a small but reliable disadvantage in the brain-injured children as a group compared with neurologically intact age-matched controls. Figure 7.7 compares results for younger versus older children on one grammatical index (mean number of errors in grammatical morphology per proposition), divided into children with left temporal involvement (+LTemp), focal lesion cases without left temporal involving (−Ltemp, combining all RHD cases and all LHD cases with temporal sparing), and neurologically intact normal controls. Although we must remember that these are cross-sectional findings, they suggest that a sub-

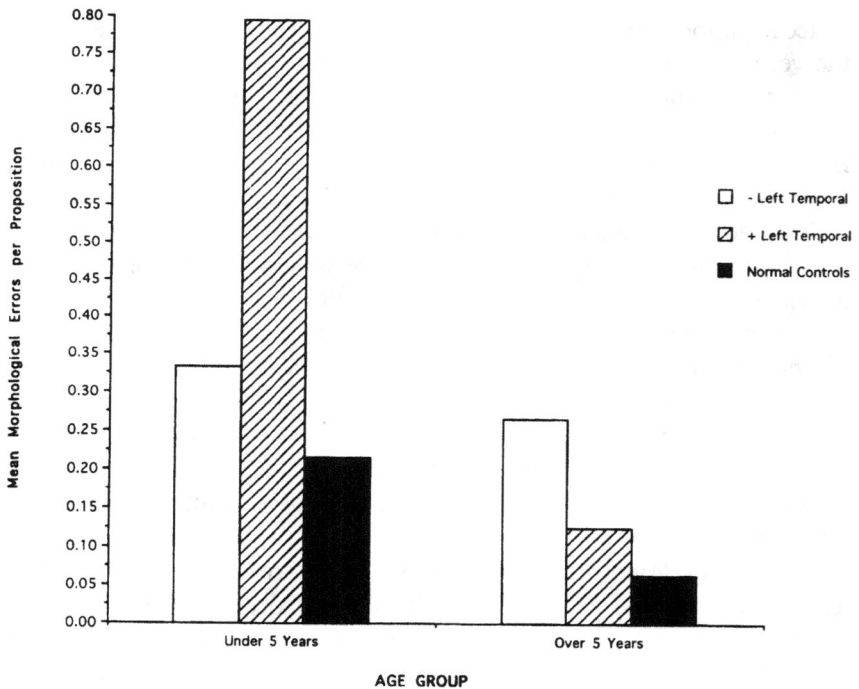

FIG. 7.7. Morphological errors as a function of age and presence/absence of left temporal damage. (Adapted from Reilly et al., 1998.)

stantial degree of recovery takes place in the LHD group during the first few years of life. In subsequent longitudinal studies, Reilly and her colleagues have followed a smaller group of children across this period of development. These longitudinal findings are compatible with the cross-sectional evidence in Figure 7.7, suggesting that the crucial period of recovery takes place before the age range covered by most of the retrospective studies in the literature on cognitive and linguistic outcomes in children with focal brain injury.

To summarize, our prospective studies of language development in children with early focal brain injury have provided evidence for specific delays, correlated with specific lesion sites. However, the nature of these lesion–symptom correlations differs markedly from those that we would expect based on the adult aphasia literature. Furthermore, these correlations are only observed within specific windows of development, followed by evidence for recovery and (by implication) reorganization. None of these results are evident in retrospective studies (including our own), where children are tested beyond the point at which this presumed reorganization has taken place.

We are occasionally asked why our results appear to be incompatible with an earlier literature on the effect of hemispherectomy (e.g., Dennis and Whitaker, 1976; but see Bishop, 1983) and/or effects of early stroke (e.g., Aram, 1988; Aram et al., 1985a,b; Woods and Teuber, 1978). Our first answer is that our results are *not* incompatible with the vast majority of studies. However, they do *appear* to be incompatible with a handful of studies that were cited (usually in secondary sources) as evidence in favor of an innate and irreversible role for the left hemisphere in some aspects of language processing. As we have noted elsewhere (Bates et al., 1999; see also Vargha-Khadem et al., 1994), apparent inconsistencies between the earlier studies and our more recent work disappear when one looks carefully at the fine print.

First, many of the earlier studies combined data for children whose injuries occurred at different points in development, and they also combined results (usually on rather global measures) for children at widely different ages at time of testing. As we saw in the previous section, there may not be a monotonic relation between age of injury and language outcomes, and the nature of the lesion–symptom mappings that we observe may be quite different depending on the age at which children are tested and the developmental events that are most prominent at that time.

Second, some of the earlier studies had methodological limitations that we have been able to overcome in the studies described above. In particular, a number of well-known studies could not perform direct comparisons of children with LHD versus RHD because of uncontrolled differences in age, education, and/or etiology. Instead, the RHD and LHD groups were each compared with a separate group of matched controls. For example, Dennis and Whitaker (1976) report that their left-hemispherectomized children performed below normal controls on subtle and specific aspects of grammatical processing; no such difference was observed between right-hemispherectomized children and their controls. These results were interpreted as though they constituted a significant difference between the LHD and RHD groups even though the latter two groups were never compared directly. As Bishop (1983) has pointed out in her well-known critique, a careful examination of results for the two lesion groups suggest that this interpretation is not warranted. The general problem that one encounters with the separate control group approach is illustrated in Figure 7.8, which compares hypothetical data for an LHD group, an RHD group, and their respective controls. As we can see, performance by the LHD group does fall reliably below performance by their controls (albeit just barely); performance by the RHD does not fall outside the confidence intervals for their control group. And yet, in this

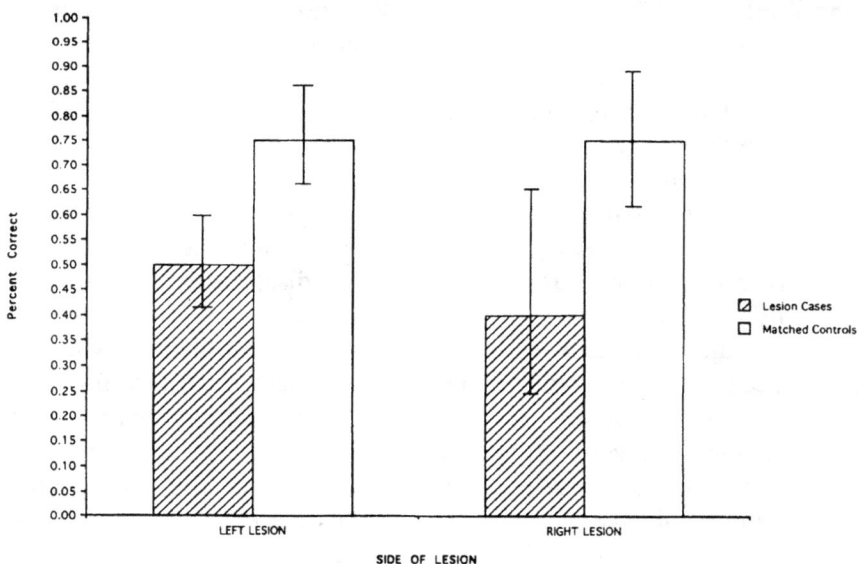

FIG. 7.8.   A hypothetical example of comparisons between left (LHD) and right (RHD) hemisphere groups with their respective controls (LHD, control difference is reliable; RHD, control difference is not).

hypothetical example, performance is actually better in the LHD cases! The key to this conundrum lies in the standard deviations for each control group: The standard deviation is larger for the RHD controls, which means that a larger difference between RHD and controls is required to reach statistical significance. Clearly, it would be unwise to draw strong conclusions about left–right differences from a data set of this kind.

Finally, some of the better-known claims in favor of an early and irreversible effect of LHD have been based on single-case studies or very small samples (including the hemispherectomy studies cited above). This fact limits the generalizability of results, and the same result is often contradicted by other individual-case or small-group studies.

For example, Stark and McGregor (1997) have recently described an interesting contrast between one child with a left hemispherectomy (seizure onset at 1;6, surgery at 4;0) and another with a right hemispherectomy (seizure onset at 2;0, surgery at 5;8). Both children were followed longitudinally with testing at 1–2-year intervals through 9;0 and 9;6 years of age, respectively. Although both children did show substantial development in language and cognition across the course of the study, they fell behind age-matched normal controls at every point. At the end of the study, the LHD case had a full-scale IQ of 71 and the RHD case had a full-scale IQ of 81, well behind the norms for development in children who are neurologically intact. For Stark and McGregor (1997), the most interesting findings lie in the contrasting patterns observed for each child for performance IQ, verbal IQ, and a series of more specific language tests. For the LHD case, verbal and performance IQ were both quite low (separated by only four points). However, performance on the specific language tasks followed a profile typical of the pattern observed in children with Specific Language Impairment, i.e., greater impairment in language measures (especially morphosyntax) than we would expect for her mental age. In contrast, the RHD case displayed a sharp dissociation at the end of the study between Verbal IQ (95) and Performance IQ (70), with scores on most of the specific language measures that were appropriate for her mental age.

This is an interesting and provocative result, and it might indeed reflect evidence for the emergence of some kind of left hemisphere specialization for language prior to the age at which the surgery occurred. However, our own experience with a relatively large focal lesion sample has made us wary of basing strong results on case studies. Individual differences in language and cognitive ability are immense, even in perfectly normal children with no history of brain injury (Bates et al., 1995; Fenson et al.,

1994). A similar degree of variation is observed even within the small cadre of cases that have undergone hemispherectomy.

Evidence for such variation comes from the case of Alex, recently reported by Vargha-Khadem et al. (1997). Alex was nearly mute prior to his surgery between 8 and 9 years of age and (to the extent that he could be tested at all) demonstrated levels of language comprehension similar to those of a normal 3-year-old. Soon after his surgery, he demonstrated remarkable recovery in both receptive and expressive language and continued to make progress into adolescence. Although Alex did suffer some degree of mental retardation (as an adolescent, he has the mental age of a 10–12-year-old on most measures), his language abilities are entirely commensurate with his mental age. In fact, his level of performance on language measures is superior to both of the cases reported by Stark and McGregor (1997), even though his surgery took place several years later. The contrast between this study and that of Stark and McGregor (1997) underscores two important points. First, it provides further evidence against the assumption that plasticity drops monotonically across a supposed critical period for language. Second, it reminds us that the effects of brain injury are superimposed on the vast landscape of individual variation observed in normally developing children (for an elaboration of this point, see Bates et al., 1995). Because there is so much variation in the normal population, it is difficult to know in a single-case or small-sample study whether the cognitive profiles we observe are statistically reliable. Indeed, they may be no different from the patterns that would be observed if brain damage were imposed randomly on cases selected from the population at large (Bates et al., 1991a; Bishop, 1997; see also Basser, 1962, for evidence that the vast majority of cases in a large sample of hemispherectomized children show no evidence at all of a speech–language impairment, regardless of side of surgery).

Despite these concerns, our results for older children are largely compatible with the retrospective literature on language development in the focal lesion population: Children with early injuries to one side of the brain usually acquire language abilities within the normal or low-normal range, with little evidence for effects of lesion side or lesion site (as reviewed by Bates et al., 1999; Eisele and Aram, 1995; Vargha-Khadem et al., 1994). Our prospective findings for children under 5 years of age are qualitatively different, but they are also so new that there is little or no comparable information in the literature, aside from a few single-case or small-sample studies with very different goals (e.g., Dall' Oglio et al., 1994; Feldman et al., 1992). Of course it will be important to replicate all these prospective findings with other samples of children, and in other laboratories. In the meantime, we can take some comfort in the fact that these results are based on the largest and most homogeneous sample of children with focal brain injury that has ever been studied in a prospective framework. Although in some cases the same children participate in more than one prospective study, the full sample across our two largest studies (Bates et al., 1997; Reilly et al., 1998) includes 72 cases of children with focal brain injury from three different laboratories. With sample sizes of 26 or more from one substudy to another, we have been able to use experimental designs and inferential statistics that would not be appropriate in a single-case or small-sample study, revealing new information about the changing nature of lesion–symptom correlations. In short, the findings are solid enough to justify some speculation about the development of brain organization for language under normal and pathological conditions.

HOW BRAIN ORGANIZATION FOR LANGUAGE EMERGES ACROSS THE COURSE OF DEVELOPMENT

The literature on language outcomes in human children with early unilateral brain injury is quite compatible with the burgeoning literature on neural plasticity in other species. Many of the human results are new, but the information from developmental neurobiology is now well established. Although few neurobiologists would argue in favor of *equipotentiality*, that is, the idea that all areas of cortex are created equal (Lenneberg, 1967), there is now overwhelming evidence in favor of *pluripotentiality*—the idea that cortical tissue is capable of taking on a wide array of representations, with varying degrees of success, depending on the timing, nature, and extent of the input to which that tissue is exposed (Elman et al., 1996; Johnson, 1997).

This conclusion is well attested in the developmental neurobiology literature, but it has had surprisingly little impact in linguistics, cognitive science, and cognitive neuroscience. In fact, the old phrenological approach to brain organization has found new life in the last two decades in various proposals that language is an "instinct" (Pinker, 1994), a "mental organ" (Chomsky, 1980a,b, 1995), or an "innate module" (Fodor, 1983; Pinker, 1997a), with its own neural architecture and its own highly specific genetic base (see also Gopnik, 1990; Pinker, 1991; Rice, 1996; Van der Lely, 1994). Indeed, Fodor's 1983 monograph celebrates the contributions of Franz Gall, the original phrenologist, and proudly bears a classic drawing of Gall's subdivided and numbered brain on its cover. The only real surprise is how little the claims have changed across the last 200 years.

Phrenology in all its reincarnations can be characterized as the belief that the brain is organized into spatially and functionally distinct faculties, each dedicated to and defined by a different kind of intellectual, emotional, or moral content. In some of the proposals put forward by Gall, Spurzheim, and others in the eighteenth century, these included areas for hope, combativeness, conjugal love, veneration, cautiousness, calculation, tune, memory, and, of course, language. A modern variant of phrenology is represented in cartoon form in Figure 7.9, which differs from the old version in at least two respects. First, the content of the proposed modules has changed a great deal in the last two centuries: With some exceptions, most of the ethical content is gone (but see Ramachandran et al., 1997, for a proposed "religiosity module"), replaced by a smaller set of species-specific cognitive and linguistic domains (e.g., music, faces,

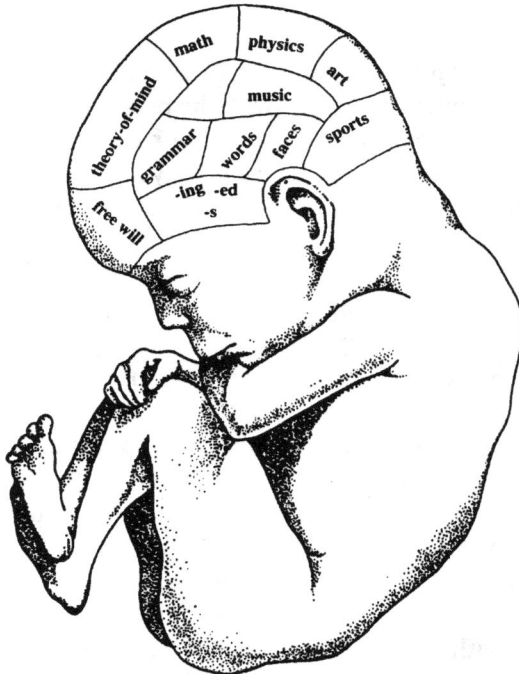

FIG. 7.9. The phrenological approach to development.

mathematics, grammar, the lexicon). To be sure, the particular entries and placements in Figure 7.9 are of my own making, but each one represents explicit claims that have been made in the last 5–10 years in the *New York Times* and other public outlets. Second, and most important for our purposes here, the modern version of phrenology has a strong nativist component. In contrast with the nineteenth century phrenologists (some of whom underscored the role of experience in setting up the functional organization of the brain—see especially Wernicke, 1977), twentieth century champions like Fodor and Pinker have wedded their theory of modular localization to the doctrine of innateness. In this variant of phrenology, the adult brain is organized along modular lines because the brain came packaged that way, in its fetal form, with specific functions assigned to specific regions by a genetic program (see also Gopnik and Crago, 1991; Rice, 1996; Van der Lely, 1994).

In part, the phrenological approach may persist because alternative accounts are difficult to understand. The adult brain is a highly differentiated organ, and the infant brain (though underspecified in comparison to the adult brain) is certainly not a *tabula rasa*. And yet efforts to reintroduce experiential effects on this brain organization (e.g., Bates and Elman, 1996; Elman and Bates, 1997) have been met with great suspicion by those who fear a reintroduction of old behaviorist accounts (Clark et al., 1997; Jenkins and Maxam, 1997; Pesetsky et al., 1997; Pinker, 1997b). Some of the heat in this exchange comes from the fact that several logically and empirically distinct issues are conflated in the argument about mental organs for language. As a result, anyone who opposes the modern doctrine of phrenology in its full-blown form is accused of (gasp!) behaviorism. To clarify the difference between old-fashioned *tabula rasa* behaviorism and the emergentist perspective that I am espousing here, we need to break the mental organ doctrine down into a series of separate and separable assumptions about (*1*) *innate representations* (i.e., synaptic connections are determined by a genetic program), (*2*) *domain-specific processing* (each region of the brain is designed to handle a specific kind of content), and three corollaries about localization, (*3*) *compact location*, (*4*) *fixed location*, and (*5*) *universal location*. Table 7.1 summarizes the five claims of modern phrenology, together with a characterization of the emergentist alternative on each of these five counts.

Consider first the assumption of innate representations. As my colleagues and I have acknowledged repeatedly, throughout this chapter and elsewhere (Bates et al., in press; Elman et al., 1996), cortex is not equipotential. There are powerful endogenous constraints in the infant

TABLE 7.1
Two Views of Brain Organization for Language

Phrenological View	Emergentist View
Innate representations	Emergent representations
Dedicated, domain-specific neural mechanisms for learning and processing	Domain-general neural mechanisms for learning and processing
Fixed localization	Plastic localization
Compact localization	Distributed localization
Universal localization	Variable localization

brain that bias the way that brain organization will proceed under normal circumstances. However, claims about the nature of these innate constraints can be made on several different levels: *innate representations* (where "representations" are operationally defined as the patterns of cortical connectivity that comprise knowledge), *innate architecture* (defined in terms of the global input–output architecture of the brain and local variations in density, speed, and style of information processing), and *innate timing* (including variations in length of neurogenesis and the onset and offset of neurotrophic factors). The mental organ doctrine is deeply committed to the existence of innate representations. The emergentist alternative is committed to the idea that knowledge itself is not innate, but emerges across the course of development, through the interaction of innate architecture, innate timing, and input to the cortex.

In fact, the case for innate representations looks very bad right now. Thirty years ago, representational nativism was a perfectly plausible hypothesis. That is, it was reasonable to suppose that knowledge is built into the infant cortex in the form of detailed and well-specified synaptic connections, independent of and prior to the effects of input to the cortex (what Pinker [1997a] refers to as an innate "wiring diagram"). Indeed, such an assumption is critical for strong forms of linguistic nativism (i.e., the idea that children are born with Universal Grammar; Chomsky, 1980a,b; Pinker, 1994; Rice, 1996) because synaptic connectivity is the only level of brain organization with the necessary coding power for complex and domain-specific representations of the sort that would be required to support an innate grammar. However, this particular form of innateness is difficult to defend in the face of mounting information on the activity-dependent nature of synaptic connectivity at the cortical level. Of course the infant brain is certainly not a *tabula rasa*. At other levels of organization, we have ample evidence for endogenous effects that bias the learning game in significant ways. These include constraints on the global input–output architecture of the brain (e.g., the fact that information from the eye usually does end up in visual cortex, in the absence of wicked interventions by Sur and his colleagues [1990]), local variations in architecture and style of computation (e.g., primary visual cortex starts out with roughly twice as many neurons as any other area), and variations in timing (e.g., variations from one region to another in the length of neurogenesis) and in the availability of nerve growth factor. It now seems that the difference between the human brain and that of other primates must be determined primarily by nonrepresentational variations of this kind, controlled by a genetic program small enough to fit into the mere 1%–2% difference between the human genome and the genome of a chimpanzee (King and Wilson, 1975; Wilson, 1985).

The second assumption in Table 7.1, domain-specific processing, is a key component of the mental organ doctrine, i.e., that distinct regions of the brain have evolved to deal with particular kinds of content of compelling interest to our species (Barkow et al., 1992; Pinker, 1997a). In addition to language (and perhaps to distinct subcomponents of language, e.g., a distinction between grammar and the lexicon), proposed modules or mental organs include a face detector, a theory-of-mind module (that contains algorithms for detecting dishonest behavior by other members of the species), a mathematics module, a music module, and so forth. These systems have presumably evolved to deal optimally with their assigned content and only with that content. Indeed, Pinker (1997a) has proposed that diverse and specific forms of psychopathology may result if a module is applied to the wrong domain (although it is not entirely clear how this might occur, given the perceptual biases that define a mental organ).

The emergentist alternative to domain-specific processing is that domain-specific knowledge can be acquired and processed by domain-general mechanisms, that is, by mechanisms of attention, perception, memory, emotion and motor planning that are involved in many different aspects of learning, thought, and behavior. In other words, the cognitive machinery that makes us human can be viewed as a new machine constructed out of old parts (Bates et al., 1979). All of the component parts that participate in language are based on phylogenetically ancient mechanisms, with homologues up and down the vertebrate line. The specific functions that make humans different from other species are superimposed on this Basic Vertebrate Brain Plan. Of course it is likely that some and perhaps all of the neural components that participate in human activity have undergone quantitative changes that permit new behaviors like language to emerge, but these components still continue to carry out older and more general functions of object detection, shifting attention, formation of new memories, motor planning, and so forth (i.e., they have kept their day jobs.).

To help us think about the kind of adaptation that would permit the construction of a new machine from old parts, consider the metaphor of the giraffe's neck. Giraffes have the same number of neckbones that you and I have, but these bones are elongated to solve the peculiar problems that giraffes are specialized for (i.e., eating leaves high up in the tree). As a result of this particular adaptation, other adaptations were necessary as well, including cardiovascular changes (to pump blood all the way up to the giraffe's brain), shortening of the hindlegs relative to the forelegs (to ensure that the giraffe does not topple over), and so on. Should we conclude that the giraffe's neck is a "high-leaf-reaching organ"? Not exactly. The giraffe's neck is still a neck, built out of the same basic blueprint that is used over and over in vertebrates, but with some quantitative adjustments. It still does other kinds of "neck work," just like the work that necks do in less specialized species, but it has some extra potential for reaching up high in the tree that other necks do not provide. If we insist that the neck is a leaf-reaching organ, then we have to include the rest of the giraffe in that category, including the cardiovascular changes, adjustments in leg length, and so on.

In the same vein, our "language organ" can be viewed as the result of quantitative adjustments in neural mechanisms that exist in other mammals, permitting us to walk into a problem space that other animals cannot perceive much less solve. Of course, once language finally appeared on the planet, it is quite likely that it began to apply its own adaptive pressures to the organization of the human brain, just as the leaf-reaching adaptation of the giraffe's neck applied adaptive pressure to other parts of the giraffe. Hence the neural mechanisms that participate in language still do other kinds of work, but they have also grown to meet the language task. In fact, it seems increasingly unlikely that we will ever be in a position to explain human language in terms of clear and well-bounded differences between our brain and that of other primates. Consider, for example, the infamous case of the planum temporale (i.e., the superior gyrus of the temporal lobe reaching back to the temporal–parietal–occipital juncture). It was noted many years ago that the planum temporale is longer on the left side of the brain in the majority of normal, right-handed human adults. Because the temporal lobe clearly does play a special role in language processing, it was argued that the asymmetry of the planum may play a key role in brain organization for language. However, surprising new evidence has just emerged showing that the same asymmetry is also observed in chimpanzees (Hollaway et al., 1997). In fact, the asymmetry is actually larger and more consistent in chimpanzees that it is in humans! I do not doubt for a moment that humans use this stretch of tissue in a quantitatively and qualitatively different way, but simple differences in size and shape may not be sufficient or even relevant to the critical difference between us and our nearest relatives in the primate line. In response to findings of this sort, Pinker (1997a) has insisted that the answer lies in the cortical microcircuitry within relevant areas. And yet, as we have seen over and over, developmental neurobiologists have abandoned the idea that detailed aspects of synaptic connectivity are under direct genetic control, in favor of an activity-dependent account. There has to be something special about the human brain that makes language possible, but that "something" may involve highly distributed mechanisms that serve many other functions.

My own favorite candidates for this category of "language-facilitating mechanisms" are capacities that predate language phylogenetically and undoubtedly involve many different aspects of the brain. They include our rich social organization and capacity for social reasoning, our extraordinary ability to imitate the things that other people do, our excellence in the segmentation of rapid auditory and visual stimuli, and our fascination with joint attention (looking at the same events together, sharing new objects just for the fun of it; for an extended discussion, see Bates et al., 1991b). These abilities are all present in human infants within the first year, and they are all implicated in the process by which language is acquired. None of them is specific to language, but they make language possible, just as quantitative adjustments in the giraffe's neck make it possible for the giraffe to accomplish something that no other ungulate can do.

Is there any evidence in favor of this domain-general "borrowed system" view? I would put the matter somewhat differently: Despite myriad predictions that such evidence will be found, there is still no unambiguous evidence in favor of the idea that specific parts of the brain are dedicated to specific kinds of objects and *only* those objects. For example, there are cells in the brain of the adult primate that respond preferentially to a particular class of stimuli (e.g., faces). However, recent studies have shown that the same cells can also respond to other kinds of content, spontaneously and/or after an extended period of training (Das and Gilbert, 1995;

De Weerd et al., 1995; Fregnac et al., 1996; Pettet and Gilbert, 1992; Ramachandran and Gregory, 1991; Tovee et al., 1996). Similarly, certain cortical regions around the sylvian fissure are invariably active in neural imaging studies of language processing, including some of the same areas that are implicated in fluent and nonfluent aphasia. However, each of these regions can also be activated by one or more forms of nonlinguistic processing. This point was made eloquently clear in a recent study by Erhard et al. (1996), who looked at all the proposed subcomponents of Broca's area while subjects were asked to carry out (covertly) a series of verbal and nonverbal actions, including complex movements of the mouth and fingers. Every single component of the Broca complex that is active during speech is also active in at least one form of covert nonverbal activity. In short, even though there is ample evidence for stretches of tissue that participate in language, there appears to be no candidate anywhere in perisylvian cortex for a pure language organ.

This brings us to three key assumptions about the nature of localization, the final three of the five contrasting issues listed in Table 7.1. On the phrenological account, precisely because of the assumptions about (1) innate representations and (2) dedicated architecture, it is further assumed that brain organization for language involves (3) a fixed architecture that cannot be replaced and cannot be modified significantly by experience, (4) a universal architecture that admits to very little individual variability, and (5) a compact and spatially contiguous architecture that operates as a coherent and autonomous unit in neural imaging studies and creates distinct deficits in or dissociations between cognitive functions when it is lesioned ("disconnection syndromes"; Caramazza, 1986; Caramazza and Berndt, 1985; Geschwind, 1965; Shallice, 1988). By contrast, the emergentist account is more compatible with forms of localization that are (3) plastic and modifiable by experience, (4) variable in form as a result of variations in experience as well as individual differences in the initial architecture, and (5) distributed across stretches of tissue that may participate in many different tasks (including spatially discontinuous systems that can perform separately or together depending on the task). Because of these properties, the emergentist view is much more compatible with all the mounting evidence from developmental neurobiology for the plasticity and activity dependence of cortical specialization, including plasticity for language in brain-injured children.

The emergentist view is also more compatible with the complex and variable findings that have emerged in recent neural imaging studies of normal adults (Courtney and Ungerleider, 1997; Poeppel, 1996). Indeed, new areas for language are multiplying at an alarming rate in language activation studies, including studies using positron emission tomography (PET), functional magnetic resonance imaging (fMRI), magnetoencephalography (MEG), and/or event-related brain potentials (ERP). Although activation is usually larger on the left than it is on the right in language activation studies, and the familiar perisylvian regions of the left hemisphere show up in study after study, there is increasing evidence for participation of homologous regions in the right hemisphere (e.g., Just et al., 1996), although there is substantial variation over individuals, tasks, and laboratories in the extent to which this occurs. Language activation studies that involve generation and maintenance of codes and/or a decision between behavioral options seem to result in reliable activation of several different prefrontal regions that were not implicated in older studies of language breakdown in aphasia (e.g., Raichle et al., 1994; Thompson-Schill et al., 1997). New regions that appear to be especially active during language activation have also appeared in basal temporal cortex (on the underside of the brain; Nobre et al., 1994), in some portions of the basal ganglia, and in the cerebellum (especially on the right side of the cerebellum). Many different aspects of both sensory and motor cortex seem to be activated in language tasks that involve imageable stimuli. More interesting still for our purposes here, these patterns of activation vary as a function of development itself, including variations with chronological age and language level in children (Hirsch et al., 1997; Mills et al., 1997; Mueller, 1996), and varying levels of expertise in adults (Hernandez et al., 1997; Kim et al., 1997; Perani et al., 1997; Raichle et al., 1994).

The picture that has emerged is one in which most of the brain participates in linguistic activity, in varying degrees, depending on the nature of the task and the individual's expertise in that task. In many respects, this is exactly what we should expect: Language is a system for encoding meaning, and there are now good reasons to believe that the activation of meaning involves activation of the same regions that participate in the original experiences on which meanings are based. Because most of the brain participates in meaning, we should expect widely distributed and dynamically shifting patterns of participation in most language-based tasks. The fact that these patterns of activation change over time is also not surprising, reflecting changes in experience as well as changes in the level of skill that individuals attain in activation and maintenance of both meaning and form.

Clearly, however, there are some important differences in the view of language organization that emerge from neural imaging studies and lesion studies. Neural imaging techniques can tell us about the areas of the brain that *participate* in language. From this point of view, we may conclude that the participation is very broad. Lesion studies can tells us about the areas of the brain that are *necessary* for normal language. The list of areas that are necessary for language (in children or adults) appears to be much smaller than the list of areas that participate freely in a language task. Even in this case, however, improved techniques for structural imaging and lesion reconstruction have yielded more and more evidence for individual variability in lesion–symptom mapping (Goodglass, 1993; Willmes and Poeck, 1993), and for compensatory organization in patients who display full or partial recovery from aphasia (Cappa et al., 1997; Cappa and Vallar, 1992).

There are of course some clear limits on this variability. Some areas of the brain simply cannot be replaced, in children or adults. For example, Bachevalier and Mishkin (1993) have shown that infant monkeys with bilateral lesions to the medial temporal regions (including the amygdala and the hippocampus) display a dense and apparently irreversible form of amnesia that persists for the rest of the animal's life, in marked contrast to the striking recovery that follows bilateral lesions to lateral temporal cortex (Webster et al., 1995). The key lies in the global input–output architecture of those medial temporal regions, a rich and broad form of connectivity that cannot be replaced because no other candidate has that kind of communication with the rest of the cortex. Other parts of the brain cannot be replaced because they are the crucial highways and offramps for information from the periphery (e.g., the insula, which receives crucial kinaesthetic feedback from the oral articulators, or the auditory nerve, which carries irreplaceable auditory input to the waiting cortex; Dronkers, 1996; Dronkers et al., 1994, 1999). These irreplaceable regions form the anchor points, the universal starting points for brain organization in normal children, and they are difficult if not impossible to replace once all the exuberant axons of the fetal brain have been eliminated.

Within this framework, learning itself also places limits on plasticity and reorganization in the developing brain. For example, Marchman (1993) has shown that artificial neural networks engaged in a language-learning task (i.e., acquiring the past tense of English verbs) can recover from "lesions" (i.e., random removal of connections) that are imposed early in the learning process. The same lesions result in a substantially greater "language deficit" when they are imposed later in the learning process. This simulation of so-called critical period effects takes place in the absence of any extraneous change in the learning potential of the network (i.e., there is no equivalent of withdrawal of neurotrophins or reduction in the learning rate). Marchman (1993) reminds us that critical period effects can be explained in at least two ways (and these are not mutually exclusive): exogenously imposed changes in learning capacity (the usual interpretation of critical periods) or the entrenchment that results from learning itself. In other words, learning changes the nature of the brain, eliminating some connections and tuning others to values that are difficult to change. Eventually the system may reach a point of no return, a reduction in plasticity that mimics critical period effects without any change in the architecture other than the changes that result from normal processes of learning and development. Marchman (1993) does not deny the possibility of exogenous effects on plasticity, but she argues convincingly that there are other ways to explain the same result, including gradual changes in the capacity to learn (and recover what was learned before) that are the product of learning itself—changes that are more compatible with the current developmental evidence than the idea of an abrupt and discontinuous critical period (see also Bates and Carnevale, 1993; Elman et al., 1996).

Finally, the emergentist view makes room for the possibility of systematic developmental changes in localization due to a shift in the processes and operations that are required to carry out a function at different points in the learning process. On the static phrenology view, a language area is a language area, always and forever. There may be developmental changes that are due to maturation (i.e., an area that was not "ready" before suddenly "comes on-line"), but the processes involved in that content domain are always carried out in the same dedicated regions. On the emergentist account, the areas responsible for learning may be totally different from the areas involved in maintenance and use of the same function in its mature form. In fact, there are at least three reasons why we should expect differences in the patterns of brain activity associated with language processing in children versus adults.

Early Competition

We may assume (based on ample evidence from animal models) that the early stages of development involve a competition among areas for control over tasks. This competition is open to any region that can receive and process the relevant information, but that does not mean that every region has an even chance of winning. In fact, as the competition proceeds, those regions that are better equipped to deal with that task (because of differences in efficiency of access and type of processing) will gradually take more responsibility for the mediation of that function. In prospective studies of language development, we are looking at this process of competition as it unfolds. This leads to the prediction that the earlier stages of development will involve more diffuse forms of processing, a prediction that is borne out by ERP studies of changes in activation across the first 3 years of language development (from activation to known words that is bilateral but slightly larger in the right toward activation that is larger on the left and localized more focally to frontotemporal sites; Mills et al., 1997).

Expertise

We may also expect quantitative and qualitative change in the regions that participate in a given task as a function of level of expertise. These changes can take three different forms: expansion within regions, retraction within regions, and a wholesale shift in mediation from one region to another. An example of expansion comes from a recent fMRI study of skill acquisition in adults (Karni et al., 1995). In this study, the first stages of learning in a finger-movement task tend to involve smaller patches of somatosensory cortex; with increased skill in this task, the areas responsible for the motor pattern increase in size. Examples of retraction come from studies that show larger areas of activation in the early stages of second-language learning compared with activation in native speakers and in more experienced second-language learners (Hernandez et al., 1997; Perani et al., 1997). Presumably this is because the novice speaker has to recruit more neural resources to achieve a goal that was far easier for a more advanced speaker (equivalent to the amount of muscle a child versus an adult must use to lift a heavy box). The third possibility may be the most interesting, and the one with greatest significance for our focal injury results. In the earliest stages, areas involved in attention, perceptual analysis, and formation of new memories may be particularly important. As the task becomes better learned and more automatic, the baton may pass to regions that are responsible for the reactivation of over-learned patterns, with less attention and less perceptual analysis. A recent example of this kind of qualitative shift is reported by Raichle et al. (1994), who observed strong frontocerebellar activation in the early stages of learning, replaced by activation in perisylvian cortex after the task is mastered.

Maturation and "Readiness"

Finally, the emergentist approach does not preclude the possibility of maturational change. Examples might include differential growth gradients for the right versus left hemisphere (Chiron et al., 1997), differential rates of synaptogenesis ("synaptic sprouting") from one region to another within the two hemispheres (Huttenlocher et al., 1982), changes from region to region in the overall amount of neural activity (as indexed by positron emission tomography; Chugani et al., 1987), variation in rates of myelination, and so forth. As a result of changes of this kind (together with the effects of learning itself in reshaping the brain; Marchman, 1993), we should expect to find marked shifts in the patterns of activity associated with language processing at different points in early childhood.

Based on these assumptions, let us return to our findings on the early stages of language development in children with early focal brain injury to see what these results suggest about the emergence of brain organization for language in normal children.

Right Hemisphere Advantage for Word Comprehension and Gesture From 10 to 17 Months

Contrary to expectations based on the adult aphasia literature, we found evidence for greater delays in word comprehension and gesture in children with RHD. This is exactly the opposite of the pattern observed in adults, where deficits in word comprehension and in production of symbolic gesture are both associated with LHD, suggesting that some kind of shift takes place between infancy and adulthood, with control over these two skills passed from the right hemisphere to the left. This result is (as we noted) compatible with observations by Mills et al. (1997) on the patterns of activation observed in response to familiar words from infancy to adulthood. There are at least two possible explanations for a developmental change, and they are not mutually exclusive.

On the one hand, the early right hemisphere advantage could be explained by hard maturational changes that are exogenous to the learning process itself. For example, Chiron and his colleagues (1997) have provided evidence from PET for a change in resting-state activation across the first 2 years, from bilateral activation that is larger on the right to greater activation on the left. Based on these findings, they suggest that the right hemisphere may mature faster than the left in the first year of development. As it turns out, this is the period in which word comprehension and gesture first emerge in normally developing children. By contrast, word production emerges in the second year and grows dramatically through 30 to 36 months, the period in which (according to Chiron et al., 1997) the left hemisphere reaches the dominant state that it will maintain for years to come. Hence one might argue that the right hemisphere "grabs" control over comprehension and gesture in the first year, and the left hemisphere "grabs" control over the burgeoning capacity for production in the second year, eventually taking over the entire linguistic-symbolic system (including word comprehension and meaningful gestures).

On the other hand, it is also possible that the right-to-left shift implied by our data reflects a qualitative difference between the learning processes required for comprehension and the processes required for production. The first time that we figure out the meaning of a word (e.g., decoding the word "dog" and mapping it onto a particular class of animals), we do so by integrating the phonetic input with information from many different sources, including visual, tactile, and auditory context ("fuzzy brown thing that moves and barks"). It has been argued that the right hemisphere plays a privileged role in multimodal integration and processing of large patterns (Stiles, 1995; more on this below), and for this reason we may expect the right hemisphere to play a more important role when children are learning to comprehend words for the first time. Presumably, this right hemisphere advantage will disappear when words are fully acquired, replaced by a rapid, efficient, and automatic process of mapping well-known sounds onto well-known semantic patterns (more on this below). If this hypothesis has merit, then we might also expect to find evidence for greater participation of the right hemisphere in the early stages of second-language learning in adults, a testable hypothesis and one that has some (limited) support.

It is much less obvious how this shift-in-strategy hypothesis might account for the early right hemisphere advantage in symbolic gesture. Although this is admittedly a speculative answer, this finding may be related to results for normal children showing that comprehension and gesture are highly correlated between approximately 9 and 20 months of age (Fenson et al., 1994). One possible explanation for this correlation may lie in the fact that symbolic gestures are acquired in the context of auditory comprehension (e.g., "Wave bye-bye to grandma," "Hug the baby!"). Hence the two skills may come in together in very small children because they are acquired together in real life.

Deficits in Expressive Vocabulary and Grammar
With Frontal Lesions to Either Hemisphere
From 19 to 31 Months

We observed specific effects of lesions involving the frontal lobes in children between 19 and 31 months of age, a brief but dramatic period of development that includes the vocabulary burst and the first flowering of grammar. Contrary to expectations based on the adult aphasia literature, the delays in expressive language associated with frontal lesions were symmetrical, that is, there was no difference between frontal lesions on the left and frontal lesions on the right. There are a number of reasons why we would expect to find specific effects of frontal involvement during this important period in the development of expressive language, including contributions to the planning and execution of motor patterns and contributions from working memory and/or the fashionable array of skills referred to by the term *executive function* (Pennington and Ozonoff, 1996). However, the absence of a left–right asymmetry is more surprising. Nor have we found any evidence for a specific effect of left frontal injury in any of our studies to date, at any age. This difference between infants and adults suggests to us that Broca's area is not innately specialized for language. It becomes specialized across the course of development, after an initial period in which frontal cortex makes a symmetrical contribution to language learning.

Deficits in Expressive Vocabulary and Grammar
With Left Temporal Injuries From 10 Months
to 5 Years of Age

This is the most robust and protracted finding in our prospective studies, and it is the only evidence we have for an asymmetry that might be systematically related to a left hemisphere advantage for language in the adult brain. Note, however, that the effect only pertains to *expressive* language (contrary to the expectation that temporal cortex is specialized for comprehension), and it applies equally to *both* vocabulary and grammar (contrary to the expectation that temporal cortex is associated with semantics while frontal cortex handles grammar; Zurif, 1980).

We have proposed that a relatively simple bias in style of computation may underlie this left temporal effect, reflecting architectural differences between left and right temporal cortex that are only indirectly related to the functional and representational specializations that are evident in adult language processing. Following a proposal by Stiles and Thal (1993), we note that left and right temporal cortex differ at birth in their capacity to support perceptual detail (enhanced on the left) and perceptual integration (enhanced on the right; see above). These differences are evident in nonverbal processing, but they may have particularly important consequences for language. For example, a number of recent studies have shown that lesions to the right hemisphere lead to problems in the integration of elements in a perceptual array, while lesions to the left hemisphere create problems in the analysis of perceptual details in the same array (e.g., Robertson and Lamb, 1991). Asked to reproduce a triangle made up of many small squares, adult patients with LHD tend to reproduce the global figure (i.e., the triangle) while ignoring information at the local level. Adult patients with RHD display the opposite profile, reproducing local detail (i.e., a host of small squares) but failing to integrate these features into a coherent whole. Stiles and Thal (1993) report that children with focal brain injury behave very much like their adult counterparts on the local–global task, suggesting that the differential contribution of left and right hemisphere processes on this task may be a developmental constant. Interestingly, this double dissociation is most evident in patients with temporal involvement, and the special role of left temporal cortex in processing of perceptual details has also been confirmed in an fMRI study of normal adults engaged in the same local–global task (Martinez et al., 1997).

The same left–right difference may be responsible for the lesion–symptom correlations that we observe in early language development. As I noted earlier, the ability to integrate information within and across modalities may be particularly helpful and important during the first stages of word comprehension and (perhaps) recognition and reproduction of familiar gestures. However, the learning task changes markedly when children have to convert the same sound patterns into motor output. At this point, perceptual detail may be of paramount importance (i.e., it is one thing to recognize the word "dog," but quite another thing to pull out each phonetic detail and construct a motor template). If it is the case that left temporal cortex plays a critical role in the extraction, storage, and reproduction of perceptual detail (visual and/or acoustic), then children with left temporal injuries will be at a greater disadvantage in this phase of learning (see also Galaburda and Livingstone, 1993; Galaburda et al., 1994; Tallal et al., 1991). However, once the requisite patterns are finally constructed and set into well-learned routines, the left temporal disadvantage may be much less evident.

No Evidence of Lesion–Symptom Correlations After 5–7 Years of Age

All of the above lesion–symptom mappings seem to have disappeared when we test children with the same congenital etiology after 5–7 years of age. Although this conclusion is based primarily on cross-sectional findings, the few cases that we have been able to study longitudinally across these periods of development are compatible with the cross-sectional results, providing further evidence for plasticity and compensatory organization across the course of language development. Of course it is entirely possible that we will find a new and improved index of efficiency in language processing that yields information about the subtle deficits that remain, for example, a residual effect of left temporal involvement that shows up in real-time sentence processing and/or in production of complex syntax under certain laboratory conditions. At the very least, however, we may conclude with some confidence that these children have found a form of brain organization for language that works very well, certainly well enough for everyday language use. As a group, children with focal brain injury do tend to perform below neurologically intact age-matched controls. But these differences also tend to disappear when the small group difference in full-scale IQ is taken into account (Bates et al., 1999).

If the familiar pattern of left hemisphere organization for language is not critical for normal language functioning, why does it develop in the first place? To answer this question, we have put forth a "modest proposal" based on the developmental findings and developmental principles listed above, as follows.

Prior to the onset of language development, the infant brain has no innate representations for language, nor does it have a "dedicated language processor" of any kind. However, the initial (prelinguistic) architecture of the infant brain is highly differentiated. The global input–output structure of the brain is well specified (e.g., the retina reports to visual cortex, the cochlea reports to auditory cortex), although there may still be a number of exuberant axons that could (if they are not eliminated in the normal course of development) sustain an alternative form of global architecture if they are needed. There are also innate (experience-independent) variations from region to region in cell density, synaptic density, speed of processing, and the kinds of neurotransmitters that are expressed (Hutsler and Gazzaniga, 1996). Furthermore, even though the infant has little experience in the world, the infant cortex has been inundated with information from the body itself. As Damasio (1994) has noted, the brain is the captive audience of the body, and the body provides the earliest and most reliable input that the growing cortex will ever receive. This includes sensory impressions from the body surface, kinaesthetic feedback from the infant's own movements, and reliable waves of activity from lower brain centers (e.g., bilateral and competitive input from lower-level visual nuclei that, we now know, is critical for the establishment of ocular dominance patterns; Miller et al., 1989; Shatz, 1992). Hence, even though there may be no direct genetic control over synaptic connectivity at the cortical level, the newborn infant starts life with a brain that has been colonized by sensorimotor input from its own body, setting down the basic parameters within which all the rest of behavioral development must take place. These facts combine with the regional differences in cortical architecture described above, setting the stage for the postnatal development of cognition and communication, including the development of grammar (Mac-Whinney, 1999).

As a result of all these forces, the infant comes to the task of language learning with a heavy set of biases about how information should be processed. Some of these biases are symmetrical (e.g., the role of frontal cortex in control of voluntary movements), and others are asymmetrical (e.g., the local–global biases described above). Following early focal brain injury, these biases show up in the early lesion–symptom mappings that we have described above, but they are eventually overcome by the competitive pressures that define plasticity and development in both the normal and the abnormal case. However, in healthy children without focal brain injury, these biases shape the development of brain organization for language in some highly predictable directions. In particular, left temporal cortex comes to play an increasingly important role in the extraction of the rapid and evanescent linguistic signal—first in the construction of motor templates to match slow and dependable inputs and later in the construction of complex meanings for both comprehension and production (events that we would expect to see in both signed and spoken language; Petitto et al., 1997). In short, under normal conditions (i.e., in the absence of focal brain injury), left temporal cortex wins the language contract. Although there is no asymmetrical bias in favor of left frontal cortex in the early stages of development, the left temporal "winner" recruits its partners in the front of the brain, setting up the familiar ipsilateral circuit that characterizes left hemisphere mediation of language in neurologically intact adults. At this point (and not before), Broca's area has a special job.

This is our proposal for the cascade of events that are responsible for the patterns of brain organization for language that lie behind 200 years of research on adult aphasia and hundreds (going on thousands) of neural imaging studies of language activation in normal adults. No doubt this proposal will have to undergo considerable revision as more information becomes available, but we are convinced that the final story will have to be one in which development and experience play a crucial role. Plasticity is not a civil defense system, a set of emergency procedures that are only invoked when something goes wrong. Rather, the processes responsible for reorganization of the brain following early focal brain injury are the same processes that organize the brain under normal conditions. It is time to exercise the ghost of Franz Gall, trading in the static phrenological view of brain organization for a dynamic approach that reconciles linguistics and cognitive science with developmental neurobiology.

REFERENCES

Support for the work described here was provided by NIH-NIDCD P50 DC1289-9351 ("Origins of Communication Disorders") and NIH/NINDS P50 NS22343 ("Center for the Study of the Neural Bases of Language and Learning").

Aram, D. M. (1988). Language sequelae of unilateral brain lesions in children. In F. Plum (ed.). Language, Communication, and the Brain. New York: Raven Press, pp. 171–197.

Aram, D. M., Ekelman, B., Rose, D., and Whitaker, H. (1985a). Verbal and cognitive sequelae following unilateral lesions acquired in early childhood. *J. Clin. Exp. Neuropsychol., 7*, 55–78.

Aram, D. M., Ekelman, B., and Whitaker, H. (1985b). Lexical retrieval in left and right brain-lesioned children. *Brain Lang., 28*, 61–87.

Bachevalier, J., and Mishkin, M. (1993). An early and a late developing system for learning and retention in infant monkeys. In M. Johnson (ed.): *Brain Development and Cognition: A Reader.* Oxford: Blackwell, pp. 195–207.

Barkow, J. H., Cosmides, L., and Tooby, J. (eds.) (1992). *The Adapted Mind: Evolutionary Psychology and the Generation of Culture.* New York: Oxford University Press.

Basser, L. (1962). Hemiplegia of early onset and the faculty of speech with special reference to the effects of hemispherectomy. *Brain, 85*, 427–460.

Bates, E., Appelbaum, M., and Allard, L. (1991a). Statistical constraints on the use of single cases in neuropsychological research. *Brain Lang., 40*, 295–329.

Bates, E., Benigni, L., Bretherton, I., Camaioni, L., and Volterra, V. (1979). *The Emergence of Symbols: Cognition and Communication in Infancy.* New York: Academic Press.

Bates, E., and Carnevale, G. F. (1993). New directions in research on language development. *Dev. Rev., 13,* 436–470.

Bates, E., Dale, P. S., and Thal, D. (1995). Individual differences and their implications for theories of language development. In P. Fletcher and B. MacWhinney (eds.): *Handbook of Child Language.* Oxford: Basil Blackwell, pp. 96–151.

Bates, E., and Elman, J. L. (1996). Learning rediscovered. *Science, 274,* 1849–1850.

Bates, E., Elman, J., Johnson, M. C., Karmiloff-Smith, A., Parisi, D., and Plunkett, K. (1998). Innateness and emergentism. In W. Bechtel and G. Graham (eds.): *A Companion to Cognitive Science.* Oxford: Basil Blackwell, pp. 590–601.

Bates, E., Thal, D., and Janowsky, J. (1992). Early language development and its neural correlates. In I. Rapin and S. Segalowitz (eds.): *Handbook of Neuropsychology, vol. 7, Child Neuropsychology.* Amsterdam: Elsevier, pp. 69–110.

Bates, E., Thal, D., and Marchman, V. (1991b). Symbols and syntax: A Darwinian approach to language development. In N. Krasnegor, D. Rumbaugh, R. Schiefelbusch, and M. Studdert-Kennedy (eds.): *Biological and Behavioral Determinants of Language Development.* Hillsdale, NJ: Erlbaum, pp. 29–65.

Bates, E., Thal, D., Trauner, D., Fenson, J., Aram, D., Eisele, J., and Nass, R. (1997). From first words to grammar in children with focal brain injury. *Dev. Neuropsychol. (Spec. Iss.), 13,* 447–476.

Bates, E., Vicari, S., and Trauner, D. (1999). Neural mediation of language development: Perspectives from lesion studies of infants and children. In H. Tager-Flusberg (ed.): *Neurodevelopmental Disorders: Contributions to a New Framework From the Cognitive Neurosciences.* Cambridge, MA: MIT Press (in press).

Bialystok, E., and Hakuta, K. (1994). *In Other Words: The Science and Psychology of Second-Language Acquisition.* New York: Basic Books.

Bishop, D. V. M. (1983). Linguistic impairment after left hemidecortication for infantile hemiplegia? A reappraisal. *Q. J. Exp. Psychol., 35A,* 199–207.

Bishop, D. V. M. (1997). Cognitive neuropsychology and developmental disorders: Uncomfortable bedfellows. *Q. J. Exp. Psychol., 50,* 899–923.

Cappa, S. F., Perani, D., Grassi, F., Bressi, S., Alberoni, M., Franceschi, M., Bettinardi, V., Todde, S., and Fazio, F. (1997). A PET follow-up study of recovery after stroke in acute aphasics. *Brain Lang., 56,* 55–67.

Cappa, S. F., and Vallar, G. (1992). The role of the left and right hemispheres in recovery from aphasia. *Aphasiology, 6,* 359–372.

Caramazza, A. (1986). On drawing inferences about the structure of normal cognitive systems from the analysis of patterns of impaired performance: The case for single-patient studies. *Brain Cog., 5,* 41–66.

Caramazza, A., and Berndt, R. (1985). A multicomponent view of agrammatic Broca's aphasia. In M.-L. Kean (ed.): *Agrammatism.* Orlando: Academic Press, pp. 27–63.

Chiron, C., Jambaque, I., Nabbout, R., Lounes, R., Syrota, A., and Dulac, O. (1997). The right brain hemisphere is dominant in human infants. *Brain, 120,* 1057–1065.

Chomsky, N. (1980a). On cognitive structures and their development: A reply to Piaget. In M. Piattelli-Palmarini (ed.): *Language and Learning.* Cambridge: Harvard University Press, pp. 35–54.

Chomsky, N. (1980b). *Rules and Representations.* New York: Columbia University Press.

Chomsky, N. (1995). *The Minimalist Program.* Cambridge: MIT Press.

Chugani, H. T., Phelps, M. E., and Mazziotta, J. C. (1987). Positron emission tomography study of human brain functional development. *Ann. Neurol., 22,* 487–497.

Clark, R., Gleitman, L., and Kroch, A. (1997). *Science, 276,* 1179.

Courtney, S. M., and Ungerleider, L. G. (1997). What fMRI has taught us about human vision. *Curr. Opin. Neurobiol., 7,* 554–561.

Curtiss, S. (1988). Abnormal language acquisition and the modularity of language. In F. J. Newmeyer (ed.): *Linguistics: The Cambridge Survey: vol. II, Linguistic Theory: Extensions and Implications.* Cambridge, UK: Cambridge University Press, pp. 96–116.

Dall' Oglio, A. M., Bates, E., Volterra, V., Di Capua, M., and Pezzini, G. (1994). Early cognition, communication and language in children with focal brain injury. *Dev. Med. Child Neurol., 36,* 1076–1098.

Damasio, A. R. (1994). *Descartes' Error: Emotion, Reason, and the Human Brain.* New York: G. P. Putnam.

Das, A., and Gilbert, C. D. (1995). Receptive field expansion in adult visual cortex is linked to dynamic changes in strength of cortical connections. *J. Neurophysiol., 74,* 779–792.

Deacon, T. (1997). *The Symbolic Species: The Co-Evolution of Language and the Brain.* New York: Norton.

Dennis, M., and Whitaker, H. (1976). Language acquisition following hemidecortication: Linguistic superiority of the left over the right hemisphere. *Brain Lang., 3,* 404–433.

De Weerd, P., Gattass, R., Desimone, R., and Ungerleider, L. (1995). Responses of cells in monkey visual cortex during perceptual filling in of an artificial scotoma. *Nature, 377,* 731–734.

Dronkers, N. F. (1996). A new brain region for coordinating speech articulation. *Nature, 384*, 159–161.

Dronkers, N. F., Redfern, B. B., and Ludy, C. A. (1999). Lesion localization in chronic Wernicke's aphasia. *Brain Lang.* (in press).

Dronkers, N. F., Wilkins, D. P., Van Valin, Jr., R. D., Redfern, B. B., and Jaeger, J. J. (1994). A reconsideration of the brain areas involved in the disruption of morphosyntactic comprehension. *Brain Lang., 47*, 461–463.

Eisele, J., and Aram, D. (1995). Lexical and grammatical development in children with early hemisphere damage: A cross-sectional view from birth to adolescence. In P. Fletcher and B. MacWhinney (eds.): *The Handbook of Child Language*. Oxford: Basil Blackwell, pp. 664–689.

Elman, J. L., and Bates, E. (1997). *Science, 276*, 1180.

Elman, J. L., Bates, E., Johnson, M., Karmiloff-Smith, A., Parisi, D., and Plunkett, K. (1996). *Rethinking Innateness: A Connectionist Perspective on Development*. Cambridge: MIT Press/Bradford Books.

Erhard, P., Kato, T., Strick, P., and Ugurbil, K. (1996). Functional MRI Activation pattern of motor and language tasks in Broca's area. *Soc. Neurosci. Abstr., 22*, 260.

Feldman, H., Holland, A., Kemp, S., and Janosky, J. (1992). Language development after unilateral brain injury. *Brain Lang., 42*, 89–102.

Fenson, L., Dale, P., Reznick, J. S., Thal, D., Bates, E., Hartung, J., Pethick, S., and Reilly, J. (1993). *The MacArthur Communicative Development Inventories: User's Guide and Technical Manual*. San Diego: Singular Publishing Group.

Fenson, L., Dale, P. A., Reznick, J. S., Bates, E., Thal, D., and Pethick, S. J. (1994). Variability in early communicative development. *Monogr. Soc. Res. Child Dev., 59*(5).

Fodor, J. A. (1983). *The Modularity of Mind: An Essay on Faculty Psychology*. Cambridge: MIT Press.

Fregnac, Y., Bringuier, V., and Chavane, F. (1996). Synaptic integration fields and associative plasticity of visual cortical cells in vivo. *J. Physiol., 90*, 367–372.

Galaburda, A. M., and Livingstone, M. (1993). Evidence for a magnocellular defect in neurodevelopmental dyslexia. *Ann. N.Y. Acad. Sci., 682*, 70–82.

Galaburda, A. M., Menard, M. T., and Rosen, G. D. (1994). Evidence for aberrant auditory anatomy in developmental dyslexia. *Proc. Natl. Acad. Sci. U.S.A., 91*, 8010–8013.

Geschwind, N. (1965). Disconnexion syndromes in animals and man. *Brain, 88*, 585–644.

Goodglass, H. (1993). *Understanding Aphasia*. San Diego: Academic Press.

Goodman, R., and Yude, C. (1996). IQ and its predictors in childhood hemiplegia. *Dev. Med. Child Neurol., 38*, 881–890.

Gopnik, M. (1990). Feature-blind grammar and dysphasia. *Nature, 344*, 715.

Hernandez, A. E., Martinez, A., Wong, E. C., Frank, L. R., and Buxton, R. B. (1997). *Neuroanatomical Correlates of Single- and Dual-Language Picture Naming in Spanish-English Bilinguals.* Poster presented at the fourth annual meeting of the Cognitive Neuroscience Society, March 1997.

Hirsch, J., Kim, K., Souweidane, M., McDowall, R., Ruge, M., Correa, D., and Krol, G. (1997). FMRI reveals a developing language system in a 15-month-old sedated infant. *Soc. Neurosci. Abstr., 23*, 2227.

Holloway, R., Broadfield, D., Kheck, N., and Braun, A. (1997). Chimpanzee brain: Left–right asymmetries in temporal speech region homolog. *Soc. Neurosci. Abstr., 23*, 2228.

Hutsler, J. J., and Gazzaniga, M. S. (1996). Acetylcholinesterase staining in human auditory and language cortices: Regional variation of structural features. *Cereb. Cortex, 6*, 260–270.

Huttenlocher, P. R., de Courten, C., Garey, L., and van der Loos, H. (1982). Synaptogenesis in human visual cortex synapse elimination during normal development. *Neurosci. Lett., 33*, 247–252.

Isacson, O., and Deacon, T. W. (1996). Presence and specificity of axon guidance cues in the adult brain demonstrated by pig neuroblasts transplanted into rats. *Neuroscience, 75*, 827–837.

Janowsky, J. S., and Finlay, B. L. (1986). The outcome of perinatal brain damage: The role of normal neuron loss and axon retraction. *Dev. Med., 28*, 375–389.

Jenkins, L., and Maxam, A. (1997). *Science, 276*, 1178–1179.

Johnson, J., and Newport, E. (1989). Critical period effects in second language learning: The influence of maturational state on the acquisition of English as a second language. *Cog. Psychol., 21*, 60–99.

Johnson, M. H. (1997). *Developmental Cognitive Neuroscience.* Oxford: Blackwell Publishers.

Just, M. A., Carpenter, P. A., Keller, T. A., Eddy, W. F., et al. (1996). Brain activation modulated by sentence comprehension. *Science, 275*, 114–116.

Karmiloff-Smith, A. (1992). *Beyond Modularity: A Developmental Perspective on Cognitive Science.* Cambridge: MIT Press.

Karni, A., Meyer, G., Jezzard, P., Adams, M. M., et al. (1995). Functional fMRI evidence for adult motor cortex plasticity during motor skill learning. *Nature, 377*, 155–158.

Kempler, D., van Lancker, D., Marchman, V., and Bates, E. (1999). Idiom comprehension in children and adults with unilateral brain damage. *Dev. Neuropsychol.* (in press).

Killackey, H. P. (1990). Neocortical expansion: An attempt toward relating phylogeny and ontogeny. *J. Cog. Neurosci., 2*, 1–17.

Killackey, H. P., Chiaia, N. L., Bennett-Clarke, C. A., Eck, M., and Rhoades, R. (1994). Peripheral influences on the size and organization of somatotopic representations in the fetal rate cortex. *J. Neurosci., 14*, 1496–1506.

Kim, K. H. S., Relkin, N. R., Lee, K. M., and Hirsch, J. (1997). Distinct cortical areas associated with native and second languages. *Nature, 388*, 171–174.

King, M. C., and Wilson, A. C. (1975). Evolution at two levels in humans and chimpanzees. *Science, 188*, 107–116.

Lenneberg, B. H. (1967). *Biological Foundations of Language*. New York: Wiley.

Liu, H., Bates, E., and Li, P. (1992). Sentence interpretation in bilingual speakers of English and Chinese. *Appl. Psycholing., 13*, 451–484.

MacWhinney, E. (1999). The emergence of language from embodiment In B. MacWhinney (ed.): *The Emergence of Language*. Mahwah, NJ: Lawrence Erlbaum Associates.

Marchman, V. (1993). Constraints on plasticity in a connectionist model of the English past tense. *J. Cog. Neurosci., 5*, 215–234.

Marchman, V., Miller, R., and Bates, E. (1991). Babble and first words in children with focal brain injury. *Appl. Psycholing., 12*, 1–22.

Martinez, A., Mosses, P., Frank, L., Buxton, R., Wong, E., and Stiles, J. (1997). Hemispheric asymmetries in global and local processing: Evidence from fMRI. *NeuroReport, 8*, 1685–1689.

Miller, K. D., Keller, J. B., and Stryker, M. P. (1989). Ocular dominance column development: Analysis and simulation. *Science, 245*, 605–615.

Mills, D. L., Coffey-Corina, S. A., and Neville, H. J. (1997). Language comprehension and cerebral specialization from 13 to 20 months. *Dev. Neuropsychol. (Spec. Iss.), 13*, 397–445.

Mueller, R.-A. (1996). Innateness, autonomy, universality? Neurobiological approaches to language. *Behav. Brain Sci., 19*, 611.

Nobre, A. C., Allison, T., and McCarthy, G. (1994). Word recognition in the human inferior temporal lobe. *Nature, 372*, 260–263.

Oyama, S. (1993). The problem of change. In M. Johnson (ed.): *Brain Development and Cognition: A Reader*. Oxford: Blackwell Publishers, pp. 19–30.

Pallas, S. L., and Sur, M. (1993). Visual projections induced into the auditory pathway of ferrets: II. Corticocortical connections of primary auditory cortex. *J. Comp. Neurol., 337*, 317–333.

Pennington, B. F., and Ozonoff, S. (1996). Executive functions and developmental psychopathology. *J. Child Psychol. Psychiatry Allied Disciplines, 37*, 51–87.

Perani, D., Paulesu, N., Sebastian, E., Dupoix, S., Dehaene, S., Schnur, T., Cappa, S., Mehler, J., and Fazio, E. (1997). Plasticity of brain language regions revealed by bilinguals. *Soc. Neurosci. Abstr., 23*, 2228.

Pesetsky, D., Wexler, K., and Fromkin, V. (1997). *Science, 276*, 1177.

Petitto, L., Zattore, R., Nikelski, E., Gauna, K., Dostie, D., and Evans, A. (1997). Cerebral organization for language in the absence of sound: A PET study of deaf signers processing signed languages. *Soc. Neurosci. Abstr., 23*, 2228.

Pettet, M. W., and Gilbert, C. (1992). Dynamic changes in receptive field size in cat primary visual cortex. *Proc. Nat. Acad. Sci. U.S.A., 89*, 8366–8370.

Pinker, S. (1991). Rules of language. *Science, 253*, 530–535.

Pinker, S. (1994). *The Language Instincts: How the Mind Creates Language*. New York: William Morrow.

Pinker, S. (1997a). *How the Mind Works*. New York: Norton.

Pinker, S. (1997b). Acquiring language. *Science, 276*, 1178.

Poeppel, D. (1996). A critical review of PET studies of phonological processing. *Brain Lang., 55*, 317–351.

Quartz, S. R., and Sejnowski, T. J. (1997). The neural basis of cognitive development: A constructivist manifesto. *Behav. Brain Sci., 20*, 537.

Raichle, M. E., Fiez, J. A., Videen, T. O., MacLeod, A. M., Pardo, J. V., Fox, P. T., and Petersen, S. E. (1994). Practice-related changes in human brain functional anatomy during non-motor learning. *Cereb. Cortex, 4*, 8–26.

Ramachandran, V. S., and Gregory, R. L. (1991). Perceptual filling in of artificial scotomas in human vision. *Nature, 350*, 699–702.

Ramachandran, V. S., Hirstein, W. F., Armel, K. C., Tecoma, E., and Iraqui, V. (1997). The neural basis of religious experience. *Soc. Neurosci. Abstr., 23*, 519.1.

Reilly, J. S., Bates, E., and Marchman, V. (1998). Narrative discourse in children with early focal brain injury. *Brain Lang. (Spec. Iss.), 61*, 335–375.

Reilly, J. S., Stiles, J., Larsen, J., and Trauner, D. (1995). Affective facial expression in infants with focal brain damage. *Neuropsychologia, 1*, 83–99.

Rice, M. (ed.). (1996). *Toward a Genetics of Language*. Mahwah, NJ: Erlbaum.

Robertson, L. C., and Lamb, M. R. (1991). Neuropsychological contributions to theories of part whole organization. *Cog. Psychol., 23*, 299–330.

Shallice, T. (1988). *From Neuropsychology to Mental Structure*. New York: Cambridge University Press.

Shatz, C. J. (1992). The developing brain. *Sci. Am., 267*, 60–67.

Stanfield, B. B., and O'Leary, D. D. (1985). Fetal occipital cortical neurones transplanted to the rostral cortex can extend and maintain a pyramidal tract axon. *Nature, 313*, 135–137.

Stark, R., and McGregor, K. (1997). Follow-up study of a right- and left-hemispherectomized child: Implications for localization and impairment of language in children. *Brain Lang., 60*, 222–242.

Stiles, J. (1995). Plasticity and development: Evidence from children with early focal brain injury. In B. Julesz and I. Kovacs (eds.): *Maturational Windows and Adult Cortical Plasticity. Proceedings of the Santa Fe Institute Studies in the Sciences of Complexity*, vol. 23. Reading, MA: Addison-Wesley, pp. 217–237.

Stiles, J., Bates, E., Thal, D., Trauner, D., and Reilly, J. (1998). Linguistic, cognitive and affective development in children with pre- and perinatal focal brain injury: A ten-year overview from the San Diego Longitudinal Project. In C. Rovee-Collier, L. Lipsitt, and H. Hayne (eds.): *Advances in Infancy Research*. Norwood, NJ: Ablex, pp. 131–163.

Stiles, J., and Thal, D. (1993). Linguistic and spatial cognitive development following early focal brain injury: Patterns of deficit and recovery. In M. Johnson (ed.): *Brain Development and Cognition: A Reader*. Oxford: Blackwell Publishers, pp. 643–664.

Stromswold, K. (1995). The cognitive and neural bases of language acquisition. In M. S. Gazzaniga (ed.): *The Cognitive Neurosciences*. Cambridge: MIT Press.

Sur, M., Pallas, S. L., and Roe, A. W. (1990). Cross-modal plasticity in cortical development: Differentiation and specification of sensory neocortex. *Trends Neurosci., 13*, 227–233.

Tallal, P., Sainburg, R. L., and Jernigan, T. (1991). The neuropathology of developmental dysphasia: Behavioral, morphological, and physiological evidence for a pervasive temporal processing disorder. *Reading Writing, 3*, 363–377.

Thal, D., Marchman, V., Stiles, J., Aram, D., Trauner, D., Nass, R., and Bates, E. (1991). Early lexical development in children with focal brain injury. *Brain Lang., 40*, 491–527.

Thompson-Schill, S., D'Esposito, M., Aguirre, G., and Farah, M. (1997). Role of left prefrontal cortex in retrieval of semantic knowledge: A re-evaluation. *Soc. Neurosci. Abstr., 23*, 2227.

Tovee, M. J., Rolls, E. T., and Ramachandran, V. S. (1996). Rapid visual learning in neurons in the primate temporal cortex. *NeuroReport, 7*, 2757–2760.

Van der Lely, H. K. J. (1994). Canonical linking rules: Forward versus reverse linking in normally developing and specifically language-impaired children. *Cognition, 51*, 29–72.

Vargha-Khadem, F., Carr, L. J., Isaacs, E., Brett, E., Adams, C., and Mishkin, M. Onset of speech after left hemispherectomy in a nine-year-old boy. *Brain, 120*, 159–182.

Vargha-Khadem, F., Isaacs, E., and Muter, V. (1994). A review of cognitive outcome after unilateral lesions sustained during childhood. *J. Child Neurol. 9(Suppl.)*, 2S67–2S73.

Weber-Fox, C. M., and Neville, H. J. (1996). Maturational constraints on functional specializations for language processing: ERP and behavioral evidence in bilingual speakers. *J. Cog. Neurosci., 8*, 231–256.

Webster, M. J., Bachevalier, J., and Ungerleider, L. G. (1995). Development of plasticity of visual memory circuits. In B. Julesz and I. Kovacs (eds.): *Maturational Windows and Adult Cortical Plasticity. Proceedings of the Santa Fe Institute Studies in the Sciences of Complexity*, vol. 23. Reading, MA: Addison-Wesley, pp. 73–86.

Wernicke, C. (1977). The aphasia symptom complex: A psychological study on an anatomic basis. (N. Geschwind, Trans.) In G. H. Eggert (ed.): *Wernicke's Works on Aphasia: A Sourcebook and Review*. The Hague: Mouton. (Original work published 1874.)

Willmes, K., and Poeck, K. (1993). To what extent can aphasic syndromes be localized? *Brain, 116*, 1527–1540.

Wilson, A. C. (1985). The molecular basis of evolution. In J. Piel (ed.): *The Molecules of Life*. New York: W. H. Freeman, pp. 120–129.

Woods, B. T., and Teuber, H. L. (1978). Changing patterns of childhood aphasia. *Ann. Neurol., 3*, 272–280.

Zurif, E. (1980). Language mechanisms: A neuropsychological perspective. *Am. Sci., 68*, 305–311.

From Ontogenesis to Phylogenesis: What Can Child Language Tell Us About Language Evolution?

Dan I. Slobin
Department of Psychology
University of California, Berkeley

In thinking about human origins, there has always been a tendency to take the child as a model of the primordial state of the species or its ancestors. For the past several centuries, philosophers and psychologists and anthropologists have made analogies between psychological characteristics of children and animals, children and "primitive" peoples, and, inevitably, children and our proto-hominid ancestors. Advances in developmental and comparative psychology, along with anthropology, have made the first two analogies untenable. Human children are not the same as mature monkeys and apes, and preliterate societies are not childlike. But in the current scientific fascination with the origin of the species, it has become fashionable again to propose that human children are in some ways models of mature proto- or prehominids. Nowhere has this proposal received more circulation than in discussions about the evolution of language (e.g., Bickerton, 1990; Givón, 1998, 2002). Most recently, Givón stated that "an analogical, recapitulationist perspective on language evolution is both useful and legitimate" (Givón, 2002, p. 35). I suggest that this recent form of the recapitulationist argument will fail. In the global classical version of Haeckel's biogenetic law, the proposal was abandoned on the basis of evidence from embryology and physiological development. By contrast, the current proposal—especially in the version proposed by Bickerton—is not compatible with what we know about the psycholinguistic development of human children and the processes of historical development of existing human languages.

There are three longstanding questions about the role of the child in language evolution and diachrony—that is, the processes whereby language emerged and developed in the genus, *Homo*, and the ceaseless changes of human language once it is present in our species, *Homo sapiens*. Briefly, the questions are: (a) Does linguistic ontogeny recapitulate phylogeny? (b) Does linguistic diachrony recapitulate ontogeny? (c) Do children create grammatical forms? To anticipate my conclusion: The answer to all three questions is mainly negative. This conclusion is supported by several types of evidence:

1. Linguistic ontogeny is shaped by the particular language being acquired. That is, there is no universal form of early child language that clearly reflects a biologically specified proto-language.

2. In historical change of existing languages, it appears that lasting innovations do not come from preschoolers but from older speakers. That is, language changes more *in use* than it does in the process of being learned.

3. Languages are sociocultural as well as individual products. Therefore, we can't expect to discover the phylogenetic origins of human language by studying the individual alone, as is implied by the neglect of social interactive factors in innatist formulations.

DOES ONTOGENY OF LANGUAGE RECAPITULATE ITS PHYLOGENY?

The *Homo sapiens* child is different from a prehuman hominid in two critically important ways. The child is exposed to some already evolved human language and is equipped with a brain that evolved to make use of such a language. This situation was already pointed out early in the last century by a leading linguist of the times, Otto Jespersen: "Manifestly, the modern learner is in quite a different position to primitive man, and has quite a different task set him . . . the task of the child is to learn an existing language . . . but not in the least to frame anything anew" (Jespersen, 1921/1964, p. 417). Furthermore, the "linguistic niche" in which our children develop has, itself, been shaped by the cognitive and social activities of our ancestors. Niche construction is one of the themes of this volume. Parker points out in her introduction: "As niche construction theory emphasizes, phenotypes progressively transform themselves by constructing and transforming a series of environments in which they participate." The children that we study today are such "transforming phenotypes." It is, therefore, not at all obvious that they might provide clues to the "co-evolving" processes out of which they arose. (See Deacon, 1997, for a plausible co-evolutionary scenario for the emergence of language.)

Nevertheless, might it be that the earliest periods of child language reveal the workings of a cognitive and linguistic core that we might share with our hominid, and even prehominid ancestors? This is a tempting possibility—especially because relevant linguistic data from all other hominid species are permanently unavailable. And so, in a search for potentially useful data, it has been suggested that early child language may serve as a plausible model for prehuman language. A contemporary linguist, Derek Bickerton, has been explicit about this parallel, on the basis of two sorts of claims: The first is that there is an identifiable "proto-language" that is shared by symbol-trained apes and toddlers. This proto-language is equated with a traditional (but inaccurate) conception of the language of "under-twos": a telegraphic code lacking in grammatical morphemes, with reliance on word order as a basic grammatical device, and expressing a collection of core prelinguistic concepts. Bickerton (1990) presents the parallel in the following terms:

> We may conclude that there are no substantive formal differences between the utterances of trained apes and the utterances of children under two. The evidence of children's speech could thus be treated as consistent with the hypothesis that the ontogenetic development of language partially replicates its phylogenetic development. The speech of under-twos would then resemble a stage in the development of the hominid line between remote, speechless ancestors and ancestors with languages much like those of today. (p. 115)

To this proposal, Bickerton adds an argument based on the nature of human postnatal brain growth:

> Haeckel's claim that ontogeny repeats phylogeny has had a checkered career in the history of biology, and certainly cannot stand as a general law of development. However, it may have application in limited domains. In particular, no one should be surprised if it applies to evolutionary developments that are quite recent and that occur in a species whose brain growth is only 70 percent complete at birth and is not completed until two or more years afterwards. (Bickerton, 1990, p. 115)

There are, however, problems in making analogies from child language to simpler ancestral languages. And there are problems in accounting for the emergence of complex capacities on the basis of brain growth. I point out some of the most salient issues, beginning with the nature of 2-year-old language.

SYMBOL-TRAINED APES, HUMAN TODDLERS, AND PROTO-LANGUAGE

Cross-species comparisons are difficult, but the temptations to see bits of ourselves in our cousins—or to deny such similarities—are strong and enduring. On the language comprehension side, Savage-Rumbaugh's reports of the accomplishments of bonobos (2001; Savage-Rumbaugh, Shankar, & Tylor, 1998) made it clear that many of the prerequisites for human language were already present before the emergence of the hominid line. Our close living relatives, bonobos, can comprehend spoken English sentences—without instruction. The capacities for acoustic segmentation of speech, lexical mapping, and some levels of syntax are thus ancient. Savage-Rumbaugh (2001) even presents evidence for English-based vocal production and writing in bonobos. All of this raises fascinating questions about the evolution and functions of these capacities; but such questions lie outside of the search for parallels in human ontogeny. Certainly, as Savage-Rumbaugh (2001) points out: "These findings render mute old questions regarding the innate limits of the ape brain" (p. 24). They also make it clear that additional factors—both cognitive and social—must have been necessary for the emergence of human language.

"Proto-Grammar" and Early Child Language

We cannot predict what new surprises will come from bonobos or other great apes, but for the moment at least, they have not been given the opportunity to acquire a rich morphological language such as Turkish (agglutinative) or Inuktitut (polysynthetic). Children under 2 who are exposed to such languages do not exhibit the sort of "pregrammatical" speech described by Bickerton, Givón, and others, such as absence of grammatical morphology and reliance on topic–comment word order. Turkish toddlers show productive use of case inflections on nouns as early as 15 months of age—that is, productive morphology at the one-word stage (Aksu-Koç & Slobin, 1985; Küntay & Slobin, 1999). For example, the direct object of a verb (accusative case) is marked by a suffixed vowel in Turkish. Thus, if I see or kiss a girl called *Deniz*, I use the form *Deniz-i*. But if the noun ends with *–k*, the final consonant is not pronounced. For example, the accusative form of *bebek* 'baby' is not *bebek-i* but *bebe-i* (written *bebeği*). An error produced by a child of 15 months (Ekmekçi, 1979) indicates that already at this age—still in the one-word stage—Turkish children use grammatical morphemes. This is shown by the report that the child said *bebek-i* in an appropriate context. This is a form that she couldn't have heard, yet matches the morphological patterns of the language. Beyond this early precocity, Turkish children quickly come to use multiple suffixes on nouns, and by the age of 24 months or younger, demonstrate full mastery of the nominal inflectional system and much of the verbal paradigm. For example, a child of 18 months (Aksu-Koç & Slobin, 1985) produced the following two-word utterance consisting of six morphemes:

kazağ -ım -ı   *at   -tı   -m*
sweater -my -ACCUSATIVE throw -PAST -1ST PERSON
'I threw my sweater.'

Similar productive use of grammatical morphology in the period of one- and two-word utterances has been documented for Inuktitut, a quite different type of highly inflected language (Allen, 1996; Fortescue & Lennert Olsen, 1992). For example, an Eskimo child of 2;6 produced a five-morpheme verb that represents an entire proposition (Allen, 2000, p. 4):

ma -una   -aq -si   *-junga*
here -VIALIS -go -PROSPECTIVE.ASPECT -PARTICIPIAL.1SG
'I'm going through here.'

Such examples are hardly possible in a "proto-language" that consists of short strings of words with no grammatical morphemes ("telegraphic speech"), yet they are typical of early utterances in highly inflected languages.

  Early child speech is also not always characterized by the use of fixed word order to express semantic relations between elements. That is, not all languages use word order to distinguish the meanings of *dog bite man* and *man bite dog*. Where these relations are marked by case inflections, as in Turkish, word-order variation is used for other functions. At the beginning of the two-word period, Turkish children are able to appropriately vary the orders of words. For example, Aksu-Koç and Slobin (1985) provide the following summary of the child studied by Ekmekçi (1979, 1986):

> Early control of the functions of word order is reflected in a number of contrastive uses, including the following: (1) Preposed adjectives are used in attributive expressions (e.g. *soğuk su* 'cold water', said at 1;7 when asking for cold water), whereas postposed adjectives are used in predicative expressions (e.g., *çorba sıcak* 'soup hot', said at 2;0 as a complaint). (2) Indefinite or nonreferential direct objects always directly precede the verb (e.g. *kalem getir* 'bring (a) pencil'), whereas definite direct objects (marked by the accusative inflection) can also follow the verb (e.g. both *kalem-i getir* 'pencil-ACCUSATIVE bring' and *getir kalem-i* 'bring pencil-ACCUSATIVE' = 'bring the pencil') [age 1;10]. (p. 856)

Because case inflections, rather than word order, are used to indicate who did what to whom, Turkish children do not make use of word-order information in comprehension in the ways that English-speaking children do. Slobin and Bever (1982) carried out a study in which children were asked to act out the meanings of sentences containing two nouns and a verb, such as *horse kick cow*, in all possible orders of subject, verb, and object: SVO, OVS, SOV, OSV, VSO, VOS. It is not until age 2;6 that English-speaking children reliably understood SVO sentences such as *the horse is kicking the cow* in Slobin and Bever's experimental task; however, Turkish children as young as 2;0 correctly understood all six orders of S, V, and O, relying on the ACCUSATIVE suffix on one of the nouns to indicate that it designated the patient of the action. Reliance on word order, therefore, is not a universal of early child language, although it has been proposed as characteristic of the "proto-language."

In sum, early telegraphic speech and reliance on fixed word-order patterns—the prototype of "pregrammar"—are characteristic of child language in only certain types of languages. And even in those languages, like English, that seem to fit the characteristic, it is not clear that children begin with broad-based rules of word combination. Research on detailed corpora of very early child speech in English (Lieven, Pine, & Baldwin, 1997; Tomasello, 1992, 1999) makes it clear that much of early language is item-based rather than reflecting productive combinations of the telegraphic or pregrammatical type.

The influence of environmental language is especially evident in the case of bilingual children. A number of investigators report that such children—as soon as they begin to produce two-word and multiword combinations—differentiate the word-order patterns of their two languages (e.g., Meisel [1989] for French–German bilinguals; de Houwer [1990, 1995] for Dutch–English bilinguals; Deuchar [1992] for Spanish–English bilinguals). These children do not show a standard pregrammar or proto-language in which the two languages are differentiated only by choice of lexical items; rather, they are differentially shaped by each of the exposure languages from very early on.

Early learners are good at extracting salient grammatical devices in the exposure language, as demonstrated, for example, in my work on Operating Principles (Slobin, 1973, 1985). But this, of course, requires a human brain in an environment of already established human languages.

"Proto-Language" and Early Child Semantics

Another part of the recapitulationist scenario assumes that the semantic concepts expressed in early child speech, across languages, represent some sort of conceptually basic core of human notions that we may have shared with our hominid ancestors. However, if we look across the languages of the world, we find unexpected diversity in the expression of "basic" notions. For example, in the domain of spatial relations, Bowerman and Choi have compared English with Korean (Bowerman & Choi, 2001; Choi & Bowerman, 1991). Developmental psychologists have assumed that infants begin with sensitivity to basic relations such as *containment* and *support*, as expressed by the English prepositions *in* and *on*. However, Korean makes a different distinction: What is important in Korean is not whether one thing is supported by another or is contained by another, but rather whether the relation between the two things is one of *tight fit* or *loose fit*. Consider, for example, the scenes represented in Fig. 8.1 and Fig. 8.2.

These figures show part of a larger set of contrasts between English and Korean. English distinguishes *containment*—using *put in* regardless of tightness of fit, and *support*—using *put on* regardless of tightness of fit. Korean uses *nehta* for *loose fit*, *kkita* for *tight fit*—whether containment or support, and *nohta* for putting something *loosely on a horizontal surface*. In a preferential looking experiment with American and Korean infants aged 18 to 23 months, Choi, McDonough, Bowerman, and Mandler (1999) found that American babies, when looking at pairs of videos and hearing *put in*, preferred to look at scenes depicting containment, whether the fit was loose or tight. Korean babies in the same task, when hearing *kkita*, preferred to look at scenes depicting tight fit, whether the fit was one of containment or support. That is, the two groups oriented to lan-

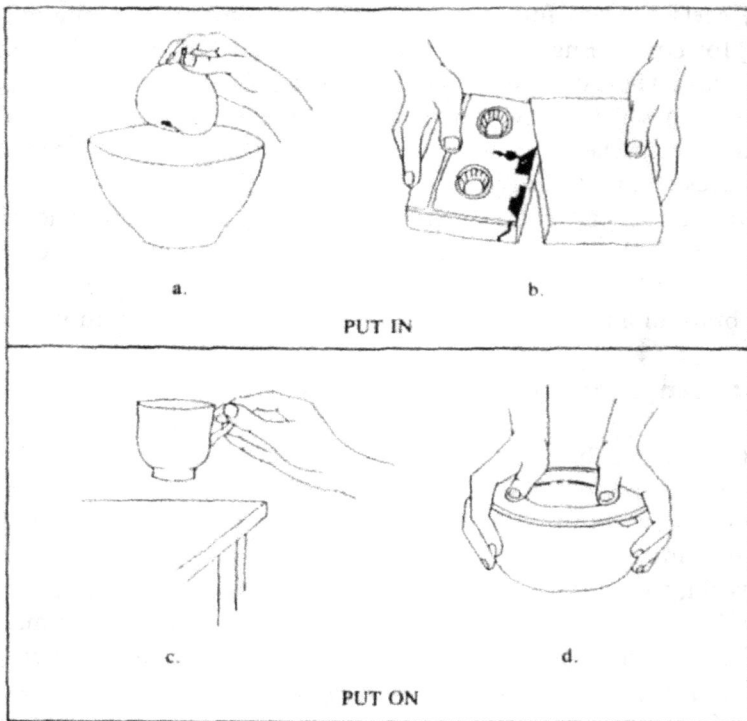

FIG. 8.1. Classification of four actions as instances of containment (a, b) versus support (c, d). From Bowerman (1996, p. 152). Reprinted with permission.

guage-specific categories in comprehension, early in the one-word period. In one- and two-word speech in the two languages, there were comparable differences in the semantic categories encoded by early words. For example, Fig. 8.3 schematizes part of the domain of spatial relations expressed by children of 16 to 20 months of age in the two languages (Choi & Bowerman, 1991). The core notions that receive early expression do not line up between the two languages. Bowerman (1996) concludes: "[I]t is striking how quickly and easily children adopted language-specific principles of semantic categorization. There was little evidence that they had strong prelinguistic biases for classifying space differently from the way introduced by their language" (pp. 169–170).

I would conclude that continuing research on both chimpanzees and human children casts doubt on Bickerton's characterizations of "proto-language" or "proto-grammar" as a sketch of the linguistic capacities of our ancestors. To be sure, chimps and human infants use reduced varie-

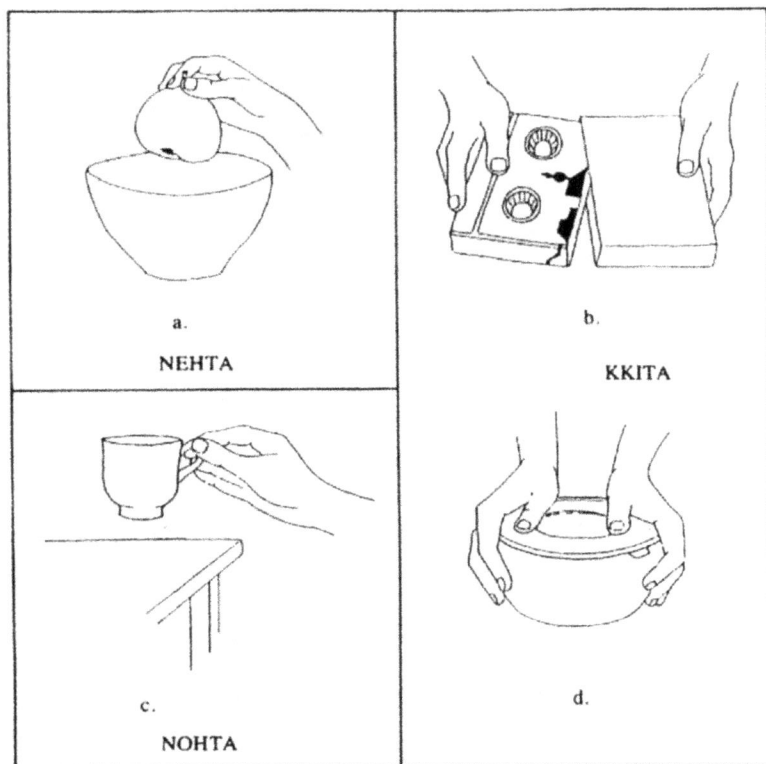

FIG. 8.2.   Classification of four actions as instances of loose fit (a) versus tight
fit (b, d) versus loose surface contact (c). From Bowerman (1996, p. 153). Reprinted with permission.

FIG. 8.3.   Early semantic categories in English and Korean child speech, 16 to
20 months. Data from Choi and Bowerman (1991).

ties of full human languages. But, as Jespersen emphasized long ago, all such reduced varieties are derived from an already developed exposure language. The child or symbol-trained chimpanzee is sampling from an existing language, and not creating without input. The structures of early language production are not independent of the structures of the exposure language. (Proposals about creation without or beyond input are taken up with regard to the third question, below.) Although the ways in which children sample from existing languages tells us a great deal about the workings of the human mind, it is not evident that any generalizations can be drawn about pre-human minds from such evidence.

Heterochrony

Another obstacle to comparing chimps and human babies—despite the huge genetic overlap between the species—is the difference in *timing* of onset and offset of abilities. A quarter-century ago, the late Steven Jay Gould, in *Ontogeny and Phylogeny*, argued for "the evolutionary importance of *heterochrony*—changes in the relative time of appearance and rate of development for characters already present in ancestors" (Gould, 1977, p. 2). In his conception, human development is retarded in relation to other primates. However, more recently, Jonas Langer (2000, 2001) has come to a somewhat different conclusion, examining heterochrony with regard to several dimensions of cognitive development that are critical for our topic. He applied comparable tests of physical cognition (causality) and logicomathematical cognition (classification) to human infants, two sister species of great apes (common chimpanzees, bonobos), and two species of monkeys (cebus, macaques). There are two important findings for our purposes: (a) Human cognitive development is accelerated in comparison to the other species (that is, we are not simply neotenized primates); (b) The two sorts of cognition are dissociated, developing in parallel in humans, but asynchronously in apes and monkeys. These heterochronic relations are evident in Fig. 8.4.

The consequence of heterochrony is that physical and logicomathetical cognition can interact from the start in human babies, whereas logicomathematical capacities are not available to apes and monkeys during the early phases of establishing physical cognition. Note also that second-order cognition appears early and synchronously in both domains for humans, allowing for immediate interaction between two types of cognition at a higher level. By contrast, second-order cognition in chimpanzees

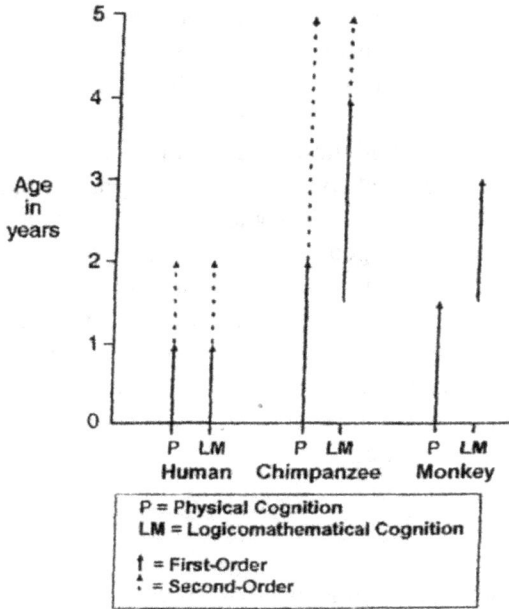

FIG. 8.4. Comparative cognitive development: Vectorial trajectories of developmental onset age, velocity, sequence, and organization (but not extent or offset age). From Langer (2000, pp. 361–388). Reprinted with permission.

emerges for physical cognition when the animals have just begun to work out first-order cognition for classification (and second-order cognition has not been observed in monkeys). The opportunities for ontogenetic interaction between cognitive capacities thus varies significantly across species, due to heterochronic effects. Langer (2000) provides a strong critique of recapitulationism:

Such phylogenetic displacements in the ontogenetic covariations between the onset, velocity and offset of cognitions in relation to each other . . . disrupt potential repetitions (i.e., recapitulation) of phylogeny in ontogeny. Thus, human, and for that matter chimpanzee, cognitive ontogeny does not simply recapitulate its phylogeny. Instead, heterochronic evolution reorganizes primate species cognitive development. Significant consequences for their respective potential growth follows.

Heterochronic evolutionary reorganization of asynchronic into descendant progressively synchronic development . . . opens up cascading possibilities for cognitions to influence each other and to be influenced by similar environmental influences. (p. 374)

Within our species, temporal covariation of cognitive and linguistic abilities shapes the emergence of language. Developing constraints of memory and attention, along with available cognitive structures of all sorts, are responsible for the nature and course of language development. These constraints and their timing vary from species to species, and we cannot know how such factors might have played themselves out in extinct ancestors. Furthermore, as argued below, I suggest that modifications of existing languages generally do not come from very young learners, but from more mature participants in social and linguistic interaction. That is, in humans, much of linguistic innovation is due to individuals who are advanced in cognitive and social development. Comparisons of human toddlers with apes and monkeys are therefore of very limited applicability to the task at hand. (And the long line of more relevant species is, alas, extinct.)

Answering Question 1

Question 1 asks whether linguistic ontogeny can be conceived of as a recapitulation of linguistic phylogeny—that is, whether phenomena of child language can provide clues about the evolutionary origins of the human language capacity. I conclude that linguistic ontogeny does not recapitulate phylogeny because the form and content of "under-two" child language is shaped by the form and content of an already existing exposure language rather than reflecting developmental trajectories that might be shared with ancestral species. Further, if we compare the rates of development of various cognitive capacities in contemporary primate species, heterochronic relations among various developing capacities indicate that any set of capacities, such as those underlying language, pattern in distinctly different ways across species. Therefore processes of human language development are not likely to mirror the phylogenetic origins of such processes.

DOES THE DIACHRONY OF LANGUAGE
RECAPITULATE ITS ONTOGENY?

Question 2 asks about a different kind of possible recapitulation: Do patterns of historical change of existing languages mirror the ways in which human children acquire existing languages? This position has been advanced repeatedly over the last several centuries, due to striking parallels between patterns of language development in individual children and repeated diachronic changes in languages. One is tempted to propose that similar cognitive processes underlie both sorts of development. For example, the linguist Paul Guillaume proposed early in the last century—about the same time that Otto Jespersen was warning against a simple answer to Question 1—that Question 2 can be answered in the affirmative:

> The facts that we cannot examine in the history of languages are available to us in the child. . . . There are certain easy routes . . . they are frequently the same ones that languages have followed in the course of their evolution and that the child, in turn, takes up in learning his language. (Guillaume, 1927/1973)

In contemporary linguistics, the child learner is seen as the source of various sorts of language change, both in generative accounts (e.g., Lightfoot, 1988) and functional approaches (e.g., Gvozdanović, 1997). Why would one propose such an explanation for historical language change? It is based on a simple set of propositions: (a) Children are imperfect learners. Their "errors" tend toward regularization. (b) Languages are imperfect systems. They tend toward regularization. (c) Child learners are responsible for changing the language. Propositions (a) and (b) are true; but the evidence suggests that (c) is false.

Most changes of the sort carried out by young children are a matter of "cleaning up" an existing grammar, rather than introducing new forms or constructions. Furthermore, for a linguistic change to have a lasting effect, it has to be maintained into adulthood; that is, the childish revisions must come to sound normal and acceptable. Sociolinguistic studies, however, show that lasting changes are more likely to be due to usage in adolescent peer groups, rather than in early childhood (e.g., Romaine, 1984).

Changes in Past-Tense Forms of the English Verb

As a small case study, consider historical changes in the past tense forms of English verbs. We now say *helped* as the past tense of *help* and *thrived* as the past tense of *thrive*. That is, these are regular verbs in contemporary English. But earlier, the past tenses of these verbs were *holp* and *throve*. This looks suspiciously similar to errors made by modern-day children, who say *telled* and *drived* instead of *told* and *drove*. Might it be that child learners have been regularizing the system over generations? To examine this possibility, Joan Bybee and I carried out a study of changes in the

English past tense, with the title: "Why Small Children Cannot Change Language on Their Own" (Bybee & Slobin, 1982). We looked for innovations in past-tense verbs forms, such as *breaked*, *hitted*, and *weeped*. The data came from three age groups: (a) spontaneous speech records of preschoolers aged 1;6–5, (b) elicited past-tense forms from school-age children aged 8;6–10, and (c) past-tense forms produced under time pressure by adults. All three age groups produced innovative forms (errors, overregularizations); however, only the forms produced by school-age and adult speakers mirrored ongoing changes in the English verb system. The preschoolers made errors on high-frequency verbs, such as *breaked*, *catched*, and *flied*—but these are not forms that are on their way to becoming standard in the language. That is, most of the preschool errors were transient. By contrast, many of the errors produced by older speakers showed some chance of becoming part of the language. They overregularized low-frequency verbs, producing forms such as *weeped* and *kneeled*, which **are** moving into standard English. Most interesting was the finding that it was only the older speakers who tended to regularize verbs that end in a final dental consonant, such as *hit–hitted*, reflecting an ongoing tendency in English to regularize verbs of this class. For example, verbs such as *started*, *lifted*, *fasted*, *roasted*, *sweated* did not use to have these overt *-ed* past tenses; the earlier past tense forms were *start*, *lift*, *fast*, *roast*, *sweat*. The study suggested that, at least in this part of the grammar, early learners are not the innovators:

> The conclusion that must be drawn from the facts is that there is nothing particularly special about the relation between small children's innovative forms and morpho-phonemic change. The innovations of older children and adults . . . may also serve as predictors of change. In fact in some cases where adult innovations differ from early child innovations, such as with the *hit*-class, the adults and older children, who are in better command of the entire system, innovate in ways that manifest more precisely the on-going changes in the system. Thus it appears that both socially and linguistically, the older children and adults are in control of morpho-phonemic changes. (Bybee & Slobin, 1982, pp. 36–37)

Answering Question 2

Question 2 asks whether historical changes in language mirror ontogenetic changes. The past-tense case study suggests that children in the early stages of acquisition are not the ones who push the language forward. Other work in developmental psycholinguistics also suggests that children are not the ones to create new grammatical forms.[1] In brief, preschoolers are at work sorting out the regularities and irregularities of subsystems of the language, on the basis of limited information. Older children, having established a working knowledge of the language and using a larger database, are able to apply patterns at the level of the language as a system. That is, they have a sufficient grasp of the overall structure of the language to allow them to adjust particular parts of the system. Again, the immature learner does not serve as an appropriate model of the processes of change.

DO CHILDREN CREATE GRAMMATICAL FORMS?

There are three sorts of situations in which we can ask whether children can create grammatical forms on their own: the emergence of creole languages, the invention of "homesigns" by deaf children with hearing parents, and the emergence of a new sign language in Nicaragua. With regard to each situation, it has been proposed that children have the capacity to innovate structure, suggesting that this is an innate capacity that arose in the evolution of our species. Again, I suggest caution in evaluating these proposals.

From Pidgin to Creole

The classic definition of a creole is "a pidgin with native speakers." Pidgin languages arise in contact situations between speakers of different languages, most typically in situations of slave labor or immigrant labor. A pidgin is no one's native language; rather, it is a limited and ad hoc composition of elements from two or more component languages. It has long been claimed that children can take such an imperfect input language and "nativize" it. In the process, it is proposed, grammatical structures emerge that were not in the pidgin input. Therefore the structures must result from an innate language-making capacity or "bioprogram" for language. I will not attempt to summarize the large and contentious literature on this topic (see, for example, Bickerton, 1981; DeGraff, 1999;

---

[1] See Slobin (1994) for a discussion of illusory parallels between the development of the PRESENT PERFECT in the history of English and in contemporary English-speaking children. See Slobin (1997) for a similar discussion of the historical development of direct-object markers in various languages and false parallels with starting points in children's cognitive and language development.

Foley, 1988; McWhorter, 1997; Muysken, 1988; Romaine, 1988; Thomason & Kaufman, 1988). Arguments in the pages of the *Journal of Pidgin and Creole Languages*, going on since its inception in 1986, provide evidence and counterevidence for a variety of theories of creole genesis. As I read the literature, there is evidence for considerable influence of substratum languages on emerging creoles, particularly influences of various African languages on Caribbean creoles. This is not surprising when one considers that African slave mothers and caretakers would probably have spoken an African native language to their infants, rather than the colonial pidgin. Thus the input must have been richer than simply a pidgin. Furthermore, demographic data strongly suggest that languages that are considerably more complex than pidgins can arise in interaction *among adults*, before there are native speakers. That is, adult pidgin speakers can produce grammatical innovations. Bickerton's proposal that creole genesis reveals an innate bioprogram for language seems far less plausible than when it was introduced 20 years ago. (Personally, I am not convinced by any of the evidence or arguments for the bioprogram.)

Most of the world's creole languages arose in the past, under linguistic and social circumstances that will always lack full documentation. But we have at least one contemporary example: the emergence of Tok Pisin as a developed language in New Guinea. Tok Pisin arose out of Melanesian pidgin, and in the course of some 150 years of use developed a number of grammatical features before it became anyone's first language (Keesing, 1991).[2] Much of this development can be attributed to the fact that Tok Pisin was called on to serve as a standard language of public communication, business, and education, as well as serving as the official language of government proceedings in Papua New Guinea after 1964 (Romaine, 1988). Gillian Sankoff was on hand to study the first native speakers of Tok Pisin—a process that she and Suzanne Laberge have aptly referred to as "the acquisition of native speakers by a language" (Sankoff & Laberge,

---

[2]John McWhorter (1995) has made a similar suggestion about Atlantic English-based creoles, suggesting that they derive from an established and elaborated West African pidgin used early in the 17th century. He concludes that this ancestor language

was by no means a rudimentary pidgin, but was, on the contrary, already relatively elaborated by the time of its exportation to the New World. . . . Hence, this contact language exhibited a structural expansion analogous to that of Tok Pisin before creolization, as opposed to the rudimentary structure documented in pidgins of limited social function . . . (p. 325)

1973, p. 32). There are two important findings for the purposes of the present argument: (a) A pidgin language can evolve into something like a creole without requiring the hypothesized special intervention of child learners. (b) The first generation of native speakers "smoothes out" the language, rather than innovating new forms. Let us briefly consider the second finding.

The children studied by Sankoff in the 1970s were learning Tok Pisin in families and social situations in which it served as a second language, spoken with some fluency, but also with some grammatical fluidity. The child learners apparently did what children are good at: making a system regular and automatic (what John Haiman [1994] has referred to as "ritualization"). This is evident on the level of speech production, as in the following example:

> The children speak with much greater speed and fluency, involving a number of morphophonemic reductions as well as reduction in the number of syllables characteristically receiving primary stress. Whereas an adult will say, for the sentence "I am going home,"
>
> (1) Mi gó long háus;
>
> a child will often say
>
> (2) Mi gò l:áus;
>
> three syllables rather than four, with one primary stress rather than two. (Sankoff & Laberge, 1973, pp. 35–36)

Grammatical morphology also changes with native speakers—but, again, they are not the innovators. For example, there was a well-established future marker, *baimbai* (from English *by and by*), which began as an optional adverbial to establish the time frame of a stretch of discourse. But long before there were native speakers, the form was reduced to *bai* and moved to preverbal position within the clause, where it tended to be used redundantly in a series of future predications. What the children did was to make the future marker obligatory, while also reducing it in substance and stress. That is, the child learners played a significant role in accelerating an ongoing process of grammaticalization, in which a preverbal clitic, *bə*, moves along a well-established path from a particle to an inflectional prefix (see Hopper & Traugott [1993] for discussion of grammaticalization in historical language development).[3] Thomason and Kaufman (1988) provide an apt reformulation of the role of the child:

> When Bickerton poses the question of how a child can "produce a rule for which he has no evidence" (1981: 6), he is, in our view, asking the wrong question. We prefer to ask how the child can create grammatical rules on the basis of input data which is much more variable than the input data received by a child in a monolingual environment. (p. 164)

Given the available evidence, I conclude that learning processes of this sort are normal, and do not reveal special capacities of the language-learning child beyond what is already known about the acquisition of "full-fledged" languages. A creole language develops over time, in contexts of expanding communicative use of a limited pidgin language. Child learners help to push the process forward, arriving at a grammar that is more regular and automated—but they do not appear to be the innovators.

The Creation of Homesigns by Deaf Children

Most deaf children are born to hearing parents and, unfortunately, most hearing parents do not learn a sign language in order to communicate with a deaf child. Such children create their own systems of gestural communication, called homesigns. Over many years, Susan Goldin-Meadow and her colleagues have documented the systematicity of homesigns in a number of deaf children, growing up in several countries (see Goldin-Meadow, 2003; Goldin-Meadow & Mylander, 1984, 1990; Goldin-Meadow, Butcher, Mylander, & Dodge, 1994; Goldin-Meadow, Mylander, & Butcher, 1995). Here we have a real opportunity to observe the language-making capacity of the child. The studies demonstrate that individual deaf children systematically use a limited set of handshapes, combined with motion, to refer to objects on the basis of specific physical properties. For example, Goldin-Meadow et al. (1995) carried out a detailed analysis of com-

---

[3]Recently, Hudson (2002) demonstrated—in a series of long-term experiments in which children or adults acquire artificial languages—that child learners regularize variable grammatical patterns to a considerably greater extent than adult learners. She proposes that general processes of probability learning can account for the emergence of consistent patterns. Because children apparently disprefer unpredictable variability, regular patterns might emerge in the course of "nativization" of a grammatically inconsistent language such as a pidgin or emerging creole. Note, however, that the regularizations are based on already existing but partial consistency in the language.

ponents of handshapes in four homesign systems created by children between the ages of 2;10 and 4;11. All four children used a set of basic handshapes, described by the researchers as Fist, O, C, Palm, Point, Thumb, V, and L. Components of hand breadth and finger curvature systematically mapped onto features of the referenced objects: Point and Thumb handshapes referred to manipulation of very narrow objects, Fist and O referred to wider objects, and C and Palm were used for the widest objects. For example, all four children used a large C-handshape to represent handling an object greater than 2 inches/5 cm in width. All of the children used Point (index finger) for straight thin objects, such as straws, candles, and pencils. Three of the children used a flat palm for vehicles. Overall, handshapes could be placed in systematic paradigms or matrices of contrasts for each child. In addition, most handshapes were combined with one or more type of motion. Goldin-Meadow et al. (1995) conclude:

> Thus, the gesture systems of the deaf children in our study appear to contain a subset of the handshape and motion components found in ASL. The similarities between sign forms in ASL and gesture forms in our subjects' gesture systems suggest that our subjects' set may reflect the units that are "natural" to a language in the manual modality—units that may form part of the basic framework not only for ASL morphology but also for the morphologies of other sign languages. . . . Whatever the details of the gesture systems, the fact that the gesture systems of all of the deaf children in our study could be characterized as having a morphological structure suggests that such structure is essential to the young communicator—so essential that it will evolve even in the absence of conventional linguistic input. (pp. 243–244)

Homesigners also uses consistent orders of signs, thus indicating a sort of beginning grammar. The following three orders appear to be typical across homesign systems:

- patient + act (e.g., CHEESE EAT)
- actor + act (e.g., YOU MOVE)
- patient + act + agent (e.g., SNACK EAT YOU)

In the case of homesign it appears that we do, at last, have evidence for a primordial human language-making capacity. Homesign systems have some of the characteristics proposed for a proto-language: referential symbols and meaningful symbol order. They also go beyond the proto-language proposals in that they appear to have morphological structure, that is, a level of meaning that is smaller than the "word." It is striking that very young children can create such systems—though we do not know if childhood is a prerequisite for the accomplishment. (It's hard to imagine a scenario in which a language-deprived deaf adult, with no prior communicative experience, would invent a homesign system.) In any event, with regard to Question 3, we can conclude that—without a language model—children can create a gestural language that has systematic patterns of reference and sign order.

However, homesign systems stagnate; they do not develop further into full human languages. Apparently more is needed—and this additional factor seems to be an interacting community of signers. The opportunity to study this factor has been made available by the emergence of a new sign language in Nicaragua, allowing us to ask a final part of Question 3: Can a group of children using homesign arrive at a common grammar?

From Homesign to Nicaraguan Sign Language

A new sign language emerged in Nicaragua in the 1980s, when deaf children were first gathered together into schools. Before that, deaf children were isolated from each other, each using some kind of homesign. What happened when they came together was remarkable: From the collection of homesigns, a common language was formed (Kegl & McWhorter, 1997; Morford & Kegl, 2000; Senghas, 1995; Senghas & Coppola, 2001; R. Senghas, A. Senghas, & Pyers, chap. 9, this volume). Nicaraguan Sign Language has attracted much attention, leading to claims such as Pinker's (1999) *New York Times* assertion: "The Nicaraguan case is absolutely unique in history. We've been able to see how it is that children—not adults—generate language . . . And it's the first and only time that we've actually seen a language being created out of thin air." Careful examination of the facts, however, leads to a conclusion that is very much like my evaluation of Tok Pisin: Linguistic structure emerges when people interact with one another and begin to communicate about a range of topics, using limited resources.

Documentation of the emergence of grammatical forms in Nicaraguan Sign Language can be found in Ann Senghas' dissertation (Senghas, 1995); for an up-to-date critical analysis, see Senghas, Senghas, and Pyers (chap. 9, this volume). The critical comparisons are between the "first cohort"—that is, the original group of deaf students who were brought together in a school in Managua, and the "second cohort" who entered later and joined an existing community of signers. (Defined in terms of year of entry, the first cohort entered in the period 1978 to 1983 and the second cohort entered after 1983.) The situation for the second cohort was thus similar to that described for the nativization of Tok Pisin: learners exposed to a language that is not yet fully developed. And, again, we can ask if new grammatical forms arise in the process of "nativization." Senghas and Coppola (2001) summarized the development of the language from the initial resources provided by homesigns and gestures:

> These initial resources were evidently insufficient for the first-cohort children to stabilize a fully developed language before entering adulthood. Nevertheless, over their first several years together, the first cohort, as children, systematized these resources in certain ways, converting raw gestures and homesigns into a partially systematized system. This early work evidently provided adequate raw materials for the second-cohort children to continue to build the grammar. (p. 328)

What would be relevant to the present discussion about language origins would be evidence that it was the *second* cohort—that is the first-language learners—that was responsible for the creation of grammar. However, Senghas' published data show that all of the grammatical innovations that she studied were already present in the first cohort, though not used consistently. Consider three types of grammatical forms that emerged in Nicaraguan Sign Language:

1. The use of space to indicate *person*: person inflection on verbs to indicate SUBJECT, DIRECT OBJECT, and INDIRECT OBJECT;

2. the use of space to indicate *coreference*: same locus to refer to a person or object in successive utterances;

3. *aspectual modulation* of verbs: movement patterns superimposed on verbs to indicate such aspects as CONSTANTLY, REPEATEDLY, or RANDOMLY.

TABLE 8.1
Use of Grammatical Forms by Nicaraguan Signers by Cohort
(Age of Entry; Senghas, 1995)

	Cohort	
Grammatical Form	First Cohort (1978–83)	Second Cohort (1983–)
Mean number of person inflections per verb	.50	.56
Mean number of person coreferences per sign	.215	.292
Mean number of aspectual modulations per verb	.332	.457

Table 8.1 presents figures on the use of these grammatical forms by the two cohorts based on narrative data gathered by Senghas (1995). In each instance, the second cohort uses the forms more frequently than the first. But note that, for each grammatical issue, the forms were already present in the first cohort; that is, there is no evidence here that innovations arise in the process of early language acquisition.[4]

Senghas carried out a more detailed analysis, considering both year of entry (cohort) and age of entry. This make it possible to separate signers both by the amount of time they have been using the language and the age at which they were first exposed. There are three categories of signers in each cohort, according to age of entry: young: age 0–6;6, medium: age 6;7–10, and old: age 10;1–27;5. The findings are especially interesting, as shown in Fig. 8.5. As in Table 8.1, we see that all of the forms in question are already attested in the signing of the first cohort. At the same time, we see an age effect: For the young and medium groups, members of the second cohort use the forms with greater frequency. For the old group, it doesn't matter whether they entered with the first or second cohort.

How are these findings to be interpreted? To begin with—contrary to the more extreme claims in the literature—linguistic structure was not the invention of the second cohort. Rather, what seems to have happened was that younger signers—that is, those who entered a community that already had a developing communication system—used the existing

---

[4]It may be necessary to modify this claim in the light of recent evidence with regard to the second form, spatial coreference. Senghas has recently presented evidence that systematic use of this form may have emerged among children of the second cohort. She provides the following summary (e-mail, Feb. 27, 2003):

First-cohort signers produce spatially nonneutral signs, no doubt, but they (a) don't consistently place coreferent referents in a common location, (b) nor do they comprehend co-spatial signs as being necessarily coreferent, (c) nor do they see a violation of co-spatial coreference as an ungrammatical utterance. Second-cohort signers do all these things (although c is variable, perhaps dependent on metalinguistic ability). . . . What I think happened in the case of spatial coreference is that second-cohort signers observed spatially modulated signs in the signing of the first cohort and reanalyzed them in such a way as to apply the form to the function of spatial coreference. . . . In this way, a "rule" was born that did not exist before.

What is at issue is whether the first-cohort uses of spatial coreference were meaningful or not. This is a difficult question to resolve with available data, but it is quite likely—as Senghas has suggested—that here we have evidence for the emergence of a new grammatical constraint, due primarily to the systematic use of space by *younger* learners. See R. Senghas, A. Senghas, and Pyers (chap. 9, this volume) for details.

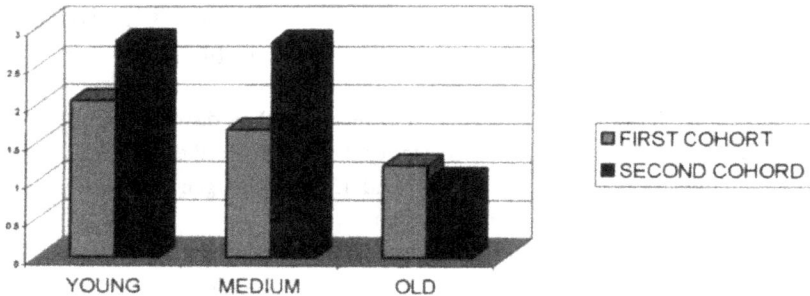

FIG. 8.5.   Mean number of inflections per verb (person, number, position, aspect) by cohort and age at entry. Data from Senghas (1995).

grammatical elements more frequently and fluently (perhaps innovating at least one systematicity, spatial coreference, as suggested in footnote 4). This account of the Nicaraguan situation matches Thomason and Kaufman's (1988) response to Bickerton with regard to the emergence of creole languages (with the exception, of course, that the starting point in Nicaragua was a collection of individual homesign systems, rather than two or more existing languages):

> [A]n entirely new language—without genetic affiliation—is created by the first members of the new multilingual community, and further developed and stabilized by later members, both children born into the community and (in many or most cases) newcomers brought in from outside. (p. 166)

In short, as regular forms begin to develop in a group, younger learners *automate* the language. Morford's (2002) discussion of the Nicaraguan situation, as well as her work with late learners of ASL, shows that efficiency in online processing is a critical factor in language mastery. And it is on this dimension that early learners have an advantage, rather than having a special, age-linked capacity "to create language." (Senghas & Coppola [2001] report that children who acquire Nicaraguan Sign Language before the age of 10 sign at a faster rate and are more skilled in comprehending grammatical forms.)

It seems clear from the evidence available thus far that individual deaf children can innovate linguistic forms, but that it takes an interacting community to push those innovations towards automated, efficient linguistic systems. Therefore, as Morford (2002) points out, the emergence of Nicaraguan Sign Language "is better described as a process of *grammaticization* than of innovation" (p. 333). We would do well to pay close attention to the stages that Morford (2002) proposes:

> Thus, the implication of this work is that there may be three distinctive stages in the emergence of language: (1) the emergence of the lexicon, (2) the emergence of system-internal grammatical properties, and (3) the emergence of processing-dependent grammatical properties. (p. 338)

This schematized formulation gives us a promising way of thinking about the emergence of language in evolutionary time, without making false analogies from the capacities and activities of already-evolved *Homo sapiens* children. It also requires us to pay as much attention to the emergence of structure in communicative practice as to the cognitive capacities of the individual—whether innate or developing, whether language-specific or general (Slobin, 1997). Senghas (1995) has made this point forcefully:

> Homesigners develop little more than a small lexicon and basic word ordering strategies. An important component missing in these cases is the dynamic interaction of a peer group whose constant attunement allows the members to converge upon a new grammar. Without a peer group of language users, a rich language does not emerge. (p. 160)

Answering Question 3

So do children create grammatical forms? To some extent they do, but within limits. At least in the gestural modality, deaf children with no sign language input can create a gestural language that has systematic patterns of reference and sign order. Children who acquire a partially structured language—either a pidgin language or an incipient sign language based on homesign—are skilled at making the language into a more efficient and regular system. But these processes go beyond the individual. On the plane of evolution, whatever scenario one might be attracted to, complex social products such as language can be allowed to emerge, in part, in processes of interpersonal use.

Attention to these two factors—individual and social—gives us a way out of the apparent insolvable problem that led Chomsky and his followers to appeal to an innate syntactic module. Consider, for example, a typical formulation of the nativist program:

The claim, then, is that some aspects of our language capacity are not the result of learning from environmental evidence. Aside from divine intervention, the only other way we know of to get them into the mind is biologically: genetic information determining brain architecture, which in turn determines the form of possible computations. In other words, certain aspects of the structure of language are *inherited*. (Jackendoff, 1987, p. 87)

There is a jump in the nativist argument (and the quote from Jackendoff is but one of hundreds that could have been chosen): The claim begins with a discussion of language *capacity* but ends up with a claim about language *structure*. There can be no disagreement that aspects of the capacity to acquire and use language are inherited; this is a general truth about species-specific behavior. (And the ongoing debate about domain-specific and domain-general capacities remains open.) The evidence considered in this chapter repeatedly points to an interaction between the emergence of linguistic structures in the processes of communication and the capacities of human individuals who can learn and use such structures. This conclusion echoes the themes of niche construction and co-evolution that are prominent in this volume. As I have argued previously:

[The] *structure* of language arises in *two* diachronic processes: biological evolution and the ever-changing processes of communicative interaction. The structure of language could not have arisen in the genetically determined brain architecture of an individual ancestor alone, because language arises only in communication between individuals. That is, after all, what language is for. As soon as we free ourselves of this confusion of levels of analysis—the individual and the social—many of the puzzles of language structure appear to have solutions beyond divine intervention or genetic determinism. The traditional attempt to account for linguistic structure is rather like trying to locate the law of supply and demand in the minds of the individual producer and consumer, or the shape of a honeycomb in the genetic structure of the individual bee. (Slobin, 1997, p. 297)

The present chapter appears in a publication dedicated to Jean Piaget's (1971) *Biology and Knowledge*, therefore it is worth remembering that Piaget, too, was well aware of these two levels of analysis. For example, in that book, he pointed explicitly to the role of social factors in genetic epistemology:

[F]rom the psychogenetic point of view . . . interindividual or social (and nonhereditary) regulations constitute a new fact in relation to the thought processes of the individual . . . (p. 361)

Society is the supreme unit, and the individual can only achieve his intentions and
intellectual constructions insofar as he is the seat of collective interactions that are
naturally dependent on society as a whole. (p. 368)

AN INTERIM CONCLUSION

I have briefly examined three longstanding proposals about possible
contributions of child language study to questions of linguistic
diachrony and evolution, with mainly negative conclusions. My field of
developmental psycholinguistics provides insights into the capacities for
language, thought, and communication in our species. Children's early
formulations of grammar and semantics provide a window into basic
operating principles and organizing factors of the human mind. Therefore, ontogenetic theory and data are useful in pinpointing some of the
basic concepts and processes that are needed in order to evaluate neo-
recapitulationist proposals of language development. In addition, comparisons with other surviving primate species—their capacities and developmental patterns—give clues about the road that had to be traversed by
our ancestors. All of this growing information provides material for a
range of speculative scenarios. At best, close attention to biology, development, and linguistic behavior can heighten the plausibility of those
scenarios. Children's creation of homesign systems suggests a human
capacity to create something like a proto-language (of course, using a
human brain). However, for such a language to develop further, a community of users is needed. This would have existed for prehumans, of
course. And more complex structures could have emerged as a social
product, like so many other achievements of human social and material
technology. Language is like other sorts of human technology; once it is
present, it provides a "niche"—a modified environment—creating new
pressures for the refinement of that technology and the human capacities that are adaptive for its acquisition and use. At issue here are the
particular structures of human language. I suggest that such structures
are emergent and are not prespecified. They can be learned and refined,
using various capacities—not necessarily language-specific. But here I
am launching into another sort of scenario building, beyond the aims of
this chapter. In any event, we can never have sufficient evidence to scien-
tifically evaluate such narratives, however appealing and plausible some
of them may be. I hope that my largely negative conclusions can at least
serve to reasonably constrain our irresistible speculations about who we
are and how we got here.

ACKNOWLEDGMENTS

An earlier version of this chapter was presented at a conference at the University of Oregon, "The Rise of Language Out of Pre-Language: An Interdisciplinary Symposium," organized by T. Givón, B. Malle, and J. Bybee. That paper, which contains more linguistics and less developmental psychology than the present chapter, is published as Slobin (2002). I am grateful to Talmy Givón and Derek Bickerton for vigorous discussion of my ideas and hope that their disagreements have helped me to clarify the arguments presented here, albeit without dissuading them of their positions. Ann Senghas and Jennie Pyers have helped me better understand the processes of the formation of Nicaraguan Sign Language. I am grateful to Sue Parker's careful reading and thoughtful suggestions for revision of this chapter.

REFERENCES

Aksu-Koç, A. A., & Slobin, D. I. (1985). The acquisition of Turkish. In D. I. Slobin (Ed.), *The crosslinguistic study of language acquisition. Vol. 1: The data* (pp. 839–878). Hillsdale, NJ: Lawrence Erlbaum Associates.
Allen, S. E. M. (1996). *Aspects of argument structure acquisition in Inuktitut.* Amsterdam/Philadelphia: John Benjamins.
Allen, S. E. M. (2000). A discourse-pragmatic explanation for argument representation in child Inuktitut. *Linguistics, 38*, 483–521.
Andersen, R. W. (1983). *Pidginization and creolization as language acquisition.* Rowley, MA: Newbury House.
Bickerton, D. (1981). *Roots of language.* Ann Arbor, MI: Karoma.
Bickerton, D. (1990). *Language and species.* Chicago: University of Chicago Press.
Bowerman, M. (1996). The origins of children's spatial semantic categories: Cognitive versus linguistic determinants. In J. J. Gumperz & S. C. Levinson (Eds.), *Rethinking linguistic relativity* (pp. 145–176). Cambridge, UK: Cambridge University Press.
Bowerman, M., & Choi, S. (2001). Shaping meanings for language: Universal and language-specific in the acquisition of spatial semantic categories. In M. Bowerman & S. C. Levinson (Eds.), *Language acquisition and conceptual development* (pp. 475–511). Cambridge, UK: Cambridge University Press.
Bybee, J. L., & Slobin, D. I. (1982). Why small children cannot change language on their own: Suggestions from the English past tense. In A. Ahlqvist (Ed.), *Papers from the 5th International Conference on Historical Linguistics* (pp. 29–37). Amsterdam/Philadelphia: John Benjamins.

Choi, S. (1997). Language-specific input and early semantic development: Evidence from children learning Korean. In D. I. Slobin (Ed.), *The crosslinguistic study of language acquisition. Vol. 5: Expanding the contexts* (pp. 111–133). Mahwah, NJ: Lawrence Erlbaum Associates.

Choi, S., & Bowerman, M. (1991). Learning to express motion events in English and Korean: The influence of language-specific lexicalization patterns. *Cognition, 41*, 83–121.

Choi, S., McDonough, L., Bowerman, M., & Mandler, J. (1999). Early sensitivity to language-specific spatial categories in English and Korean. *Cognitive Development, 14*, 241–268.

Deacon, T. (1997). *The symbolic species.* New York: Norton.

DeGraff, M. (1999). *Language creation and language change: Creolization, diachrony, and development.* Cambridge, MA: MIT Press.

De Houwer, A. (1990). *The acquisition of two languages from birth: A case study.* Cambridge, UK: Cambridge University Press.

De Houwer, A. (1995). Bilingual language acquisition. In P. Fletcher & B. MacWhinney (Eds.), *The handbook of child language* (pp. 219–250). Oxford: Blackwell.

Deuchar, M. (1992). Can government and binding theory account for language acquisition? In C. Vide (Ed.), *Lenguajes naturales y lenguajes formales VIII* [Natural languages and formal languages] (pp. 273–279). Barcelona: Universitat de Barcelona.

Ekmekçi, Ö. (1979). *Acquisition of Turkish: A longitudinal study on the early language development of a Turkish child.* Unpublished doctoral dissertation, University of Texas, Austin.

Ekmekçi, Ö. (1986). The significance of word order in the acquisition of Turkish. In D. I. Slobin & K. Zimmer (Eds.), *Studies in Turkish linguistics* (pp. 265–272). Amsterdam/Philadelphia: John Benjamins.

Foley, W. (1988). Language birth: The process of pidginization and creolization. In F. Newmeyer (Ed.), *Linguistics: The Cambridge Survey, Vol. 1: Language: The sociocultural context* (pp. 162–183). Cambridge, UK: Cambridge University Press.

Fortescue, M., & Lennert Olsen, L. (1992). The acquisition of West-Greenlandic. In D. I. Slobin (Ed.), *The crosslinguistic study of language acquisition: Vol. 3* (pp. 111–219). Hillsdale, NJ: Lawrence Erlbaum Associates.

Givón, T. (1998). On the co-evolution of language, mind and brain. *Evolution of Communication, 2*, 45–116.

Givón, T. (2002). The visual information-processing system as an evolutionary precursor of human language. In T. Givón & B. F. Malle (Eds.), *The evolution of language out of pre-language* (pp. 3–50). Amsterdam/Philadelphia: John Benjamins.

Goldin-Meadow, S. (2003). *The resilience of language: What gesture creation in deaf children can tell us about how all children learn language.* New York/Hove: Psychology Press.

Goldin-Meadow, S., Butcher, C., Mylander, C., & Dodge, M. (1994). Nouns and verbs in a self-styled gesture system: What's in a name? *Cognitive Psychology, 27,* 259–319.

Goldin-Meadow, S., & Mylander, C. (1984). Gestural communication in deaf children: The effects and non-effects of parental input on early language development. *Monographs of the Society for Research in Child Development, 49,* 1–121.

Goldin-Meadow, S., & Mylander, C. (1990). Beyond the input given: The child's role in the acquisition of language. *Language, 66,* 323–355.

Goldin-Meadow, S., Mylander, C., & Butcher, C. (1995). The resilience of combinatorial structure at the word level: Morphology in self-styled gesture systems. *Cognition, 56,* 195–262.

Gould, S. J. (1977). *Ontogeny and phylogeny.* Cambridge, MA: Harvard University Press.

Guillaume, P. (1973). First stages of sentence formation in children's speech. In C. A. Ferguson & D. I. Slobin (Eds.), *Studies of child language development* (pp. 522–541). New York: Holt, Rinehart & Winston. [translation of *"Les débuts de la frase dans le langage de l'enfant." Journal de Psychologie, 24,* 1–25] (Original work published 1927)

Gvozdanović, J. (Ed.). (1997). *Language change and functional explanations.* Berlin/New York: Mouton de Gruyter.

Haiman, J. (1994). Ritualization and the development of language. In W. Pagliuca (Ed.), *Perspectives on grammaticalization* (pp. 3–28). Amsterdam/Philadelphia: John Benjamins.

Hopper, P. J., & Traugott, E. (1993). *Grammaticalization.* Cambridge: Cambridge University Press.

Hudson, C. L. (2002). *Pidgins, creoles, and learners: How children and adults contribute to the formation of creole languages.* Unpublished doctoral dissertation, University of Rochester, New York.

Jackendoff, R. (1987). *Consciousness and the computational mind.* Cambridge, MA: MIT Press.

Jespersen, O. (1921/1964). *Language: Its nature, development and origin.* New York: Norton.

Keesing, R. M. (1991). Substrates, calquing and grammaticalization in Melanesian Pidgin. In E. C. Traugott & B. Heine (Eds.), *Approaches to grammaticalization. Vol. 1: Focus on theoretical and methodological issues* (pp. 315–342). Amsterdam/Philadelphia: John Benjamins.

Kegl, J., & McWhorter, J. (1997). Perspectives on an emerging language. In E. V. Clark (Ed.), *The Proceedings of the Twenty-eighth Annual Child Research Fo-*

rum (pp. 15–38). Stanford, CA: Center for the Study of Language and Information.

Küntay, A., & Slobin, D. I. (1999). The acquisition of Turkish as a native language. A research review. *Turkic Languages, 3,* 151–188.

Langer, J. (2000). The descent of cognitive development. *Developmental Science, 3,* 361–388.

Langer, J. (2001). The mosaic evolution of cognitive and linguistic ontogeny. In M. Bowerman & S. C. Levinson (Eds.), *Language acquisition and conceptual development* (pp. 19–44). Cambridge, UK: Cambridge University Press.

Lieven, E., Pine, J. M., & Baldwin, G. (1997). Lexically-based learning and the development of grammar in early multi-word speech. *Journal of Child Language, 24,* 187–219.

Lightfoot, D. (1988). Syntactic change. In F. J. Newmeyer (Ed.), *Linguistics: The Cambridge Survey I. Linguistic theory: Foundations* (pp. 303–323). Cambridge, UK: Cambridge University Press.

McWhorter, J. (1995). Sisters under the skin: A case for the genetic relationship between the Atlantic English-based creoles. *Journal of Pidgin and Creole Languages, 10,* 289–333.

McWhorter, J. (1997). *Towards a new model of creole genesis.* New York: Peter Lang.

Meisel, J. M. (1989). Early differentiation of languages in bilingual children. In K. Hyltenstam & L. Obler (Eds.), *Bilingualism across the lifespan: Aspects of acquisition, maturity and loss* (pp. 13–40). Cambridge, UK: Cambridge University Press.

Morford, J. P. (2002). Why does exposure to language matter? In T. Givón & B. Malle (Eds.), *The evolution of language out of pre-language* (pp. 329–342). Amsterdam/Philadelphia: John Benjamins.

Morford, J. P., & Kegl, J. (2000). Gestural precursors of linguistic constructs: How input shapes the form of language. In D. McNeill (Ed.), *Language and gesture* (pp. 358–387). Cambridge, UK: Cambridge University Press.

Muysken, P. (1988). Are creoles a special type of language? In F. Newmeyer (Ed.), *Linguistics: The Cambridge Survey II. Linguistic theory: Extensions and implications* (pp. 285–301). Cambridge, UK: Cambridge University Press.

Piaget, J. (1971). *Biology and knowledge: An essay on the relations between organic regulations and cognitive processes.* Chicago: University of Chicago Press. [Translation of (1967) *Biologie et connaissance: Essai sur les relations entre les régulations organiques et les processus cognitifs.* Paris: Gallimard.]

Pinker, S. (1999, October 24). A linguistic big bang. [http:www.indiana.edu/~langacq/E105/Nicaragua.html]

Romaine, S. (1984). *The language of children and adolescents: The acquisition of communicative competence.* Oxford: Blackwell.

Romaine, S. (1988). *Pidgin and creole languages.* London: Longman.

Sankoff, G. (1979). The genesis of a language. In K. C. Hill (Ed.), *The genesis of language* (pp. 23–47). Ann Arbor, MI: Karoma.

Sankoff, G., & Laberge, S. (1973). On the acquisition of native speakers by a language. *Kivung, 6*(1), 32–47.

Savage-Rumbaugh, S., Shankar, S. G., & Tylor, T. J. (1998). *Apes, language, and the human mind.* Oxford: Oxford University Press.

Savage-Rumbaugh, S., & Taglialatela, J. (2001, May). *Language, apes, and understanding speech.* Paper presented at conference, "The Rise of Language out of Pre-Language," University of Oregon, Eugene.

Senghas, A. (1995). *Children's contribution to the birth of Nicaraguan Sign Language.* Unpublished doctoral dissertation, MIT.

Senghas, A., & Coppola, M. (2001). Children creating language: How Nicaraguan Sign Language acquired a special grammar. *Psychological Science, 12,* 323–328.

Slobin, D. I. (1973). Cognitive prerequisites for the development of grammar. In C. A. Ferguson & D. I. Slobin (Eds.), *Studies of child language development* (pp. 175–208). New York: Holt, Rinehart & Winston.

Slobin, D. I. (1985). Crosslinguistic evidence for the language-making capacity. In D. I. Slobin (Ed.), *The crosslinguistic study of language acquisition. Vol. 2: Theoretical issues* (pp. 1157–1256). Hillsdale, NJ: Lawrence Erlbaum Associates.

Slobin, D. I. (1994). Talking perfectly: Discourse origins of the present perfect. In W. Pagliuca (Ed.), *Perspectives on grammaticalization* (pp. 119–133). Amsterdam/Philadelphia: John Benjamins.

Slobin, D. I. (1997). The origins of grammaticizable notions: Beyond the individual mind. In D. I. Slobin (Ed.), *The crosslinguistic study of language acquisition. Vol. 5: Expanding the contexts* (pp. 265–323). Mahwah, NJ: Lawrence Erlbaum Associates.

Slobin, D. I. (2002). Language evolution, acquisition, diachrony: Probing the parallels. In T. Givón & B. Malle (Eds.), *The evolution of language from prelanguage* (pp. 375–392). Amsterdam/Philadelphia: John Benjamins.

Slobin, D. I., & Bever, T. G. (1982). Children use canonical sentence schemas: A crosslinguistic study of word order and inflections. *Cognition, 12,* 229–265.

Thomason, S., & Kaufman, T. (1988). *Language contact, creolization, and genetic linguistics.* Berkeley: University of California Press.

Tomasello, M. (1992). *First verbs: A case study in early grammatical development.* Cambridge, UK: Cambridge University Press.

Tomasello, M. (1999). *The cultural origins of human cognition.* Cambridge, MA: Harvard University Press.

9

The Emergence of Nicaraguan Sign Language: Questions of Development, Acquisition, and Evolution

Richard J. Senghas
Sonoma State University, Rohnert Park, California

Ann Senghas
Barnard College of Columbia University

Jennie E. Pyers
University of California, Berkeley

The emergence of a new sign language in Nicaragua over the past 25 years provides an opportunity to examine the relationship between intercohort contact and individual development in their link to historical language change. This chapter examines these forces in contemporary circumstances as they set a new language in motion. In Nicaragua, we have observed that a new sign language emerged only after a potential speech community (Gumperz, 1968) of older and younger members was brought together. The resulting development of the new language suggests that the interactions across age cohorts are crucial in language emergence. We propose, then, that language genesis requires at least two age cohorts of a community in sequence, the first providing the shared symbolic environment upon which the later cohorts can build. It requires the capacities of both children and adults to create a viable new language.

In this chapter, we consider specific changes in linguistic patterns of Nicaraguan Sign Language (NSL).[1] We identify which members of the

---

[1]Natural sign languages are used throughout the world by communities of deaf people. These languages (e.g., American Sign Language, British Sign Language, Swedish Sign Language) have their own complex grammars, just as do spoken languages, and natural sign languages should not be seen as codes of spoken languages. (Some artificial sign languages

speech community employ specific linguistic forms, and under what conditions. These findings suggest the necessary factors that individuals bring to this process: cultural and social (e.g., age cohort or generation), and developmental (e.g., ontogenetic stage). These factors have an effect on the lexicon, the syntax, and the use of this new language.

Rather than ask which of these factors created NSL, we ask how all these factors interact to create a language. We find that linguistic innovation involves not only the creation of new linguistic forms, but also the selection (i.e., the continued, regular use) of the constructions in which they appear. The case of the recent emergence of Nicaraguan Sign Language highlights the complex interrelationships between culture and individuals, and their respective development, and convinces us that theories of language change must be informed by both sociocultural and psychological principles. During the creation of a language, the sociocultural impact of any given individual changes with age, with individuals having their greatest effect on their environment after entering adolescence. The influence of psychological capacities also changes with age, as certain language-learning capacities are available only in childhood. The nature of the linguistic changes that result from intercohort (or intergenerational) contact, therefore, depends on the age of the individuals involved, as age determines the type of influence they exert in each of the sociological and linguistic domains.

Language is an inherently social phenomenon, and must be studied as part of larger, sociocultural systems (Duranti, 2001; Gumperz, 1968; Hymes, 1972). The perpetuation of linguistic changes is dependent on the suitability of those changes to the sociocultural environment and the learning capacities of individuals. Circumstances will favor some changes over others, and thus selection will occur (Mufwene, 2001). However, although the general process of natural selection is common to both sociocultural and linguistic domains, the mechanisms of information change, transmission and selection differ. Crucially, the effects that individuals have on their environment, and the subsequent effects of that changed environment, depend on individuals' cohort (or generation) and age. To accurately account for language emergence and change, we must examine both individual- and group-level phenomena, in both sociocultural and linguistic domains.

[e.g. Signing Exact English; Gustason, Pfetzing, & Zawolkow, 1980] have been invented to aid the acquisition of spoken/written languages.) See Senghas and Monaghan (2002) for a brief summary of these distinctions.

Keeping these domains in mind, we begin with an account of the sociocultural circumstances of the emergence of NSL. We identify distinct periods within the historical development of this language, and discuss the qualitative differences between them. Next, we discuss linguistic variation among deaf Nicaraguan signers. In doing so, we consider various characteristics associated with each of the historical periods described in this chapter. We also examine developmental factors within individual signers. In combining these factors, we demonstrate how environmental and ontogenetic factors interact. More specifically, we conclude that the linguistic environments surrounding Nicaraguan signers, combined with changing first-language acquisition capabilities of individuals, explain the linguistic changes observed. With this approach, this case uncovers important principles that apply to all cases of language emergence and change.

THE SOCIOCULTURAL HISTORY OF NSL

Let us place the individuals involved in the emergence of NSL within their recent and current historical settings. By doing so, we address the ways they might have regulated or directed changes in their communicative circumstances. The sociocultural environment that surrounded deaf Nicaraguans in the period prior to the emergence of a sign language differed in crucial ways from the environment present during the beginning and later phases of the emergence of NSL.

Establishing a Durable Speech Community

Until relatively recently, deaf individuals in Nicaragua had minimal contact with other deaf people. Ethnographic fieldwork (R. J. Senghas, 1997, 2003; Polich, 1998) and archival research (Polich, 1998) indicate that prior to the 1970s, there was no Deaf[2] community in Nicaragua, nor any established sign language. Although Deaf communities have existed in many parts of the world since as far back as the 19th century (cf. Erting,

---

[2]Consistent with other literature on deafness and Deaf communities, *Deaf* in this chapter is written with an upper-case *D* to signify cultural Deafness, that is, membership within a self-identified Deaf community, generally one that uses a sign language as its primary language. The term *deaf* written with a lowercase *d* refers to hearing loss, without necessarily denoting membership in a cultural and linguistic Deaf community. (See R. J. Senghas & Monaghan, 2002, for a discussion of these distinctions.)

Johnson, Smith, & Snyder, 1994; Monaghan, Schmaling, Nakamura, & Turner, 2003; Plann, 1997), circumstances in Nicaragua apparently prevented any such communities from forming. With no special schools available to deaf students until at least 1946, and no widely accessible special education available until 1977, a critical factor for the formation of a Deaf community was absent (cf. Schein, 1989).[3]

Typically, deaf individuals are the only deaf members in their immediate families, and usually they are the only deaf members of their extended families (Schein, 1989). In such situations, deaf individuals often develop *homesign* systems, that is, idiosyncratic and rudimentary gestural systems used to communicate within the family (Goldin-Meadow & Mylander, 1984, 1998; cf. Morford, 1996, for a review of homesign studies). For those few deaf Nicaraguans who had access to tutoring or special education clinics, the methods were *oralist*, that is, the emphasis was on speaking and understanding spoken Spanish, occasionally supplemented with a few signs to support Spanish acquisition. In other words, signing and gesture were discouraged, whereas spoken and written forms of language were encouraged.

In 1946, the government established the first special education school in Managua, Nicaragua's capital city. Initially, ten deaf and hard-of-hearing students enrolled in a program that covered primary and elementary grades. By the early 1970s, enrollments rose to approximately 50 (Polich, 1998). The school's pedagogy was oralist, and students with residual hearing sometimes practiced articulation with the aid of microphones and headphones (R. J. Senghas, 1997). Alumni of the first governmental schools report today that they did not socialize with one another outside of school, and they lost touch with one another once they no longer attended.

In 1977, the Nicaraguan government established a larger special education center in Barrio San Judas, Managua. This school's deaf program covered preschool through grade 6, and used oralist pedagogy. Initially, approximately 25 deaf and hard-of-hearing students were enrolled, within just a few years rising to over 100 (Polich, 1998). In 1980, the then new Sandinista government established a special education vocational school for adolescents. Many of the graduates from the elementary school entered this vocational program, where they attended classes in

---

[3]Schein's (1989) theory on the formation of Deaf communities includes issues of absolute size of the deaf population (or "critical mass"), relative population size, issues of inclusion and exclusion in the larger society, and the roles of schools and education.

carpentry, hairdressing, tailoring, and other vocations (R. J. Senghas, 1997). The students of the vocational school often rode public buses to the school, and having gained familiarity with public transportation, began to date and socialize with each other outside of school hours (R. J. Senghas, 1997, 2003; R. J. Senghas & Kegl, 1994). They would meet at ice cream shops, at each other's houses, or go to markets and public places together. In 1986, a hearing teacher, with assistance from other hearing adults, established a club in Managua providing assistance and opportunities for social interaction to deaf adolescents and adults. By 1990, this club had become a national association for Deaf Nicaraguans, and was directed by the Deaf members themselves.

Interviews with older deaf people today suggest that initially there was often confusion about what signs referred to which referent. As in other such cases, as the lexicon became more and more conventionalized within this emerging linguistic community (Gumperz, 1962), certain linguistic forms were left behind, for reasons varying from efficiency and ease of production to the charismatic nature of a particular signer.

Not surprisingly, in this community of adolescents and adults, signers occasionally disagreed over the appropriate use and meanings of particular signs. As a result, many of the Deaf Nicaraguans (and their teachers and parents) soon felt that a dictionary would be a useful tool to standardize the lexicon. After one earlier effort at compiling a dictionary in the late 1980s, the Deaf association in Managua launched a more concerted project with considerable support from the Swedish Federation of the Deaf (SDR). This dictionary was published in 1997 (ANSNIC [*Asociación Nacional de Sordos de Nicaragua*], 1997). The development of this dictionary coincided with instruction in NSL and written Spanish, also offered by the Deaf association. Throughout the dictionary project, most explicit discussion of the sign language by Deaf Nicaraguans focused exclusively on lexical signs and their meanings.

In an effort to make this dictionary an authentic *Nicaraguan* dictionary, the project team put a good deal of effort into excluding signs borrowed from other languages, despite the fact that these signs were commonly used in Managua. The explicit effort to avoid borrowed signs confirms that borrowing was indeed already occurring as a consequence of contact with signing visitors from other countries. Despite these efforts, some borrowed signs do appear in the dictionary. One example is the sign for *association*, often used to refer to the Deaf association. This sign, apparently borrowed directly from American Sign Language (ASL), did make it into the NSL dictionary (ANSNIC, 1997, p. 48; cf., for comparison, the ASL sign for *association* that appears in Sternberg, 1987, p. 24).

Evidence indicates that borrowing went beyond appropriation of particular signs, to incorporation of lexical principles. For example, initialized signs (signs that incorporate a handshape representing the first letter of a corresponding Spanish word) began to appear in patterns analogous to those found in other sign languages. Accordingly, the sign for *clean* adopted an L handshape, representing the first letter of the corresponding Spanish word *limpiar* (ANSNIC, 1997, p. 162). In this case, what is borrowed is a convention for generating and modifying lexical items, rather than lexical items themselves. Adults also develop new lexical conventions that spread throughout the lexicon, for example: the BUENO handshape in BIEN, BONITA, and SEGURO.

Increasingly through the 1990s and into the current decade, the Deaf association, with the support of SDR, has been advocating the training of sign language interpreters, as well as Deaf teachers' aides. The presence of Deaf adults as assistants in the classrooms not only allows for the course content presented by hearing teachers to be made much more accessible to deaf students, but, more importantly, also provides fluent signers as linguistic models for the younger students who are still in their early stages of first-language acquisition.

Thus, once adolescents were provided the opportunity to socialize and interact with each other, a Deaf community was formed. The durable nature of this community, unlike the earlier situations when deaf individuals were isolated from each other, provided a potential speech community in which a language could emerge. This Deaf community actively supported its members, created a dictionary of standardized signs, and ensured that later cohorts of deaf children would receive what they had not—early exposure to sign language in the classroom and other social fora.

Periods in NSL History

The emergence of NSL can be divided into three distinct periods:

1. Pre-Emergence Period (up to mid-20th century)—pre-contact
   a. Prior to interaction among deaf individuals in Nicaragua;
   b. Use of isolated homesign systems in families with deaf members;
   c. Earliest (small) schools with oralist programs for deaf students.

2. Initial Contact Period (~1977 through mid-1980s)—contact and consolidation

   a. Establishment of a larger program in San Judas;

   b. Establishment of vocational program for adolescents.

3. Sustained Contact Period (mid-1980s to present)—beginnings of an established linguistic community

   a. Establishment of a Deaf association;

   b. Control and direction of Deaf association assumed by Deaf members;

   c. Dictionary projects;

   d. Deaf individuals as linguistic models in schools.

Each of these historical periods has distinctive qualities. The transitions between these periods correlate with the changes in the types of influence the deaf individuals and groups have on their environments. Increasingly, over time, deaf individuals influence the structure of their social organization, and as a result the cultural forms produced (including language) increasingly bear their mark.

The first (Pre-Emergence) period has no established beginning date, and covers the period when deaf individuals in Nicaragua were not in contact with other deaf individuals. As mentioned, deaf individuals in Nicaragua at this time only rarely interacted with other deaf individuals. The communicative patterns of deaf individuals of this period would have been highly idiosyncratic (Coppola, 2002). With no linguistic community of deaf signers, no conventionalized sign language could develop or be maintained.

The social situations of deaf individuals of the Pre-Emergence period were structured by hearing people, primarily family members. Polich (1998) proposed the concept of the *Eternal Child* to characterize the type of dependent status of deaf individuals of this era. Even in the earliest deaf education programs, students did not interact with one another outside of school, instead returning to their homes once classes had ended.

The date for transition from the Pre-Emergence period to the Contact period could be reasonably assigned to several candidate dates. The scale of the San Judas program, and the fact that its deaf students continued to interact with one another after leaving this school, suggest that the establishment of the 1977 San Judas program is the most significant historical event, and is therefore our preferred choice for marking the transition from the Pre-Emergence period to the Initial Contact period. In any case, the period of deaf Nicaraguans' isolation from one another ends with students interacting in special education schools that continue to bring them together even into adolescence.

The Initial Contact period is characterized by the creation—primarily by hearing people—of circumstances that enabled deaf individuals to interact socially, providing an opportunity for the homesigns that each brought into these situations to be shared and modified. At this time, signers began to converge on a common lexicon and develop common linguistic structure. Already a new language was being born. Signed conversations from this period were characterized by frequent redundant phrases for clarification of reference. By the 1990s, especially among the younger signers, the frequency of such redundancy had noticeably diminished. The Initial Contact period was relatively short, less than a decade, which ended with a linguistic community of adolescent and adult signers supplying a progressively richer linguistic environment to younger members—an environment markedly different from that of the Pre-Emergence period.

The Sustained Contact period is distinguished by the conscious choices by deaf individuals to form enduring formal and informal relationships, including the establishment and control of the Managua-based Deaf association. These enduring relationships include friendships, participation in the Deaf association, marriage and domestic partnerships, sometimes despite opposing pressure from individuals, families or institutions to do otherwise. In the Sustained Contact period, which continues to the present day, there is frequent contact not only among Deaf adults, but also between Deaf adults and children. Deaf adolescents and adults repeatedly return to the school to participate in events involving young Deaf children. They attend school promotion and graduation ceremonies at the end of the academic year. The larger Managuan Deaf community often comes together for the fiestas and social gatherings at the Deaf association. The association's center offers more than simply a place for socializing; Deaf adolescents and adults also attend seminars there on subjects ranging from elementary Spanish to vocational training, usually offered by other Deaf adults. Deaf adults can be trained at the center to become teachers' aides in the deaf classrooms at the special education schools.

With this sustained Deaf community contact, there has also arisen a political consciousness about the rights and powers of Deaf people. As members of the Deaf association, some individuals have become involved in local and even national-level politics, often lobbying for the rights of deaf individuals or working for recognition of Deaf people and their sign language. Consider the ideological effects of an "official" dictionary, and the sign language seminars offered by the *de facto* national Deaf association. Recognition of NSL as a valid and effective language has reinforced the Ministry of Education's efforts to use sign language as a medium of instruction in programs for deaf students, with significant implications (R. J. Senghas, 1997).

While the dictionary and seminar projects have provided stabilizing influences on the language, they have also motivated the creation of alternate signs as part of oppositional positioning or regional identification by signers in the linguistic community. Such positioning might involve linguistic forms to assert or deny social identities and roles (cf. Schieffelin, Woolard, & Kroskrity, 1998). R. J. Senghas (1997) identifies one such event observed in an outlying town in 1993, where two deaf individuals argued about which of two signs was the "correct" one to use, a local form or the Managuan one. In times of linguistic doubt, signers can now consult sources of authority such as dictionaries and community leaders.

As we have indicated, in the beginning, hearing Nicaraguans structured the social and cultural environments of individual deaf Nicaraguans. Isolated deaf Nicaraguans responded socially and linguistically by developing homesign systems. Even in the first small oralist schools, signing remained limited and the social opportunities did not extend beyond the school grounds or school hours. Later, hearing people set up new social circumstances by establishing special education schools that brought many deaf Nicaraguan children together. Again, the deaf children responded socially and linguistically—but in these circumstances, deaf Nicaraguans began to have significant effects on the sociocultural environment of their deaf peers. They provided one another with a richer linguistic environment—one that included shared signing. Finally, with the addition of adolescents and adults to their community, Deaf Nicaraguans had significantly increasing influence over their sociocultural and linguistic environments. They could now structure intercohort social situations in which Deaf signers figure prominently, thereby providing models of sociocultural behavior, especially language use. The conditions were now in place for language emergence, change, and perpetuation.

EFFECTS OF ONTOGENETIC DEVELOPMENT ON NSL: CHILDREN'S MINDS MATTER

As already described, individuals act on their environment in sociocultural and linguistic ways, affecting themselves and the other members of their community from that day forward. However, another process is simultaneously taking place over time—the individuals themselves are getting older. As people mature, both sides of the interaction are affected; the way individuals affect the environment changes, and the way they are affected by the environment changes. Thus, to understand how language emergence and change take place, we must factor ontogenetic development into the interaction between individuals and their changing language environment.

The actions of each member of the Nicaraguan Deaf community alter the environment for signers of all ages. However, the types of activities in which individuals participate change over the course of their lifetime. It is worth examining empirically how children differ from adolescents and adults in the nature of their effects on the language they are learning. We have observed that adolescents and adults actively form social communities in which language can emerge, consciously add vocabulary to the language, and aggressively ensure that their language is passed down to younger children. We now consider effects that change the internal structure of a language, for example, when individuals apply a form to a function different from that observed, fail to adopt a form, or introduce a novel construction. In the section that follows, we identify one such measurable change, and examine how, as individuals mature, their effect on the language takes on a different nature.

We also consider how a given language environment differentially affects individuals of different ages. It has been found that the age at which learners are first exposed to a language determines their eventual linguistic abilities, with those who start younger achieving greater proficiency (Lenneberg, 1967; Newport, 1990). For example, adults who moved to the United States from Korea during early childhood have a better command of English than those who moved here in adolescence or adulthood (Johnson & Newport, 1989). Similarly, Deaf adults in the United States who entered the signing community in early childhood have a better command of ASL than those who entered in adolescence or adulthood (Mayberry & Eichen, 1991; Newport, 1990). Evidently, as learners age, it becomes more difficult to learn language natively, whether signed or spoken.

Some parts of a language will be easier to master in adulthood than others. For example, among the native Korean speakers who had learned English, all had acquired a large vocabulary, but only those who were exposed as children had mastered the complicated use of articles like *a* and *the* (Johnson & Newport, 1989).

In the case of NSL, this familiar effect of age (ontogenetic development) interacted in an unusual way with the effect of language change (historical development) over the course of the 1980s. With each passing year, individual proficiency at language learning declined, decreasing each learner's potential. At the same time, with each passing year, the ambient language became progressively richer, increasing each learner's potential. These simulatneous, opposing forces make it tricky (but not impossible) to differentiate the effects of the language on its learners from the effects of learners on their language.

We can tease apart the interaction by comparing the grammars of learners exposed to the language in different years and at different ages. Consider that all signers retain outcomes of earlier periods of their own development. Adults remember what they learned as children. For this reason, in the language of many adults, we find constructions that can be learned only in childhood, such as the native use of *a* and *the*, or the pronunciation of the English /r/ sound. The fact that they can use these constructions as adults reveals that their childhood environment included them. The fact that others (including many Korean immigrants to the United States) cannot use these constructions as adults reveals that their childhood environment did not include them, and that these particular constructions are difficult to learn in adulthood.

Of course, any constructions that are easy to learn in adulthood will be present in the language of all adults, regardless of age of exposure. For example, we can all use words we acquired only as adults, such as many of the words that appear on this page. For this reason, constructions that are learned easily by adults are not useful tools for determining the content or richness of an individual's childhood language environment.

The constructions useful for illuminating a learner's childhood linguistic environment will be those that are not easily learned in adolescence and adulthood. Such constructions will be present only in the language of those who were exposed to them as children. If an element is missing from an individual's version of NSL, we can conclude that it emerged after that individual had already reached adolescence. Conversely, the set of such constructions present in the language of each individual represents the total contributions of that person's age cohort and its predecessors. Constructions are distributed across cohorts today like rings on a tree, enabling us to date when each one entered the language.

Following this logic, we have examined the emergence of *spatial coreference* in the grammar of Nicaraguan signing (A. Senghas & Coppola, 2001). Most signs can be produced in a neutral location in front of the signer's body. However, a signer can choose to spatially modify a sign, producing it with a movement toward or away from a particular location.

These modifications, or *spatial modulations*, can serve various grammatical functions. In NSL (as in many other sign languages) they are often used for co-reference; that is, to indicate that several signs are associated with a common referent. Figure 9.1 presents the verbs *see* and *pay* in their neutral form and spatially modulated. In the spatially modulated versions, the signs' shared spatial modulation would indicate their link to a single person who was both seen and paid.

In this analysis, we identified spatial modulations in videotaped narratives elicited from Deaf Nicaraguan signers. We then coded how often utterances that referred to the same referent used the same spatial modulation. Although a common spatial modulation on two different signs will sometimes occur by chance, signers who frequently use common spatial modulations in cases of co-reference are more likely to be using them to indicate co-reference grammatically.

In order to examine the effects of the changing language environment, subjects were divided into two groups, or cohorts, based on their initial year of exposure: the *first cohort* entered the community between 1978 and 1983, the *second cohort* entered between 1984 and 1990. To examine the effects of the age of individual learners, subjects were further divided into three groups based on the age at the time of exposure: *early-exposed* (birth to 6;6), *middle-exposed* (6;6 to 10), and *late-exposed* (after age 10). The proportion of co-referential spatial modulations per verb for each group was determined, and is presented in Fig. 9.2.

Comparing the third pair of columns with the other two reveals an effect of age: late-exposed signers of both cohorts are equally (un)likely to produce co-referential spatial modulations. Evidently, spatial co-reference is not as easily mastered once one is older than 10, and late-learners of both cohorts were already past that age when they were first exposed to NSL. We take this low frequency of common modulations to be our best approximation of how often spatial modulations will co-occur by chance, or to what degree they might be learnable after early childhood.

In contrast, for the early- and middle-exposed signers, the year of exposure made a crucial difference; members of the second cohort produced spatially co-referent forms significantly more often than the first. As children, the second-cohort signers did not replicate the pattern of signing used by the older signers from whom they were learning. Instead, the second-cohort signers were much more apt to produce common spatial modulations in contexts with potential co-reference. In this way, they were using the form with a systematic pattern that they had not observed in the signing of their first cohort models.

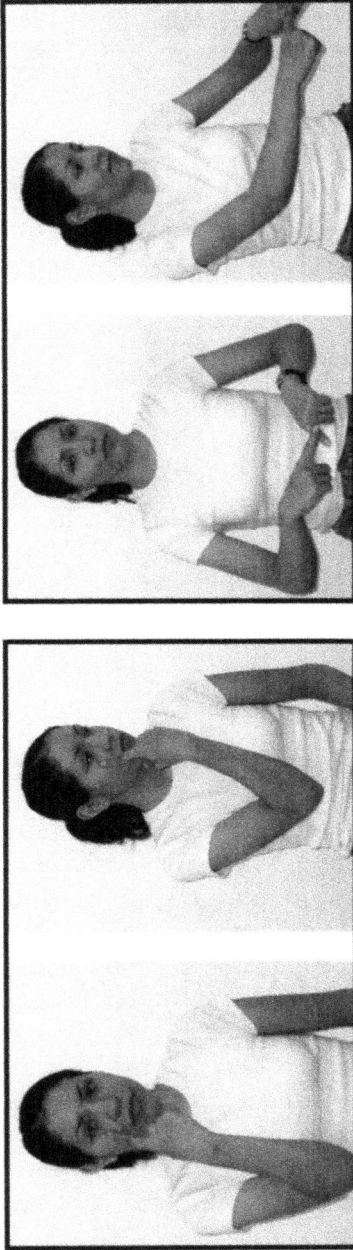

see   **pay**

FIG. 9.1.   The Nicaraguan signs *see* and *pay* produced in their neutral form, and spatially modulated to the signer's left.

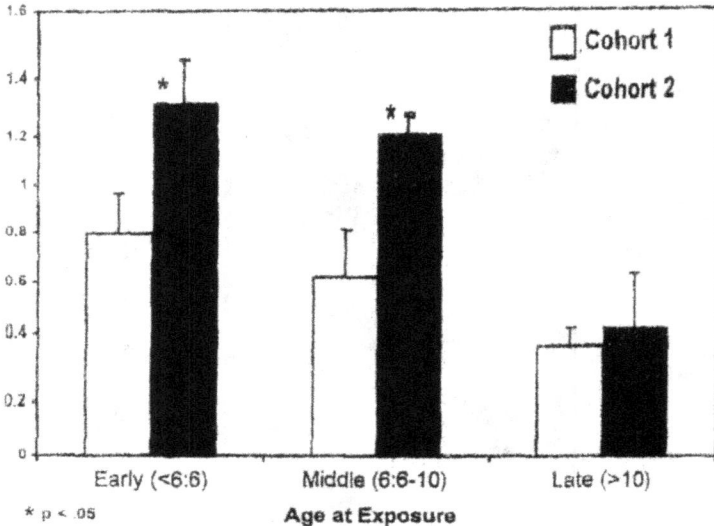

FIG. 9.2.   Spatial modulations in co-referential contexts produced per verb by early-, middle-, and late-exposed signers of the first and second cohort (from A. Senghas & Coppola, 2001).

It seemed likely that the spatial modulations produced by members of the first cohort, even if they were occasionally produced, were not ever being used to indicate co-reference. To test this, we conducted a comprehension study to determine how spatial modulations are interpreted (A. Senghas, 2000). Early-exposed signers from both cohorts watched video clips of signed sentences that included a spatially modulated sign (along with several unmodulated fillers) and indicated the meaning of each sentence by selecting from a set of pictures. The difference in usage is striking. *None* of the first-cohort signers constrained their choices based on the direction of the spatial modulation; all of the second-cohort signers did. Evidently, even though first-cohort signers occasionally produce spatially modulated forms, they do so without regard for potential co-reference. The young signers that were exposed to such utterances in the late 1980s nevertheless acquired a system that is systematic and rule-governed; accordingly, their usage is constrained in both production and comprehension.

Based on these and related analyses (A. Senghas, 1995; A. Senghas, Coppolla, Newport, & Supalla, 1997), we conclude that the present-day use of spatial modulations to indicate co-reference was developed over the course of the 1980s by sequential cohorts of child learners. Some form of *spatial modulation*, that is, modifying signs with respect to specific locations, was probably already present in the homesign systems developed by some of the children with their families before they entered school (Coppola, 2002). In the early 1980s, children of the first cohort began producing these modulations more frequently. Then, crucially, the children of the late 1980s imposed a constraint on this device. They restricted the side toward which a sign was produced in order to indicate coreference or agreement; that is, signs produced in a common location now unambiguously indicated a common referent.

At this point, the construction could be used to link a verb to its arguments, a noun to its modifiers. Now a common spatial modulation could be used to mean that a single person was both seen and paid. Because this constraint arose among the children of the late 1980s, who are today's adolescents, it can be observed in their language still, and in the signing of today's children, but not in the language of those who were already adolescents in the late 1980s, that is, today's adults.

Note that the particular innovation contributed by the second-cohort children was not the act of signing in space; it was the constraint on how space could be used. This innovation limits not only the way a set of signs can be produced; it limits what the set of utterances can mean, and in this way it makes the grammar more specific. For example, consider the sentence in which *see* and *pay* are both produced to the left. To a first cohort signer, the sentence could mean that one person was seen and another paid, or that a single person was both seen and paid. To early-exposed second-cohort signers, the first reading is not only unlikely—it is ungrammatical, even though such sentences must have been present in their environment when they were children.

ONTOGENETIC AND HISTORICAL TIME FRAMES MEET (CASCADING NICHE CONSTRUCTION)

Every cohort at every age has played an indispensable role in the emergence of NSL. Considering the community's history, together with the data on spatial modulations, it is clear that no single cohort "invented" NSL. We therefore do not propose a scenario in which the first cohort's language was agrammatical and the second cohort "innovated" a grammar.[4] We propose instead that the grammar of NSL has been developing from the Initial Contact Period onward, and every cohort since that time

---

[4]We do not find evidence to support a single-cohort view, although such a view is occasionally implied in others' discussion of our research (e.g., Slobin, chap. 8, this volume).

played a crucial role in this development. Each cohort, in turn, enriched the grammar of the language while they were children, during a period of early sensitivity to language structure. As they entered adolescence, they continued to learn the language and add to their vocabularies, but stabilized on their use of grammatical constructions such as spatial modulations. At this point, they also began to create and maintain or modify the social structures that enabled them to pass their progress on to a new cohort of children. The newer children, surrounded by a now-changed social and linguistic environment, quickly picked up the language of the day, and continued to develop it where their older peers left off.

This account is supported by the following findings: (a) the sociocultural environment of deaf Nicaraguans changed dramatically in the late 1970s and the early 1980s, from essentially no contact, to extensive peer contact, to intercohort contact among members of a new community; and (b) the linguistic environment also changed during this period, becoming grammatically richer. A close examination of spatial modulations in particular indicates that a system of spatial co-reference emerged and was available in the language environment from the mid-1980s on.

Furthermore, both the increasing intercohort contact and the linguistic enrichment stemmed from the very community that then benefited from them. In this way, at the community level, deaf Nicaraguans are constructing a niche, a new, changed linguistic environment for themselves, a niche that then provides a shaping influence on the members of the community.

Let us momentarily shift our attention away from the individuals and their linguistic community, and toward the changes in the language itself. Historical language change can be viewed as the evolutionary development of a language. This perspective is adopted by Mufwene (2001), and provides a useful approach for understanding the emergence of NSL. The concept of *natural selection* as applied to linguistic behaviors is especially relevant, because it is not only the appearance of novel linguistic forms that is of concern, but also their retention (i.e., selection for regular continued use) that marks true historical change. Novel forms would be more likely to be retained by speakers if those forms are seen as more effective at communicating, whether through increased efficiency, precision, flexibility, or compatibility with either the cognitive capacities of the speakers or the structures of the linguistic system (i.e., the language). As elements of the language and, ultimately, the language itself change, the environments of the speakers change, including the environments of those children in the process of constructing their cognitive capacities.

This brings us to the epigenetic process of cognitive constructivism that Piaget describes whereby an individual child, in the course of developing cognitive capacities, changes its environment in ways that then, through feedback, transform the developing child in return, propelling development into more advanced stages (cf. Parker, chap. 1 and chap. 2, this volume). In this case, significant transformation also happens at the community level, as the changes to the environment derive directly from community interaction.

At the individual level, some of these adaptive and transformative abilities will not be direct or immediate as they interact in an important way with individual, ontogenetic development. Although the ability to creatively build up one's own language, and the ability to shape one's sociocultural environment are available to some degree throughout the lifespan, they are each especially prominent during a particular, limited period in ontogenetic development. Constructive linguistic abilities peak early in life; constructive social abilities peak later in life. As a result, the creative influence of an individual's childhood language abilities must await adolescence to exert their full effect. Only then, together with age peers, can the individual actively serve as a language model to a new, younger cohort that can benefit from the linguistic change. Thus, there will be a lag of five to ten years from when a new construction initially emerges to when it transforms the language environment of others. As a result, each age cohort transforms the environment of the subsequent age cohorts more than the environment of their own. What we have, then, is *niche construction* (Laland, Odling-Smee, & Feldman, 2000), but with a cascading, delayed impact.

Thus, across multiple cohorts, both adults and children play crucial roles in creating a language. NSL could emerge only when a cohort of adolescents and adults provided the social and linguistic environment from which it grew, and ensured the perpetuation of its signs and conventions. The grammatical elements to be perpetuated, however, depended on a complementary role that only children are equipped to play. Their capacity to acquire grammatical systematicity (even where it is absent in the environment) is essential for the initial appearance of linguistic structure.

CONCLUDING REMARKS: THE INTEGRATION
OF CULTURE AND BIOLOGY

We argue that the emergence of NSL has been an evolutionary process, subject to evolutionary principles, including selection. This is not to imply that the appearance of this new language represents a reenactment of the original emergence of language in human societies, as the appearance of the first language was situated in a vastly different sociocultural environment from that observed in this contemporary case.[5] Rather, it is the principle of selection, as it interacts with sociocultural and psychological development, that underlies both scenarios. Within the model of niche construction (Laland et al., 2000), selection is affected by environmental factors that themselves may be modified by the biological individuals subject to the selection. In this case, culture (language) meets ontogenetic development in a reciprocally changing, at times reinforcing, process.

Certain sociocultural and psychological conditions, brought together, can trigger the creation of a language, with all of its lexicon, grammar, and conventions of use. Since the late 1970s, the sociocultural influence of Deaf Nicaraguan adolescents and adults interacted with the language-receptive and language-creative mental abilities of preadolescent children to establish, systematize, and internalize the new grammar of Nicaraguan Sign Language.

Note that an individual's potential contribution, in both psychological and sociocultural domains, changes over the lifespan. Strong language-creating abilities emerge early in life, and decrease with age. Social self-determination emerges later in life, and thus, the ability to influence the environment of others increases with age. Ironically, this ability to provide fertile sociocultural conditions, which must occur first, develops later ontogenetically. For this reason, no single age cohort can progress through the developmental stages in the order necessary to create a language in a single pass. Consequently, language genesis requires at least two cohorts of the community in sequence, the first providing the shared symbolic environment that the second can exploit. Neither children, nor adults—independent of each other—can create a language. But a community in which both are available, interacting with each other and passing developments down as they age, can provide the fertile ground out of which language grows.

REFERENCES

ANSNIC [Asociación Nacional de Sordos de Nicaragua]. (1997). *Diccionario del Idioma de Señas de Nicaragua*. Managua, Nicaragua: Author.

---

[5]And as Deacon (1997) argues, the first emergence of language is likely to have arisen in ancestors with brains and minds organized differently than the modern configuration of *Homo sapiens* after the co-evolutionary development of the human mind and human language.

Coppola, M. (2002). *The emergence of grammatical categories in homesign: Evidence from family-based gesture systems in Nicaragua.* Unpublished doctoral dissertation, University of Rochester, New York.

Deacon, T. (1997). *The symbolic species.* New York: W. W. Norton.

Duranti, A. (2001). Linguistic anthropology: History, ideas, and issues. In A. Duranti (Ed.), *Linguistic anthropology: A reader* (pp. 1–38). Malden, MA: Blackwell.

Erting, C. J., Johnson, R. C., Smith, D. L., & Snider, B. D. (1994). *The Deaf way: Perspectives from the International Conference on Deaf Culture.* Washington, DC: Gallaudet University Press.

Goldin-Meadow, S., & Mylander, C. (1984). Gestural communication in deaf children: The effects and noneffects of parental input on early language development. *Monographs of the Society for Research in Child Development, 49*(3–4, Serial No. 207).

Goldin-Meadow, S., & Mylander, C. (1998). Spontaneous sign systems created by deaf children in two cultures. *Nature, 391,* 279–281.

Gumperz, J. J. (1962). Types of linguistic communities. *Anthropological Linguistics, 4*(1), 28–40.

Gumperz, J. J. (1968). The speech community. In *International Encyclopedia of the Social Sciences* (Vol. 1, pp. 381–386). New York: Macmillan.

Gustason, G., Pfetzing, D., & Zawolkow, E. (1980). *Signing exact English.* Los Alamitos, CA: Modern Signs.

Hymes, D. (1972). On communicative competence. In J. B. Pride & J. Holmes (Eds.), *Sociolinguistics* (pp. 269–293). Harmondsworth, England: Penguin.

Johnson, J., & Newport, E. (1989). Critical period effects in second language learning: The influence of maturational state on the acquisition of English as a second language. *Cognitive Psychology, 21*(1), 60–99.

Laland, K., Odling-Smee, J., & Feldman, M. (2000). Niche construction, biological evolution and cultural change. *Behavioral and Brain Sciences, 23*(1), 131–146.

Lenneberg, E. (1967). *Biological foundations of language.* New York: John Wiley & Sons.

Mayberry, R. I., & Eichen, E. B. (1991). The long-lasting advantage of learning sign language in childhood: Another look at the critical period for language acquisition. *Journal of Memory & Language, 30*(4), 486–512.

Monaghan, L., Schmaling, C., Nakamura, K., & Turner, G. (Eds.). (2003). *Many ways to be Deaf: International variation in Deaf communities.* Washington, DC: Gallaudet University Press.

Morford, J. (1996). Insights to language from the study of gesture: A review of research on the gestural communication of non-signing deaf people. *Language and Communication, 16*(2), 165–178.

Mufwene, S. S. (2001). *The ecology of language.* Cambridge, UK: Cambridge University Press.

Newport, E. L. (1990). Maturational constraints on language learning. *Cognitive Science, 14*(1), 11–48.

Plann, S. (1997). *A silent minority: Deaf education in Spain, 1550–1835.* Berkeley, CA: University of California Press.

Polich, L. G. (1998). *Social agency and deaf communities: A Nicaraguan case study.* Unpublished doctoral dissertation, University of Texas, Austin.

Schein, J. D. (1989). *At home among strangers.* Washington, DC: Gallaudet University Press.

Schieffelin, B. B., Woolard, K., & Kroskrity, P. (Eds.). (1998). *Language ideologies: Practice and theory.* Oxford: Oxford University Press.

Senghas, A. (1995). *Children's contribution to the birth of Nicaraguan Sign Language.* Unpublished doctoral dissertation, Massachusetts Institute of Technology.

Senghas, A. (2000). The development of early spatial morphology in Nicaraguan Sign Language. In S. C. Howell, S. A. Fish, & T. Keith-Lucas (Eds.), *Proceedings of the 24th Annual Boston University Conference on Language Development* (pp. 696–707). Boston, MA: Cascadilla Press.

Senghas, A., & Coppola, M. (2001). Children creating language: How Nicaraguan Sign Language acquired a spatial grammar. *Psychological Science, 12*(4), 323–328.

Senghas, A., Coppola, M., Newport, E., & Supalla, T. (1997). Argument structure in Nicaraguan Sign Language: The emergence of grammatical devices. In E. Hughes, M. Hughes, & A. Greenhill (Eds.), *Proceedings of the 21st Annual Boston University Conference on Language Development* (Vol. 2, pp. 550–561). Somerville, MA: Cascadilla Press.

Senghas, R. J. (1997). *An 'unspeakable, unwriteable' language: Deaf identity, language and personhood among the first cohorts of Nicaraguan signers.* Unpublished doctoral dissertation, University of Rochester, New York.

Senghas, R. J. (2003). New ways to be Deaf in Nicaragua: Changes in language, personhood, and community. In L. Monaghan, C. Schmaling, K. Nakamura, & G. H. Turner (Eds.), *Many ways to be Deaf: International variation in Deaf communities* (pp. 260–282). Washington, DC: Gallaudet University Press.

Senghas, R. J., & Kegl, J. (1994). Social considerations in the emergence of *Idioma de Signos Nicaragüense* (Nicaraguan Sign Language). *Signpost, 7*(1), 40–45.

Senghas, R. J., & Monaghan, L. (2002). Signs of their times: Deaf communities and the culture of language. *Annual Review of Anthropology, 31*, 69–97.

Sternberg, M. L. A. (1987). *American Sign Language dictionary.* New York: Harper & Row.

10

Can Developmental Disorders Be Used to Bolster Claims From Evolutionary Psychology? A Neuroconstructivist Approach

Annette Karmiloff-Smith
Michael Thomas
*Neurocognitive Development Unit,
Institute of Child Health, London*

Data from adult neuropsychological patients and studies of individuals with genetic disorders are often used by evolutionary psychologists to motivate strong nativist claims about the organization of the neonate brain in terms of innately specified cognitive modules (Barkow, Cosmides, & Tooby, 1992; Duchaine, Cosmides, & Tooby, 2001; Pinker, 1997). Such hypotheses are based, in our view, on static snapshots of phenotypic outcomes in middle childhood and adulthood and tend to ignore one vital causal factor affecting disorders, that is, the actual process of ontogenetic development. In contrast to nativists, we take a truly developmental approach to both normal and atypical outcomes by focusing on the infant start state and the developmental trajectories that lead to such outcomes.

In this chapter, we discuss why it is essential to take a neuroconstructivist approach to interpreting the data from developmental disorders and why these latter cannot be used to bolster nativist claims. From our studies of older children and adults with the neurodevelopmental disorder, Williams syndrome, we show how processes that some claim to be "intact" actually display subtle impairments and cannot serve to divide the cognitive system into independent parts that develop normally from parts that develop atypically. Likewise, from our studies of infants and toddlers with developmental disorders, we identify low-level deficits in general capacities that have differential effects on the phenotypic outcome of different cognitive domains. Indeed, a tiny impairment very early in development can have a huge impact on some domains (the seemingly "selectively impaired cognitive modules") and a very subtle impact on other domains (the seemingly "intact cognitive modules"). It is thus crucial not only to focus on domains showing serious deficits in developmental disorders, but also to carry out in-depth studies of domains that seem at first blush to be unimpaired (Karmiloff-Smith, 1998). Because the brain develops as a whole system from embryogenesis onwards, we believe it to be highly unlikely that children with genetic disorders will end up with a patchwork of neatly segregated, preserved and impaired cognitive modules.

The aforementioned argumentation does not only hold for atypical development, of course. In keeping with some theorists of infant development, we also find it highly unlikely that the normal infant brain starts out with prespecified modules solely dedicated to the independent processing of specific cognitive domains. Indeed, we challenge the "Swiss Army Knife" metaphor adopted by some evolutionary psychologists for the neonate brain (Barkow et al., 1992; Duchaine et al., 2001). Rather, we argue that the infant brain is not like a Swiss Army knife, simply handed down by evolution with preformed, specialized components that form, in the case of developmental disorders, a segregated pattern of individually impaired/preserved modules at birth. Rather, as Piaget did for the normal child (Piaget, 1953, 1971), we contend that ontogenetic development itself is the clue to understanding both normal and atypical development and its relation to the structure of the resulting adult cognitive system. In a similar vein to Piaget's constructivism (1953, 1971), we embrace the notion that the child constructs his own environment and sculpts the microcircuitry of his own brain through his physical and mental actions on the world.

HOW THE INFANT BRAIN SCULPTS ITSELF THROUGH ONTOGENETIC DEVELOPMENT

Undeniably, all constructs—including nativism—impute some role to the external stimuli. However, unlike staunch *nativism* that considers environmental stimuli as mere triggers to a genetic blueprint for development, and unlike staunch *empiricism* that sees the environment as the major contributor to cognitive outcomes, we contend that gene expression and environment constantly undergo complex and dynamic interactions that only an in-depth analysis of ontogenesis can reveal.

For example, the onset of complex functions in the cerebral cortex of the infant brain can be traced to a burst in synaptic activity—the formation of rich networks of connections that allow knowledge to be encoded. This burst of synaptogenesis is under genetic control and appears to take place across the cortex relatively independently of input from the environment (Huttenlocher, 2002). However, synaptogenesis creates a surfeit of possible connections (many more than are retained in the eventual adult system), and it is this environment that selects which connections will be functionally useful. Unused connections are gradually eliminated. This pruning process continues over many years, that is, well into adolescence for the frontal regions, for example, allowing the environment to shape the raw mechanisms that genetic processes have put in place. What is included in the notion of "environment"? First, for a given cognitive system within the organism, the "internal" environment potentially includes inputs from other cognitive systems as well as sensory inputs. Environment also includes the social and physical worlds external to the organism that provide a wide variety of inputs to the different sensory systems.

To reiterate, it is in our view highly improbable that the infant starts life with independently functioning cognitive modules, simply awaiting appropriate triggers from the environment. Rather, our argument is that infant brain development is an activity-dependent process in which the environment acts not merely as a trigger but actually plays an important role in sculpting the final outcome in terms of both structure and function. In our view, initial noncognitive perceptual biases orient the infant toward certain aspects of the environment such as, for example, a sequential processor that pays particular attention to the flow of real-time speech output but less attention to, say, static spatial inputs. With repeated exposure and repeated processing of certain types of inputs (such as speech in our example), certain circuits of the brain become increasingly specialized (Elman et al., 1996; Johnson, 2001). Thus, a domain-relevant mechanism becomes a domain-specific mechanism as a function of development (Karmiloff-Smith, 1998). In other words, adult modules are, we contend, the result of a very progressive process of modularization over developmental time (Karmiloff-Smith, 1992).

NEUROCONSTRUCTIVISM

There are several competing theories about the structure of the infant brain at birth (see Johnson, 2001, for full discussion). *Maturationists* claim that different parts of the brain come on line sequentially during development as a result of genetic programming. They tend to explain the absence of a particular behavior in infancy by the hitherto absence of functioning of a specific region of the brain. *Interactionists*, by contrast, claim that at birth most parts of the brain function to some degree, but that it is the network of interactions both within and across regions that changes as a function of development. We have termed this *neuroconstructivism* (Karmiloff-Smith, 1998) or more recently, the *interactive specialization approach* (Johnson, Halit, Grice, & Karmiloff-Smith, 2002). Rather than waiting for a region to come on line maturationally, infant brain regions may initially be *more* active than the adult's until the processes of specialization and localization of function gradually stabilize. It has now been shown that seemingly identical overt behavior in infants and adults is supported by different brain regions or interactions between regions (e.g., Csibra, Davis, Spratling, & Johnson, 2002; de Haan, Pascalis, & Johnson, 2002; Neville, Mills, & Lawson, 1992). By the time we reach adulthood, our brains are indeed highly structured and functionally specialized, but this in no way entails that we started out in infancy with anything like this structure in place.

A compelling example of very progressive specialization and localization comes from the development of infant face processing. What could be more evolutionarily important than species-specific recognition? If the nativist position held, then face processing would seem to be an ideal candidate for a built-in module, ready to function independently of other brain circuits as soon as appropriate triggering stimuli were presented. Yet, although a preference for face-like stimuli seems to be present from birth (Johnson & Morton, 1991), infant face processing is very different from adult face processing in terms of both behavior and the brain circuits involved. Initially infants are just as likely to track pictures of real faces as those of very schematic faces with only three blobs in the appropriate eye and mouth regions. By 2 months of age, however, they only track real faces. But even the neonate preference is not constrained to facelike stimuli alone. Rather, the stimuli that are preferred are those that have more information in the upper region than the lower region, like a T-shape (Simion, Valenza, Macchi Cassia, Turati, & Umilta, 2002). Although this happens to coincide with the overall visual stimulus of a face, it is clearly not an innately given "face template," the brain processing not being initially dedicated to face processing alone. So it seems that evolution does not need to provide more than a domain-general kick start to face processing, with the guarantee that the external environment will furnish massive face input early in life. After huge quantities of face inputs over the first months of life, even 6-month-olds do not display the brain activity typical of both 12-month-olds and adults in terms of binding the perceptual features of a facial stimulus (Csibra et al., 2002). It is also known that early on, both hemispheres of the infant brain actively process faces. However, by the end of the first year, processing of faces shifts predominantly to the right hemisphere, the one typically more active in older children and adults (de Haan et al., 2002).

These are but a few aspects of how face processing develops during infancy, highlighting the fact that it does not come ready to display adultlike functioning once face stimuli have triggered a so-called innately specified module. On the contrary, infants seem to require hundreds of thousands of face stimuli to progressively develop their face processing expertise such that by the end of the first year of life, they start to display adultlike processing in terms of both behavior and underlying brain processes. We contend that any face processing module that ultimately exists in adults, and that could by then be selectively impaired (e.g., McNeil & Warrington, 1993), actually develops out of initial attention biases in interaction with the rich face processing experience available to the infant.

Further evidence for progressive neuroconstructivism comes from the study of infants with perinatal unilateral brain lesions to the right hemisphere. A review of their subsequent face processing abilities between 5 and 14 years of age revealed two things (de Haan, 2001). First, their impairments were mild compared to adults who had experienced similar damage—less than half the children exhibited impairments in face or object recognition compared to controls. Whatever the early damage, it had been attenuated by developmental plasticity. Second, face-processing deficits were no more common than problems identifying objects, and a face processing deficit never occurred in the absence of an object-processing deficit. The specialization of face processing and its progressive separation from object processing appears to be a product of development, with the face recognition system emerging as a gradual specialization of an initially more general-purpose system. The dissociation of face and object recognition in the adult cannot be replicated by early damage to the normal system.

Now, nativists might claim that the progressive changes in infant face processing simply constitute the unfolding of a genetic timetable. However, other work on early processing of language, for example, challenges this. Neville and her colleagues examined the brain processes of toddlers when they listened to a series of words. They found that it was the number of words that the infant could produce, and not maturational age, that predicted which brain circuits were used (Neville et al., 1992). In sum, the ball is in the court of the evolutionary psychologist to demonstrate that the infant brain is really anything like the metaphor suggested by the Swiss army knife with its highly specialized component parts in place from the outset.

A REEXAMINATION OF DATA
FROM DEVELOPMENTAL DISORDERS

Adult neuropsychological patients may in some cases display highly specific impairments in their performance, suggesting independently functioning modules and impairment to a very specialized area of the brain. It must be recalled, however, that in the adult neuropsychological case, the adult has suffered a brain insult to a hitherto normally developed and highly structured brain. Such structure, as we have consistently argued, is the result of prior development and tells us nothing about the start state. Yet, at first blush, overt behavioral outcomes in older children and adults with genetic disorders seem also to present a neat case of preserved and impaired modules. So, why do we continue to question this? People with genetic disorders do not, in our view, have normal brains with parts preserved and parts impaired. Rather, they have developed an atypical brain throughout embryogenesis and subsequent postnatal growth, so we should expect fairly widespread impairments across the brain rather than a very localized one. How can we then reconcile our theoretical assumptions with the empirical data suggesting clear-cut selective impairments?

We argue that the empirical data themselves need to be reexamined, both from the viewpoint of the overt behavior versus the underlying cognitive processes, and from the viewpoint of the control groups used to make theoretical claims about genetic disorders. To do this, we take the example of one genetic disorder, Williams syndrome, and briefly examine three domains that some researchers have claimed to be "spared" in this clinical population—face processing, language and social cognition.

WILLIAMS SYNDROME

Williams syndrome (WS) is a neurodevelopmental disorder caused by a submicroscopic deletion of some 24 genes on one copy of chromosome 7q.11.23 (Donnai & Karmiloff-Smith, 2000). It occurs in approximately 1 in 20,000 live births. Clinical features include several physical abnormalities that are accompanied by mild to moderate mental retardation and a specific personality profile. The interest of WS to neuroscience stems from its very uneven profile of cognitive abilities, with spatial and numerical cognition seriously impaired, whereas language, social interaction and face processing seem surprisingly proficient for a clinical population with IQs in the 50s to 60s range (Bellugi, Wang, & Jernigan, 1994; Udwin & Yule, 1991).

Work by Bellugi and her collaborators first drew attention to the potential theoretical interest of the seeming dissociations in the Williams syndrome cognitive phenotype (Bellugi, Marks, Bihrle, & Sabo, 1988). Surprising proficiency with language was shown to co-exist with serious problems with nonverbal tasks, in particular those calling on spatial processing. Moreover, people with WS scored at floor, for example, on the Benton Line Orientation Task, but were within the normal range on the Benton Face Processing Task (Bellugi et al., 1988). This striking contrast between facial and spatial processing led some researchers (e.g., Bellugi et al., 1988) to maintain that face processing in WS is "intact" demonstrating, together with *prosopagnosia* (the inability to identify previous known faces) in the adult neuropsychological patients, that face processing is an independently functioning module.

Face Processing in Williams Syndrome

The early claims about an intact face processing module in WS have since been challenged, not with respect to the behavioral data themselves, but targeting the underlying cognitive and brain processes involved. Several studies have now replicated Bellugi's findings showing, indeed, that older children and adults with WS achieve behavioral scores in the normal range on some face processing tasks (Grice et al., 2001; Karmiloff-Smith, 1997; Udwin & Yule, 1991). However, this behavioral success is only superficially the same as that of normal controls. Usually we process faces configurally; our brains rapidly analyze the spatial relations between facial elements. By contrast, people with WS tend to predominantly analyze faces featurally; they focus more on the separate elements of a face, and less on the second-order relations between elements (Deruelle, Mancini, Livet, Casse-Perrot, & de Schonen, 1999; Karmiloff-Smith, 1997; Rossen, Bihrle, Klima, Bellugi & Jones, 1996). So the cognitive processes underpinning the superficially successful face processing of people with WS seem to be different from the normal case (Karmiloff-Smith et al., in press).

A similar situation holds for the electrophysiology of the brain (Grice et al., 2001, in press; Mills et al., 2000). People with WS are more likely to show a predominance of the left hemisphere when processing faces in contrast to the typical right-hemisphere dominance for face processing. Furthermore, people with WS do not display the normal inversion effect, whereby upside down faces are processed differently from upright faces (Karmiloff-Smith et al., in press-b). In WS, both types of display are processed in a similar way, again suggesting that this clinical group processes face stimuli on a feature-by-feature basis. This cognitive difference does not hold only for facial stimuli. Work by Deruelle and her collaborators (1999) revealed that people with WS are more inclined to use featural than configural processing also of nonface displays. In sum, people with WS do not present with a normally developed "intact" face processing module and an impaired space processing module, as nativists would claim. Rather, from the outset, they have followed an atypical developmental trajectory such that both facial and spatial processing reveal a similar underlying impairment in second-order configural processing. It is simply that the problem space of face processing lends itself more readily to featural analysis than spatial analysis does, so that it merely seems normal in the older child and adult. In other words, a fairly low-level impairment in configural processing early on impacts differentially on face processing and space processing during development, such that one domain can call on certain compensatory processes, whereas the other cannot.

Language in Williams Syndrome

Perhaps face processing just happens not to be the right domain for the evolutionary psychologist to establish a dissociation between innate components of the cognitive system. So, let us briefly examine another domain. Early claims were made for another dissociation in WS, this time between language and cognition. Language has been argued to be an innate mental organ specific to humans, and not reliant on general cognition (Pinker, 1994). So, on this account, we might expect certain genetic disorders to allow normal language to develop even in the presence of an impairment to general cognition. Such a dissociation was initially claimed for WS. But as we have seen, such dissociations are actually highly unlikely given what we know about the processes of language development. And, in exactly the same way as our example from face processing, subsequent careful analysis of the ostensibly "intact" language capacity in WS revealed many, sometimes quite subtle, atypicalities, which suggested that WS language was learned via an atypical developmental trajectory (Karmiloff-Smith et al., 1997; Laing et al., 2002; Nazzi & Karmiloff-Smith, 2002; Nazzi, Paterson, & Karmiloff-Smith, 2002; Singer-Harris, Bellugi, Bates, Jones & Rossen, 1997; Vicari, Brizzolara, Carlesimo, Pezzini, & Volterra, 1996; Volterra, Capirci, Pezzini, Sabbadini, & Vicari, 1996).

Initial comparisons were made between the abilities of individuals with WS and those from other syndromes who present with equivalent general cognitive abilities. Certainly compared to a disorder such as Down syndrome (DS), language in WS appears strikingly more advanced. For example, although the language of individuals with DS often shows appropriate word ordering, their speech often remains telegraphic, with a reduced use of function words, poorly inflected verbs, predominant use of the present tense, and a lack of appropriate feature marking on pronouns and anaphors, a state that largely persists into the adult years (Fowler, Gelman, & Gleitman, 1994). On the other hand, the language of individuals with WS often reveals sophisticated linguistic knowledge. For instance, in an analysis of the expressive language of four children with WS, Clahsen and Almazan (1998) reported the presence of complex syntactic structures and grammatical morphemes that were almost always used correctly.

A number of studies have pursued comparisons between language in WS and DS, presumably under the view that DS can serve as a baseline of what one might expect of language development in the presence of mental retardation, against which the achievements of WS may be measured (see discussion in Karmiloff-Smith, Ansari, Campbell, Scerif, & Thomas, in press-a). Thereafter, however, detailed investigations began to demonstrate that language performance is not at normal levels in WS, and at the very least shows a developmental delay of at least 2 years (Singer-Harris et al., 1997). Most recent studies that compare the performance of individuals with WS to typically developing children now use a control group matched for mental age, to which their performance levels are more closely tied. Paradoxically, this matching procedure implicitly concedes that language development in WS is not independent of general cognitive ability.

While the language performance of individuals with WS is relatively impressive (compared to other syndromes with low IQs), evidence of atypicalities has accumulated in all areas of language, and at all stages of language development, including vocabulary, syntax, morphology, and pragmatics, as well as the precursors to language development in infants (see Thomas & Karmiloff-Smith, 2003, for a review). Moreover, comparisons with Down syndrome actually exaggerate the apparent language ability in WS, given that individuals with DS demonstrate a particular developmental deficit in phonological processing that is not found in WS. And, most crucially, when this pattern of deficits in the endstate of each disorder—better language in WS than DS—was traced back to the respective abilities in early language comprehension in infancy, the pattern did not hold. Infants with WS and DS showed equal (and very delayed) early language comprehension, implying that adult phenotypes were the product of differential atypical trajectories of development (Paterson, Brown, Gsodl, Johnson, & Karmiloff-Smith, 1999).

Social Cognition in Williams Syndrome

The story we have seen for face recognition and for language development in WS is now being repeated in the study of social cognition in this disorder. Here again, an initial claim was made that in WS, social cognition developed normally against a background of other impaired functions. Yet, subsequent detailed research has suggested that social cognition and pragmatics are atypical in WS, sometimes subtly but sometimes quite markedly (Jones et al., 2000; Tager-Flusberg & Sullivan, 2000). The study of WS illustrates that in every case that a "preserved function" has been heralded in this genetic developmental disorder; that claim did not stand up to subsequent detailed investigation. Indeed, whenever a claim has been put forward that is inconsistent with what we know about development in general, this claim has turned out to be false. And similar results have also started to emerge from other developmental disorders with a genetic basis such as in the study of Specific Language Impairment, developmental dyslexia, Fragile X syndrome and Velo-cardiofacial syndrome (see discussions in Chiat, 2001; Karmiloff-Smith, 1998; Karmiloff-Smith et al., in press-a; Thomas, 2003; Thomas & Karmiloff-Smith, 2002, 2003).

THE IMPORTANCE OF NEUROCONSTRUCTIVISM

What becomes clear from the examples given is that genetic disorders do not provide data pointing to neatly impaired and spared cognitive domains that lend themselves to the evolutionary psychology claims.

Rather, studies of developmental disorders demonstrate just how very complex and dynamic are the processes of gradual ontogenetic development and how important it is to recall that for humans, selection has favored a very lengthy period of postnatal brain development. It is one thing to spot consistency in the pattern of adult cognitive structures following development in the environments to which human adults are typically exposed. It is quite another, however, to assume—against accumulating counterevidence—that these structures are innately present in the infant brain. And it is yet a further act of faith to then argue that selection somehow favored them!

So, what is wrong with selection, one might ask. Have its mechanisms gone awry? Why is a process as crucial to recognizing conspecifics as, say, face processing, not innately specified and cordoned off to function independently from all other processes? The reason may well lie in two different types of control, and the fact that some higher level cognitive outcomes may not even be possible at all without the gradual ontogenetic process of learning (Elman et al., 1996; Piaget, 1971).

It is generally accepted that there are two forms of biological control: *mosaic control* and *regulatory control* (Elman et al., 1996). Mosaic control involves deterministic epigenesis; genes tightly control timing and outcome, the process is fast and operates independently of other processes. This form of control is fine under optimal conditions. However, it places serious limits on complexity and flexibility of the developmental process. Some parts of human development are likely to involve mosaic control, such as the very basic macrostructures of the brain and of the body. However, the other type of control, regulatory control, is much more common and involves probabilistic epigenesis. It is especially prominent in the developing microstructure of the brain. It is under broad rather than tight genetic control, is slow and progressive, with limited prespecification. In this type of control, different parts of a system develop interdependently. And, unlike in mosaic control, there are fewer constraints on complexity and plasticity. This does not mean, of course, that there are no biological constraints, as the empiricist position might claim, but it is far less constrained than mosaic control. Genes and their products are most unlikely to code for the cognitive level, but rather for differences in developmental timing, neuronal density, neuronal migration, neuronal type, firing thresholds, neurotransmitter differences, and the like.

The notion of neuroconstructivism embodies regulatory control, with ontogeny seen as the prime force for turning a number of domain-relevant learning mechanisms progressively into domain-specific outcomes in the adult. This does not imply that the infant brain is a single, homogeneous learning device; there is, no doubt, much heterogeneity in the initial gross wiring of the brain. But this heterogeneity bears little resemblance to the ultimate functional structures that can only emerge through the process of ontogeny. In other words, rather than the mosaic form of tight genetic control that some evolutionary psychology models invoke, the human brain may well have evolved to favor very progressive development and neuroconstructivist plasticity rather than prespecification. If we are to understand what it is to be human, our continuing emphasis must be on the process of development itself.

ACKNOWLEDGMENTS

The writing of this chapter was supported by grants to AK-S from the Williams Syndrome Foundation, the Medical Research Council UK (Programme Grant No. G971462), the PPP Foundation (No. 1748/961), and NIH Grant (No. R2ITW06761-01).

REFERENCES

Barkow, J. H., Cosmides, L., & Tooby, J. (Eds.). (1992). *The adapted mind: Evolutionary psychology and the generation of culture*. New York: Oxford University Press.

Bellugi, U., Marks, S., Bihrle, A. M., & Sabo, H. (1988). Dissociation between language and cognitive functions in Williams syndrome. In D. B. K. Mogsford (Ed.), *Language development in exceptional circumstances* (pp. 177–189). New York: Churchill Livingstone.

Bellugi, U., Wang, P. P., & Jernigan, T. L. (1994). Williams syndrome: An unusual neuropsychological profile. In S. H. Broman & J. Grafman (Eds.), *Atypical cognitive deficits in developmental disorders: Implications for brain function* (pp. 23–56). Hillsdale, NJ: Lawrence Erlbaum Associates.

Chiat, S. (2001). Mapping theories of developmental language impairment: Premises, predictions and evidence. *Language & Cognitive Processes, 16*, 113–142.

Clahsen, H., & Almazan, M. (1998). Syntax and morphology in Williams syndrome. *Cognition, 68*, 167–198.

Csibra, G., Davis, G., Spratling, M. W., & Johnson, M. H. (2002). *Gamma* oscillations and object processing in the infant brain. *Science, 290*, 1582–1585.

De Haan, M. (2001). The neuropsychology of face processing during infancy and childhood. In C. A. Nelson & M. Luciana (Eds.), *Handbook of developmental cognitive neuroscience* (pp. 381–398). Cambridge, MA: MIT Press.

De Haan, M., Pascalis, O., & Johnson, M. H. (2002). Specialisation of neural mechanisms underlying face recognition in human infants. *Journal of Cognitive Neuroscience, 14*, 199–209.

Deruelle, C., Mancini, J., Livet, M. O., Casse-Perrot, C., & de Schonen, S. (1999). Configural and local processing of faces in children with Williams syndrome. *Brain & Cognition, 41*, 276–298.

Donnai, D., & Karmiloff-Smith, A. (2000). Williams syndrome: From genotype through to the cognitive phenotype. *American Journal of Medical Genetics, 97*, 164–171.

Duchaine, B., Cosmides, L., & Tooby, J. (2001). Evolutionary psychology and the brain. *Current Opinion in Neurobiology, 11*, 225–230.

Elman, J. L., Bates, E., Johnson, M. H., Karmiloff-Smith, A., Parisi, D., & Plunkett, K. (1996). *Rethinking innateness: A connectionist perspective on development*. Cambridge, MA: MIT Press.

Fowler, A., Gelman, R., & Gleitman, R. (1994). The course of language learning in children with Down syndrome: Longitudinal and language level comparisons with young normally developing children. In H. Tager-Flusberg (Ed.), *Constraints on language acquisition: Studies of atypical populations* (pp. 91–140). Hillsdale, NJ: Lawrence Erlbaum Associates.

Grice, S. J., de Haan, M., Halit, H., Johnson, M., Csibra, G., Grant, J., & Karmiloff-Smith, A. (in press). ERP abnormalities of illusory contour perception in Williams syndrome. *Neuroreport*.

Grice, S. J., Spratling, M. W., Karmiloff-Smith, A., Halit, H., Csibra, G., de Haan, M., & Johnson, M. H. (2001). Disordered visual processing and oscillatory brain activity in autism and Williams syndrome. *NeuroReport, 12*, 2697–2700.

Huttenlocher, P. R. (2002). *Neural plasticity*. Cambridge, MA: Harvard University Press.

Johnson, M. H. (2001). Functional brain development in humans. *Nature Reviews Neuroscience, 2*, 475–483.

Johnson, M. H., Halit, H., Grice, S., & Karmiloff-Smith, A. (2002). ÊNeuroimaging of typical and atypical development: A perspective from multiple levels of analysis. *Development and Psychopathology, 14*, 521–536.

Johnson, M. J., & Morton, J. (1991). *Biology and cognitive development: The case of face recognition*. Oxford: Blackwell.

Jones, W., Bellugi, U., Lai, Z., Chiles, M., Reilly, J., Lincoln, A., & Ralphs, A. (2000). Hypersociability in Williams syndrome. *Journal of Cognitive Neuroscience, 12* (Suppl. 1), 30–46.

Karmiloff-Smith, A. (1992). *Beyond modularity: A developmental perspective on cognitive science.* Cambridge, MA: MIT Press/Bradford Books.

Karmiloff-Smith, A. (1997). Crucial differences between developmental cognitive neuroscience and adult neuropsychology. *Developmental Neuropsychology, 13,* 513–524.

Karmiloff-Smith, A. (1998). Development itself is the key to understanding developmental disorders. *Trends in Cognitive Sciences, 2,* 389–398.

Karmiloff-Smith, A., Ansari, D., Campbell, L., Scerif, G., & Thomas, M. S. C. (in press-a). Theoretical implications of studying genetic disorders: The case of Williams syndrome. In C. Morris, H. Lenhoff, & P. Wang (Eds.), *Williams-Beuren Syndrome: Research and clinical perspectives.* Baltimore, MD: Johns Hopkins University Press.

Karmiloff-Smith, A., Grant, J., Berthoud, I., Davies, M., Howlin, P., & Udwin, O. (1997). Language and Williams syndrome: How intact is "intact"? *Child Development, 68,* 246–262.

Karmiloff-Smith, A., Thomas, M. S. C., Annaz, D., Humphreys, K., Ewing, S., Grice, S., Brace, N., Van Duuren, M., Pike, G., & Campbell, R. (in press-b). Exploring the Williams syndrome face processing debate: The importance of building development trajectories. *Journal of Child Psychology and Psychiatry.*

Laing, E., Butterworth, G., Ansari, D., Gsödl, M., Longhi, E. Panagiotaki, G., Paterson, S., & Karmiloff-Smith, A. (2002). Atypical development of language and social communication in toddlers with Williams syndrome. *Developmental Science, 5,* 233–246.

McNeil, J. E., & Warrington, E. K. (1993). Prosopagnosia: A face-specific disorder. *Quarterly Journal of Experimental Psychology, 46A,* 1–10.

Mills, D. L., Alvarez, T. D., St. George, M., Appelbaum, L. G., Bellugi, U., & Neville, H. (2000). Electrophysiological studies of face processing in Williams syndrome. *Journal of Cognitive Neuroscience, 12,* 47–64.

Nazzi, T., & Karmiloff-Smith, A. (2002). Early categorization abilities in young children with Williams syndrome. *NeuroReport, 13,* 1259–1262.

Nazzi, T., Paterson, S., & Karmiloff-Smith, A. (2002). Early word segmentation by infants and toddlers with Williams syndrome. *Infancy, 4,* 251–271.

Neville, H., Mills, D., & Lawson, D. (1992). Fractionating language: Different neural subsystems with different sensitive periods. *Cerebral Cortex, 2,* 244–258.

Paterson, S. J., Brown, J. H., Gsödl, M. K., Johnson, M. H., & Karmiloff-Smith, A. (1999). Cognitive modularity and genetic disorders. *Science, 286,* 2355–2358.

Piaget, J. (1953). *The origins of intelligence in children.* London: Routledge & Kegan Paul.

Piaget, J. (1971). *Biology and knowledge: An essay on the relations between organic regulations and cognitive processes.* Edinburgh: Edinburgh University Press.

Pinker, S. (1994). *The language instinct.* London: Penguin Books.

Pinker, S. (1997). *How the mind works.* New York: Norton.

Rossen, M., Bihrle, A., Klima, E. S., Bellugi, U., & Jones, W. (1996). Interaction between language and cognition: Evidence from Williams syndrome. In J. H. Beitchmen, N. Cohen, M. Konstantareas, & R. Tannock (Eds.), *Language learning and behavior* (pp. 367–392). New York: Cambridge University Press.

Simion, F., Valenza, E., Macchi Cassia, V., Turati, C., & Umilta, C. (2002). Newborns' preference for up-down asymmetrical configurations. *Developmental Science, 5*(4), 427–434.

Singer Harris, N. G., Bellugi, U., Bates, E., Jones, W., & Rossen, M. (1997). Contrasting profiles of language development in children with Williams and Down syndromes. *Developmental Neuropsychology, 13*, 345–370.

Tager-Flusberg, H., & Sullivan, K. (2000). A componential view of theory of mind: Evidence from Williams syndrome. *Cognition, 76*, 59–89.

Thomas, M. S. C. (2003). Limits on plasticity. *Journal of Cognition and Development. 4*, 95–121.

Thomas, M. S. C., & Karmiloff-Smith, A. (2002). Are developmental disorders like cases of adult brain damage? Implications from connectionist modelling. *Behavioral and Brain Sciences, 25*(6), 727–758.

Thomas, M. S. C., & Karmiloff-Smith, A. (2003). Modeling language acquisition in atypical phenotypes. *Psychological Review, 110*(4), 647–682.

Udwin, O., & Yule, W. (1991). A cognitive and behavioral phenotype in Williams syndrome. *Journal of Clinical and Experimental Neuropsychology, 13*, 232–244.

Vicari, S., Brizzolara, D., Carlesimo, G., Pezzini, G., & Volterra, V. (1996). Memory abilities in children with Williams syndrome. *Cortex, 32*, 503–514.

Volterra, V., Capirci, O., Pezzini, G., Sabbadini, L., & Vicari, S. (1996). Linguistic abilities in Italian children with Williams syndrome. *Cortex, 32*, 663–677.

Author Index

A

Aakalu, G., 165, 174
Adams, C., 215, 226, 252
Adams, G., 48, 52, 81
Adams, M. M., 238, 249
Adolphs, R., 196, 197, 198
Aguirre, G., 235, 252
Akshoomoff, N. A., 151, 163, 171
Aksu-Koç, A. A., 258, 259, 281
Alberoni, M., 236, 246
Albus, J. S., 169, 170
Allada, V., 164, 173
Allard, L., 226, 245
Allen, G., 47, 82, 131, 137, 151, 163, 170
Allen, S. E. M., 259, 281
Allison, T., 235, 250
Almazan, M., 315, 318
Altman, J., 148, 165, 166, 167, 170
Alvarez, T. D., 314, 320
Ancrenaz, M., 163, 177
Andrew, P., 125, 137
Andrew, R. J., 125, 142
Andrews, P., 161, 170
Ansari, D., 314, 315, 316, 320

Antoun, N., 196, 198
Appelbaum, L. G., 314, 320
Appelbaum, M., 226, 245
Aram, D. M., 205, 206, 219, 220, 221, 223, 226, 227, 245, 246, 248, 252
Arbib, M. A., 189, 200
Armel, K. C., 228, 251
Asfaw, B., 133, 143
Austin, J., 20, 26
Avery, O., 59, 82
Avikainen, S., 189, 200
Ayala, F., 65, 82

B

Bachevalier, J., 208, 209, 212, 236, 245, 252
Baer, K. E. v., 51, 82
Bailey, P., 132, 137
Baldwin, G., 260, 284
Baldwin, J. M., xvi, xvii, 9, 10, 26, 43, 78, 82, 88, 109, 110, 121
Balota, D. A., 153, 171
Bard, K. A., 163, 176

Barkow, J. H., 24, 27, 231, 245, 307, 308, 318
Baron, G., 129, 143, 158, 174
Barthelmy, C., 189, 199
Barton, R. A., 156, 170, 177
Basser, L., 205, 226, 245
Bates, C., 126, 137
Bates, E., 206, 207, 211, 212, 213, 214, 216, 217, 219, 220, 221, 222, 223, 225, 226, 227, 229, 231, 233, 237, 240, 243, 245, 246, 247, 248, 249, 250, 251, 315, 321
Bayer, S. A., 148, 165, 166, 167, 170
Bekkering, H., 189, 200
Bellugi, U., 313, 314, 315, 316, 318, 319, 320, 321
Benefit, B. R., 160, 170
Benigni, L., 206, 231, 246
Bennett-Clarke, C. A., 208, 249
Bergson, H., 40, 82
Berndt, R., 234, 247
Berthoud, I., 315, 320
Bettinardi, V., 189, 202, 236, 246
Bever, T. G., 260, 285
Bialystok, E., 214, 246
Bickerton, D., xii, 255, 257, 269, 272, 281
Bihrle, A. M., 313, 314, 318, 321
Binkofski, F., 189, 198
Bishop, D. V. M., 223, 224, 246
Bixby, J., 186, 203
Bjaalie, J. G., 168, 171
Black, J. E., 126, 140
Bloedel, J. R., 147, 169, 170
Boccia, M., 124, 141
Boden, M. A., 134, 137
Boesch, C., 19, 29, 162, 163, 170, 177
Boesiger, E., 46, 82
Borgen, G., 163, 177
Bourgeois, J. -P., 164, 176
Bower, J. M., 149, 150, 162, 170, 172
Bower, T. G. R., 130, 137
Bowerman, M., 261, 262, 263, 281, 282
Boyd, R., 19, 27
Bracha, V., 147, 170
Brandon, R. N., 20, 27
Brass, M., 189, 200
Braun, A. R., 132, 138, 233, 249

Bressi, S., 236, 246
Bretherton, I., 206, 231, 246
Brett, E., 215, 226, 252
Bringuier, J. C., 41, 82
Bringuier, V., 234, 248
Brizzolara, D., 315, 321
Broadfield, D. C., 132, 138, 233, 249
Brodal, P., 168, 171
Bronfenbrenner, U., 21, 27
Brooks, A. S., 19, 29, 134, 137, 140
Brooks, D., 23, 27
Brooks, R., 181, 201
Brothers, L., 186, 198
Brown, J. H., 316, 320
Bruner, J., 24, 27
Buccino, G., 189, 198
Burkhardt, R. W., Jr., 45, 46, 82
Buss, L. W., 49, 50, 72, 82
Butcher, C., 272, 273, 283
Butterworth, G., 315, 320
Buxton, R. B., 131, 137, 151, 163, 170, 235, 238, 242, 249, 250
Bybee, J. L., 268, 281
Byers, J. A., 163, 171
Byrne, J. M. E., 162, 171
Byrne, R. W., 124, 136, 137, 143, 162, 171

C

Calder, A. J., 196, 198
Camaioni, L., 206, 231, 246
Camarda, R., 187, 188, 201, 202
Campbell, D. T., 91, 121
Campbell, L., 314, 315, 316, 320
Cant, J. G. H., 160, 161, 175
Cantalupo, C., 132, 137
Capirici, O., 315, 321
Cappa, S., 235, 236, 238, 246, 250
Caramazza, A., 234, 247
Carevale, G. F., 237, 246
Carey, D. P., 186, 198, 199
Carlesimo, G., 315, 321
Carpenter, P. A., 235, 249
Carr, L. J., 196, 199, 215, 226, 252
Case, R., 124, 128, 130, 131, 137

Casse-Perrot, C., 314, 319
Changeaux, J. P., 126, 137
Chan-Palay, V., 166, 171
Chavane, F., 234, 248
Chen, S., 164, 173
Cheney, M. K., 152, 171
Cherry, S. R., 164, 174
Chiaia, N. L., 208, 249
Chiat, S., 316, 318
Chiles, M., 316, 319
Chiron, C., 236, 246
Chitty, A. J., 186, 202
Choi, S., 261, 262, 263, 281, 282
Chomsky, N., 227, 230, 247
Chugani, H. T., 164, 173, 239, 247
Churchill, F., 49, 52, 82
Churchill, S. E., 133, 138
Clahsen, H., 315, 318
Clark, R., 229, 247
Clutton-Brock, T. H., 161, 171
Cochin, S., 189, 199
Coffey-Corina, S. A., 235, 238, 239, 250
Conel, J. L., 129, 138
Constantine-Paton, M., 107, 121
Cooper, G., 196, 198
Coppola, M., 274, 275, 277, 285, 293, 297, 300, 301, 305, 306
Correa, D., 235, 249
Corsi, P., 50, 82
Cosmides, L., 24, 27, 124, 138, 143, 231, 245, 307, 308, 318, 319
Costes, N., 189, 199
Courchesne, E., 131, 137, 151, 163, 165, 170, 171
Courtney, S. M., 234, 247
Couzin, J., 63, 82
Cowell, P. E., 126, 138
Cramer, J. S., 134, 137
Crawford, M. L., 127, 143
Crick, F., 59, 86
Csibra, G., 310, 311, 313, 314, 318, 319
Curtiss, S., 206, 214, 247
Cutting, J. E., 193, 199
Cuvier, G., 45, 82

D

D'Esposito, M., 235, 252
Dale, P. S., 214, 220, 225, 226, 240, 246, 248
Dall'Oglio, A. M., 226, 247
Damasio, A. R., 127, 140, 196, 198, 244, 247
Damasio, H., 196, 198
Damerose, E., 126, 138
Darlington, R. B., 156, 172
Darwin, C., 8, 10, 11, 19, 27, 43, 44, 46, 82, 83, 123, 138
Das, A., 233, 247
David, A. C., 74, 83
Davidson, I., 20, 29, 134, 141
Davies, M., 315, 320
Davis, G., 310, 311, 318
Dawkins, R., 13, 14, 17, 18, 24, 27, 110, 121
De Beer, G., 52, 83
de Courten, C., 239, 249
de Haan, M., 310, 311, 314, 314, 319
De Houwer, A., 261, 282
de Schonen, S., 314, 319
De Weerd, P., 234, 247
Deacon, T., 2, 10, 20, 27, 92, 113, 114, 115, 121, 207, 247, 249, 257, 282, 304, 305
Decety, J., 189, 199
DeGraff, M., 269, 282
DeGusta, D., 133, 143
Dehaene, S., 235, 238, 250, 311, 319
Denckla, M. B., 165, 174
Denenberg, V. H., 126, 138
Deng, C., 125, 138
Dennis, M., 223, 224, 247
Derber, M. P., 127, 142
Deruelle, C., 314, 319
Desimone, R., 185, 199, 234, 247
Deuchar, M., 261, 282
DeVries, H., 48, 83
Di Capua, M., 226, 247
Diamond, A., 154, 171
Diamond, M. C., 126, 138
Diener, H. C., 150, 173
Dodd, G., 127, 142

Dodge, M., 272, 283
Donnai, D., 313, 319
Dostie, D., 244, 250
Dow, R. S., 147, 166, 169, 173
Doyon, J., 153, 171
Dronkers, N. F., 236, 248
Dubeau, M. C., 196, 199
Dubner, R., 185, 199
Duchaine, B., 307, 308, 319
Dulac, O., 236, 246
Dupoix, S., 235, 238, 250
Duranti, A., 288, 305
Durham, W., 19, 27
Duscheck, J., 125, 138

E

Ebner, T. J., 169, 170
Eccles, J. C., 168, 171
Eck, M., 208, 249
Eckenhoff, M. F., 164, 176
Eddy, W. F., 235, 249
Edelman, G., 2, 24, 27
Eibl-Eibesfeldt, I., 12, 15, 27
Eichen, E. B., 296, 305
Eisele, J., 205, 219, 220, 221, 226, 227, 246, 248
Ekelman, B., 223, 245
Ekmekçi, O., 259, 260, 282
Elman, J. L., 206, 207, 214, 227, 229, 237, 246, 248, 309, 317, 319
Engels, F., 19, 27
Erhard, P., 234, 248
Erting, C. J., 289, 290, 305
Evans, A., 244, 250

F

Fadiga, L., 188, 189, 198, 199, 200, 202
Fairbanks, L. A., 163, 164, 171
Farah, M., 235, 252
Farver, J. A. M., 21, 27
Fazio, E., 235, 238, 250
Fazio, F., 189, 199, 236, 246
Fazio, G., 189, 202
Feirabend, H. K. P., 147, 177

Feldman, H., 205, 226, 248
Feldman, M., 13, 14, 17, 18, 19, 28, 29, 303, 304, 305
Fenson, J., 219, 220, 221, 225, 226, 227, 240, 246, 248
Fiez, J. A., 152, 153, 171, 172, 176, 235, 238, 251
Fink, G. R., 189, 198
Finlay, B. L., 156, 172, 207, 249
Fleagle, J. C., 160, 161, 172
Fodor, J. A., 227, 248
Fogassi, L., 188, 189, 190, 198, 199, 200, 201
Fogassi, M., 187, 188, 202
Foley, W., 270, 282
Forss, N., 189, 200
Forster, D., 21, 30
Fortescue, M., 259, 282
Fowler, A., 315, 319
Fox, E. A., 161, 177
Fox, P. T., 149, 152, 153, 162, 172, 175, 176, 235, 238, 251
Frahm, H. D., 129, 143, 158, 174
Franceschi, M., 236, 246
Frank, L. R., 235, 238, 242, 249, 250
Franklin, A., 134, 137
Fregnac, Y., 234, 248
Freund, H. –J., 189, 198
Fromkin, V., 229, 250
Fukuyamas, R., 113, 121
Fuster, J. M., 132, 138

G

Galaburda, A. M., 242, 248
Galdikas, B., 163, 177
Gallese, V., 182, 183, 187, 188, 189, 198, 190, 194, 195, 196, 197, 199, 200, 201, 202, 203
Gannon, P. J., 132, 138
Gao, J. –H., 149, 162, 172
Gardner, B. T., 124, 138, 162, 172
Gardner, R. A., 124, 138, 162, 172
Garey, L., 239, 249
Garraghty, P. E., 127, 142
Garufi, G., 133, 142

Gattass, R., 234, 247
Gauna, K., 244, 250
Gazzaniga, M. S., 243, 249
Gelman, R., 315, 319
Gentilucci, M., 187, 202
Gerrits, N. M., 169, 177
Gerstein, M., 60, 85
Geschwind, N., 132, 138, 234, 248
Gesell, A., 2, 27
Ghiselin, M., 44, 46, 51, 83
Gibson, K. R., 2, 20, 27, 28, 124, 128,
    129, 130, 131, 134, 136, 137, 138,
    139, 140, 141, 155, 158, 160, 161,
    162, 164, 172, 174, 175
Gilbert, C. D., 233, 234, 247, 251
Gilbert, P. F. C., 169, 172
Gilbert, S. F., 54, 56, 57, 58, 83
Gilissen, E., 132, 139
Gill, T. V., 162, 176
Givón, T., 255, 282
Gleitman, L., 229, 247
Gleitman, R., 315, 319
Glickstein, M., 166, 169, 172, 188, 201
Goldin-Meadow, S., 128, 139, 272, 273,
    283, 290, 305
Goldman, A., 195, 197, 200
Goldman-Rakic, P. S., 132, 139, 164, 176
Goodall, J., 124, 139, 163, 177
Goodglass, H., 205, 236, 248
Goodman, R., 215, 248
Gopnik, M., 227, 248
Gottlieb, G., 34, 83
Gould, J. L., 114, 121, 128, 139
Gould, S. J., 51, 83, 114, 121, 264, 283
Gowlett, J. A. J., 133, 139
Grabowski, T. J., 127, 140
Grafman, J., 127, 142
Grafton, S. T., 189, 200
Grammaldo, L. G., 153, 174
Grant, J., 314, 315, 319, 320
Grassi, F., 189, 199, 236, 246
Grauer, D., 65, 84
Greenfield, P. M., 123, 124, 131, 140,
    161, 162, 172
Greenough, W. T., 126, 140
Gregory, R. L., 234, 251
Grèzes, J., 189, 199

Grice, S. J., 310, 313, 314, 319
Griffiths, P., 9, 16, 28
Gsödl, M., 315, 316, 320
Guillaume, P., 267, 283
Gumperz, J. J., 287, 288, 291, 305
Güntürkün, O., 125, 140
Gustason, G., 288, 305
Gvozdanovic, J., 267, 283

H

Haas, R. H., 151, 163, 171
Haeckel, E., 43, 50, 83
Haiman, J., 271, 283
Hakuta, K., 214, 246
Halit, H., 310, 313, 314, 319
Hallett, M., 126, 127, 140, 142
Hamberger, V., 2, 28, 52, 83
Hamilton, W. D., 8, 28
Hari, R., 189, 200
Harries, M. H., 186, 202
Hartung, J., 220, 248
Harvey, P. H., 156, 161, 170, 171
Helgren, D. M., 134, 137
Henikoff, S., 74, 75, 85
Hepp-Raymond, M. –C., 187, 200
Hernandez, A. E., 235, 238, 249
Hess, D. T., 169, 177
Hesselink, J. R., 151, 165, 171
Hinton, G. E., 113, 121
Hirsch, J., 235, 249, 250
Hirstein, W. F., 228, 251
Holland, A., 205, 226, 248
Holloway, R. L., 132, 138, 233, 249
Holmes, G., 147, 173
Hopkins, W. D., 132, 137
Hopper, P. J., 271, 283
Hornyak, W., 134, 137
Horvasse, R., 39, 83
Hovda, D. A., 164, 174
Howell, F. C., 133, 143
Howlin, P., 315, 320
Hubel, D., 127, 140
Hudson, C. L., 272, 283
Hüsler, E. J., 187, 200
Husserl, E., 192, 198, 200

Hutchins, E., 21, 28, 30
Hutchinson, W. D., 195, 200
Hutsler, J. J., 243, 249
Huttenlocher, P. R., 239, 249, 309, 319
Huxley, J., 49, 83
Hymes, D., 288, 305

I

Iacoboni, M., 189, 196, 199, 200
Ibanez, V., 127, 142
Ingold, T., 20, 27
Inhelder, B., 1, 28, 30, 35, 85
Inhoff, A. W., 150, 173
Insel, T. R., 160, 176
Iraqui, V., 228, 251
Isaacs, E., 205, 215, 223, 226, 252
Isacson, O., 207, 249
Ito, M., 145, 168, 169, 171, 173
Iversen, S. K., 161, 175
Ivry, R. B., 150, 154, 173

J

Jablonka, E., 34, 42, 45, 50, 71, 72, 73,
    74, 83, 99, 121
Jackendof, R., 279, 283
Jacob, F., 60, 83
Jacobs, B., 164, 173
Jaeger, J. J., 236, 248
Jambaque, I., 236, 246
James, H. E., 151, 163, 171
Janosky, J., Jr., 205, 226, 248
Janowsky, J. S., 207, 246, 249
Jansen, J., 158, 173
Jeannerod, M., 189, 199
Jellema, T., 186, 200
Jenkins, L., 229, 249
Jenuwein, T., 74, 83
Jerison, H. J., 155, 173
Jernigan, T. L., 151, 165, 171, 242, 252,
    313, 318
Jesperson, O., 256, 283
Jessee, S., 124, 139
Jezzard, P., 238, 249
Johnson, J., 214, 218, 249, 296, 305

Johnson, M. C., 206, 246
Johnson, M. H., 180, 201, 206, 207, 209,
    214, 227, 229, 237, 248, 249, 309,
    310, 311, 314, 316, 317, 318, 319,
    320
Johnson, R. C., 289, 290, 305
Jones, W., 314, 315, 316, 319, 321
Judson, H., 58, 59, 60, 83
Just, M. A., 235, 249

K

Kaas, J. H., 127, 141, 142
Kalthoff, K., 62, 63, 66, 67, 68, 70, 71,
    74, 76, 83
Kano, M., 169, 173
Karmiloff-Smith, A., 206, 207, 214, 227,
    229, 246, 248, 249, 308, 309,
    310, 313, 314, 315, 316, 317, 319,
    320, 321
Karni, A., 238, 249
Kato, T., 234, 248
Kaufman, T. C., 58, 63, 77, 84, 270, 272,
    277, 285
Kawai, M., 123, 140
Kay, R. F., 161, 173
Keane, J., 196, 198
Keating, J. M., 134, 137
Keele, S. W., 154, 173
Keesing, R. M., 270, 283
Kegl, J., 274, 277, 283, 284, 291, 306
Keller, E. F., 58, 59, 60, 66, 67, 84
Keller, J. B., 244, 250
Keller, T. A., 235, 249
Kemp, S., 205, 226, 248
Kempler, D., 213, 249
Keysers, C., 190, 196, 200, 201, 203
Kheck, N., 233, 249
Killackey, H. P., 126, 137, 140, 207, 208,
    249
Killen, M., xi, xvii
Kim, J., 184, 200
Kim, K. H. S., 235, 249, 250
Kim, S. G., 151, 161, 173
Kimura, M., 65, 84
King, M. C., 231, 250

Kirveskari, S., 189, 200
Klein, R. G., 134, 137, 140
Klima, E. S., 314, 321
Knott, C., 163, 177
Kohler, E., 190, 200, 201
Kohler, R. E., 58, 59, 67, 68, 84
Kolb, B., 165, 173
Kornack, D. R., 126, 142
Kozlowski, L. T., 193, 199
Krech, D., 126, 138
Kroch, A., 229, 247
Krol, G., 235, 249
Kroskrity, P., 295, 306
Küntay, A., 258, 284
Kurata, K., 187, 201

L

Laberge, S., 270, 271, 285
Lai, Z., 316, 319
Laing, E., 315, 320
Lakoff, G., 180, 201
Laland, K., xii, xvi, xvii, 13, 14, 17, 18, 19, 28, 29, 303, 304, 305
Lamarck, J. -B., xvi, xvii, 43, 45, 84
Lamb, M., 34, 42, 45, 50, 71, 72, 73, 74, 83, 99, 121, 242, 251
Land, P. W., 126, 142
Langer, J., xi, xvii, 23, 28, 76, 85, 128, 140, 264, 265, 284
Larsell, O., 158, 173
Larsen, J., 219, 251
Larsen, S. G., 160, 173
Lau, L., 151, 163, 171
Law, M. I., 107, 121
Lawson, D., 310, 312, 320
Leaky, R., 133, 143
Lecours, A. R., 129, 143
Lee, K. M., 235, 250
Leiner, A. L., 147, 166, 169, 173
Leiner, H. C., 147, 166, 169, 173
Lejeune, B., 189, 199
Lenneberg, B. H., 205, 206, 214, 227, 250
Lenneberg, E., 296, 305
Lennert Olsen, L., 259, 282
Lenzi, G. L., 196, 199

LeVay, S., 127, 140
Lewis, E. B., 58, 84
Lewis-Williams, D., 135, 140
Lewontin, R., 14, 20, 23, 24, 28
Li, J., 149, 162, 172
Li, P., 216, 217, 220, 250
Li., W. -H., 65, 84
Lichtman, J. W., 164, 176
Lieven, E., 260, 284
Lightfoot, D., 267, 284
Lincoln, A. J., 151, 163, 171, 316, 319
Lindquist, S., 101, 121
Linquist, L., 57, 85
Lipps, T., 192, 201
Liu, H., 216, 217, 250
Livet, M. O., 314, 319
Livingstone, F. B., 19, 29
Livingstone, M., 242, 248
Longhi, E., 315, 320
Lorenz, K., 12, 29
Lounes, R., 236, 246
Ludy, C. A., 236, 248
Lumsden, C., 19, 28
Luppino, G., 187, 188, 189, 201, 202
Luria, A., 132, 140

M

Macchi Cassia, V., 310, 321
MacLeod, A. –M., 153, 176, 235, 238, 251
MacLeod, C. E., 59, 82, 131, 140, 155, 158, 160, 173, 174
MacWhinney, E., 244, 250
Maier, M. A., 187, 200
Mancini, J., 314, 319
Mandler, J., 261, 282
Manes, F., 196, 198
Marchman, V., 206, 211, 212, 213, 214, 220, 221, 222, 227, 233, 236, 237, 239, 246, 249, 250, 251
Marks, S., 313, 318
Marler, P. M., 128, 139
Marr, D., 169, 174
Marshack, A., 134, 140
Martin, L., 161, 170

Martineau, J., 189, 199
Martinez, A., 235, 238, 242, 249, 250
Matano, S., 158, 174
Matelli, M., 187, 188, 189, 201, 202
Maunsell, J., 185, 186, 201, 203
Maxam, A., 229, 249
Mayberry, R. I., 296, 305
Mayley, G., 113, 121
Mayr, E., xvi, xvii, 11, 29, 40, 46, 47, 48,
    49, 50, 72, 80, 84
Mazzacco, M. M., 165, 174
Mazziotta, J. C., 189, 196, 199, 200, 239,
    247
McBrearty, S., 19, 29, 134, 140
McCarthy, G., 235, 250
McCarty, M., 59, 82
McClean, P., 62, 84
McClintock, B., 65, 84
McCoby, E. E., 24, 29
McCulloch, W. S., 132, 137
McDonough, L., 261, 282
McDowall, R., 235, 249
McGregor, K., 225, 226, 251
McGrew, W. C., 163, 177
McKinney, M. L., 21, 23, 25, 29, 76, 85,
    129, 141, 164, 175
McKinney, N., 51, 84
McLaren, A., 57, 84
McLennan, D., 23, 27
McNamara, K., 51, 84
McNeil, J. E., 311, 320
McWhorter, J., 270, 274, 283, 284
Mehler, J., 235, 238, 250
Meisel, J. M., 261, 284
Mellars, P., 134, 135, 141
Mellors, P., 19, 20, 29
Meltzoff, A., 181, 182, 191, 201, 202
Menard, M. T., 242, 248
Mercader, J., 19, 29
Merleau-Ponty, M., 193, 202
Merrill, M., 163, 177
Merzenich, M., 127, 141
Messerly, J. G., 39, 40, 84
Meyer, G., 238, 249
Miall, R. C., 150, 166, 176
Michkin, M., 127, 142
Middleton, F. A., 151, 165, 166, 174

Milbrath, C., 161, 175
Miles, H. L., 162, 174
Miller, K. D., 244, 250
Miller, R., 220, 250
Milliken, G. W., 127, 141
Mills, D. L., 235, 238, 239, 250, 310, 312,
    314, 320
Milton, K., 161, 174
Minoshoma, S., 113, 121
Mintun, M., 152, 175
Mishkin, M., 215, 226, 236, 245, 252
Mistlin, A. J., 186, 202
Mitchell, R. W., 124, 141
Mithin, S., 20, 29, 134, 135, 141
Molinari, M., 153, 174
Monaghan, L., 290, 305
Monod, J., 60, 83
Moore, A. H., 164, 174
Moore, M. K., 181, 182, 201, 202
Morford, J. P., 274, 277, 278, 284, 290,
    305
Morgan, C. L., 88, 121, 123, 141
Morgan, T. H., 58, 84
Morton, J., 310, 319
Mosses, P., 242, 250
Mostofsky, S. H., 165, 174
Mueller, R. –A., 207, 235, 250
Mufwene, S. S., 288, 306
Muter, V., 205, 223, 226, 252
Muysken, P., 270, 284
Mylander, C., 128, 139, 272, 273, 283,
    290, 305

N

Nabbout, R., 236, 246
Nakamura, K., 290, 305
Nanetti, L., 190, 200
Nanney, D. L., 73, 84
Nass, R., 206, 219, 220, 221, 227, 246,
    252
Nazzi, T., 315, 320
Neville, H. J., 127, 141, 214, 235, 238,
    239, 250, 252, 310, 312, 314, 320
Newport, E. L., 214, 218, 249, 296, 300,
    305, 306

Nieuwenhuys, R., 148, 175
Nikelski, E., 244, 250
Nishida, T., 163, 177
Nishikimi, M., 113, 121
Noble, W., 20, 29, 134, 141
Nobre, A. C., 235, 250
Noland, S. J., 113, 121
Nudo, R. J., 127, 141
Nuñez, R., 180, 201

O

O'Leary, D. D., 207, 251
Odling-Smee, J., xii, xvi, xvii, 12, 13, 14, 17, 18, 19, 28, 29, 303, 304, 305
Odum, E. B., 16, 21, 29
Ohno, S., 65, 84
Olby, R., 48, 85
Ommaya, A. K., 127, 142
Oram, M. W., 186, 198, 199
Oyama, S., 214, 250
Ozonoff, S., 241, 250

P

Pallas, S. L., 208, 231, 250, 252
Panagiotaki, G., 315, 320
Pandya, D. N., 165, 176
Panger, M., 19, 29
Pardo, J. V., 153, 176, 235, 238, 251
Parisi, D., 206, 207, 214, 227, 229, 237, 246, 248, 309, 317, 319
Parker, S. T., 21, 23, 24, 25, 29, 76, 85, 124, 129, 141, 161, 162, 163, 164, 175, 176
Parsons, L. M., 149, 162, 172
Pascalis, O., 310, 311, 319
Pascual-Leone, A., 127, 142
Passingham, R. E., 131, 141
Paterson, S. J., 315, 316, 320
Patterson, F. G., 162, 175
Paulesu, E., 189, 202
Paulesu, N., 235, 238, 250
Pavesi, G., 189, 199
Penisi, E., 71, 85
Pennington, B. F., 241, 250

Perani, D., 189, 199, 202, 235, 236, 238, 246, 250
Perrett, D. I., 186, 198, 199, 200, 202
Pesetsky, D., 229, 250
Petersen, S. E., 152, 153, 171, 175, 176, 235, 238, 251
Peterson, A. C., 2, 28
Pethick, S., 220, 225, 226, 240, 248
Petitto, L., 244, 250
Petrides, M., 161, 175
Petrosini, L., 153, 174
Pettet, M. W., 234, 251
Pezzini, G., 226, 247, 315, 321
Pfetzing, D., 288, 305
Phelps, M. E., 164, 173, 174, 239, 247
Piaget, J., xi, xvi, xvii, 1, 2, 3, 4, 5, 8, 23, 24, 26, 28, 29, 30, 35, 36, 37, 38, 39, 40, 41, 78, 80, 81, 85, 87, 90, 91, 92, 94, 98, 101, 102, 108, 121, 124, 128, 141, 146, 175, 279, 284, 308, 317, 320, 334
Piatelli-Palmarini, M., 41, 85, 92, 121
Pine, J. M., 260, 284
Pinker, S., 24, 30, 124, 142, 227, 229, 230, 231, 233, 251, 274, 284, 307, 314, 321
Plaily, J., 196, 203
Plann, S., 290, 306
Plotkin, H. C., 12, 13, 30
Plunkett, K., 206, 207, 214, 227, 229, 237, 246, 248, 309, 317, 319
Poeck, K., 236, 253
Poeppel, D., 234, 251
Polich, L. G., 289, 290, 293, 306
Pollack, D. B., 164, 173, 174
Pons, T. P., 127, 142
Portman, A., 129, 142
Posner, M. I., 152, 175
Potts, R., 135, 142, 161, 175
Povinelli, D., 160, 161, 175
Premack, D., 162, 176
Press, G. A., 151, 165, 171
Prigman, G. W., 192, 202
Procyk, E., 189, 199
Provine, W., xvi, xvii, 11, 29, 49, 50, 72, 84
Purves, A., 164, 176
Pysh, J. J., 164, 176

Q

Qi, H. –X., 187, 200
Quartz, S. R., 207, 251

R

Rafal, R. D., 150, 173
Raff, R. A., 23, 30, 58, 63, 77, 84
Raichle, M. E., 152, 153, 171, 172, 175,
    176, 235, 238, 251
Raife, E. A., 153, 171
Rakic, P., 126, 142, 164, 176
Raleigh, M. J., 164, 173
Ralphs, A., 316, 319
Ramachandran, V. S., 228, 234, 251, 252
Redfern, B. B., 236, 248
Reilly, J. S., 206, 211, 212, 219, 220, 221,
    222, 227, 248, 251, 316, 319
Reiss, A. L., 165, 174
Relkin, N. R., 235, 250
Reynolds, V., 163, 177
Reznick, J. S., 220, 225, 226, 240, 248
Rhoades, R., 208, 249
Ricards, R. J., 6, 7, 8, 9, 10, 11, 30
Rice, M., 227, 229, 230, 251
Richards, G. D., 133, 143
Richards, R. J., 45, 85, 88, 93, 94, 95,
    104, 121
Richerson, P., 19, 27
Riley, H. A., 158, 177
Rilling, J. K., 131, 140, 155, 158, 160,
    174, 176
Ring, B., 186, 198
Rink, W. J., 134, 137
Rizzolatti, G., 187, 188, 189, 190, 196,
    198, 199, 200, 201, 202, 203
Robertson, L. C., 242, 251
Roe, A. W., 13, 30, 208, 231, 252
Rogers, L. J., 125, 138, 142
Rogoff, B., 21, 30
Rolls, E. T., 234, 252
Romaine, S., 267, 270, 284
Rose, D., 223, 245
Rosen, G. D., 242, 248
Rosensweig, M. R., 126, 138

Rossen, M., 314, 315, 321
Roux, S., 189, 199
Royet, J. –P., 196, 203
Ruge, M., 235, 249
Rumbaugh, D. M., 124, 133, 142, 162,
    176
Russon, A. E., 19, 30, 124, 137, 161, 162,
    163, 176
Rutherford, S. L., 57, 74, 75, 85, 101, 121

S

Sabbadini, L., 315, 321
Sabo, H., 313, 318
Sadato, N., 127, 142
Sainburg, R. L., 242, 252
Saito, O., 151, 163, 171
Salenius, S., 189, 200
Sankoff, G., 270, 271, 285
Savage-Rumbaugh, E. S., 123, 124, 128,
    130, 131, 133, 140, 142, 258, 285
Scandolara, C., 187, 202
Scerif, G., 314, 315, 316, 320
Schein, J. D., 290, 306
Schick, K. D., 133, 142
Schieffelin, B. B., 295, 306
Schleicher, A., 131, 140, 155, 158, 160,
    174
Schmahmann, J. D., 165, 176
Schmalhausen, I. I., 43, 53, 85, 113, 122
Schmaling, C., 290, 305
Schnur, T., 235, 238, 250
Schoen, J. H. r., 147, 177
Schreibman, L., 151, 163, 171
Schwarcz, H., 134, 137
Schwwarz, J. P., 153, 171
Sebastian, E., 235, 238, 250
Seitz, R. J., 189, 198
Sejnowski, T. J., 207, 251
Senghas, A., 274, 275, 276, 277, 278, 285,
    288, 289, 290, 291, 295, 297, 300,
    306
Senghas, R. J., 288, 289, 291, 306
Sevcik, R., 133, 142
Shallice, T., 234, 251

Shanker, S. G., 124, 128, 130, 142, 258, 285
Shannon, C., 90, 122
Shapiro, J. A., 64, 65, 76, 85
Shatz, C. J., 207, 244, 251
Shea, J., 133, 142
Sherman, J. C., 165, 176
Shimizu, N., 113, 121
Shipp, S., 185, 203
Sifuentes, F., 127, 141
Simion, F., 310, 321
Simons, D. J., 126, 142
Simpson, G. G., 13, 30, 114, 122
Singer Harris, N. G., 315, 321
Singleton, I., 163, 177
Sitompul, A. F., 161, 177
Slobin, D. I., 258, 259, 260, 261, 268, 269, 278, 279, 281, 284, 285
Smith, A., 192, 202
Smith, D. L., 289, 290, 305
Smith, J. N. L., 134, 137
Snider, B. D., 289, 290, 305
Snyder, M., 60, 85
Souweidane, M., 235, 249
Spencer, H., xvi, xviii, 94, 122
Spratling, M. W., 310, 311, 313, 314, 318, 319
St. George, M., 314, 320
Stanfield, B. B., 207, 251
Stark, R., 225, 226, 251
Stein, E., 193, 202
Stein, J. F., 150, 166, 176
Stephan, H., 129, 143, 158, 174
Stern, D. N., 181, 202
Sternberg, M. L. A., 292, 306
Stewart, K., 134, 137
Stewart, R., 62, 71, 85
Stiles, J., 206, 219, 220, 221, 240, 242, 250, 251, 252
Strick, P. L., 151, 161, 165, 166, 173, 174, 234, 248
Stringer, C., 19, 20, 29
Stromswold, K., 206, 252
Strum, S. C., 21, 30
Stryker, M. P., 244, 250
Sugiyama, Y., 163, 177
Sullivan, K., 316, 321

Supalla, T., 300, 306
Sur, M., 208, 231, 250, 252
Suwa, G., 133, 143
Suzuki, A., 163, 177
Syrota, A., 236, 246
Szentágothai, J., 168, 171

T

Tager-Flusberg, H., 316, 321
Taglialatela, J., 258, 285
Tallal, P., 242, 252
Tanji, J., 187, 201
Tattersall, I., 133, 143
Taub, E., 127, 142
Taylor, T. J., 124, 128, 130, 142
Tecoma, E., 228, 251
Teuber, H. L., 223, 253
Thach, W. T., 149, 153, 158, 169, 172, 176
Thal, D., 206, 207, 214, 219, 220, 221, 225, 226, 227, 233, 240, 242, 246, 248, 251, 252
Thomas, M. S. C., 314, 315, 316, 320, 321
Thomason, S., 270, 272, 277, 285
Thompson-Schill, S., 235, 252
Tilbert, H., 133, 143
Tilney, F., 158, 177
Tinbergen, N., 12, 30
Titchener, E. B., 192, 202
Todd, J., 163, 177
Todd, N. E., 134, 137
Todde, S., 236, 246
Tomasello, M., 163, 177, 260, 285
Tooby, J., 24, 27, 124, 138, 143, 231, 245, 307, 308, 318, 319
Toth, N., 133, 142
Tovee, M. J., 234, 252
Townsend, J., 151, 163, 171
Tranel, D., 196, 198
Traugott, E., 271, 283
Trauner, D., 206, 219, 220, 221, 223, 226, 227, 243, 246, 251
Trevarthan, C., 2, 30
Trevathan, W., 24, 30

Turati, C., 310, 321
Turner, G., 290, 305
Tutin, C. E. G., 163, 177
Tylor, T. J., 258, 285

U

Udwin, O., 313, 315, 320, 321
Ugurbil, K., 151, 161, 173, 234, 248
Umilta, C., 310, 321
Umiltà, M. A., 190, 200, 201
Ungar, P. S., 161, 173
Ungerleider, L. G., 185, 199, 208, 209,
    212, 234, 236, 247, 252
Utami, S., 163, 177

V

Valentine, J., 65, 82
Valenza, E., 310, 321
Valler, G., 236, 246
Van der Lely, H. K. J., 227, 229, 252
van der Loos, H., 239, 249
Van Essen, D., 185, 186, 201, 203
Van Huijzen, C., 148, 175
van Lancker, D., 213, 249
Van Schaik, C. P., 161, 163, 177
Van Valin, R. D., Jr., 236, 248
Vargha-Khadem, F., 205, 215, 223, 226,
    252
Vauclair, J., 126, 138
Verniers, J., 134, 137
Vicari, S., 223, 226, 243, 246, 315, 321
Vidal, F., 40, 85
Videen, T. O., 153, 176, 235, 238, 251
Volterra, V., 206, 226, 231, 246, 247, 315,
    321
Von Bertalanfy, L., 90, 122
Von Bonin, G., 132, 137
von Glasersfeld, E. C., 162, 176
Von Noorden, G. K., 127, 143
Voogd, J., 147, 148, 169, 175, 177
Vrba, E., 114, 121
Vygotsky, L. S., 24, 31

W

Waddington, C. H., xvi, xvii, 1, 8, 10, 13,
    20, 33, 36, 39, 41, 42, 43, 54, 55,
    56, 59, 81, 85, 86, 88, 100, 122,
    124, 143
Wake, D., 53, 86
Walker, A., 133, 143
Walker, C., 163, 171
Wallace, C., 126, 140
Wang, P. P., 313, 318
Warofsky, I. S., 165, 174
Warrington, E. K., 311, 320
Washburn, S. L., 19, 31
Waters, N. S., 126, 138
Watson, J. B., 59, 86
Watson, J. S., 24, 31
Weaver, W., 90, 122
Weber-Fox, C. M., 214, 252
Webster, M. J., 208, 209, 212, 236, 252
Weindling, P., 50, 51, 86
Weir, D. J., 150, 166, 176
Weismann, A., xvii, xviii, 43, 86, 96, 104,
    109, 122
Weiss, G. M., 164, 176
Welker, W., 168, 177
Wernicke, C., 229, 253
West-Eberhard, M. J., xvi, xviii, 10, 13,
    14, 15, 16, 24, 31, 34, 35, 43, 56,
    59, 66, 76, 78, 79, 80, 86
Wexler, K., 229, 250
Whishaw, I. Q., 165 173
Whitaker, H., 223, 224, 245, 247
White, T., 133, 143
Whiten, A., 124, 143, 163, 177
Whiting, B. A., 156, 177
Wicker, B., 196, 203
Wiener, N., 90, 122
Wiesel, T. N., 127, 140
Wilkins, D. P., 236, 248
Willard, K., 295, 306
Willmes, K., 236, 253
Wilson, A. C., 231, 250, 253
Wilson, E. O., 19, 28
Winther, R., 46, 47, 86
Wise, B. M., 127, 141
Wolpoff, M. H., 133, 143

Wong, E. C., 131, 137, 151, 163, 170,
    235, 238, 242, 249, 250
Woods, B. T., 223, 253
Woods, R. P., 189, 200
Wrangham, R. W., 163, 177
Wynn, T., 133, 143

X–Y

Xiong, J., 149, 162, 172
Yagi, K., 113, 121
Yakovlev, P. I., 129, 143
Yellen, J. E., 134, 137
Yeung-Courchesne, R., 151, 163, 165, 171

Young, A. W., 196, 198
Yude, C., 215, 248
Yule, W., 313, 321

Z

Zahavi, D., 198, 203
Zattore, R., 244, 250
Zawolkow, E., 288, 305
Zecevic, N., 164, 176
Zeki, S. M., 185, 186, 199, 203
Zilles, K., 131, 140, 155, 158, 160, 174,
    189, 198
Zurif, E., 241, 253

Subject Index

A

Ape-human
  ancestor, 127
Ape-human
  differences, 131
Aphasia, 205
Aphasia
  adult, 214, 221
Aphasia
  Broca's, 221
Aphasia
  fluent, 234
  nonfluent, 234
  Wernicki's, 221
Accommodation, 3, 8, 23, 36, 80, 92, 93, 109, 146
Adaptation, 6, 11, 14, 17, 91, 96, 97, 99, 232
Adaptation
  pre-adaptation, 114
  exaptation, 114
Assimilation, 1–4, 23, 36, 39, 80, 92, 93, 109, 146
Autoregulation, 3, 5, 6, 35, 53, 81, 90, 103, 108, 116, 119, 120

B

Baldwin effect
  see Organic evolution
Behavior
  role of in evolution, 3, 6, 11, 13
Behaviorism, 92, 96, 97, 229
Brain
  development, 2, 3, 24, 113, 125, 127, 128, 164, 205
  injury, 205, 206, 211, 223, 239, 242, 245
  lateralization, 125, 126
  myelination, 129, 164, 239
  organization, 206, 227, 229, 230, 233, 236, 243, 244, 245
  reorganization, 124, 206, 236, 245
  size, 132–134
  volumes, 156, 157
Broca's area, 132, 189, 241, 244

C

Cells, 2, 35, 43
  germline, 2, 15, 47, 49
  somatic, 2, 35, 43, 49

Capacities
    domain-general, 279
    domain-specific, 279, 309, 317
Cerebellar anatomy, 164–170
Cerebellum
    neurophysiological functioning of, 147,
        149–155
    dentate nucleus of, 148–151
    significance of lateral cerebellum ex-
        pansion, 158–163
Chreods, 36, 55, 94
Chromosomes, 58, 59, 62, 67
Co-evolution, 17, 20, 279
Cognitive capacities, 128, 129, 278
Cognitive development, 23, 35, 36, 90,
        103, 117, 120, 124, 154, 181, 185,
        194, 264
Cognitive construction, 4, 14, 302
Comparative studies, 20, 23, 258
Constructional capacities, 130, 133, 136
Constructivism, 1, 3, 23–25, 81, 87, 89,
        117, 308
Cortex
    frontal, 221, 235, 241
    parietal, 188, 189, 208
    somatosensory, 208, 238
    temporal, 208, 220, 232, 235, 236, 241,
        242, 244
    visual, 106, 107, 127
Creoles, 269, 270, 272, 277
Critical periods, 207, 214, 215, 216, 218,
        226, 236, 237
Cybernetics, 33, 89, 90, 94, 97, 98, 108,
        118, 119

D

Developmental biology, 43, 57, 58, 66,
        68, 88, 119
Developmental canalization, 10, 41,
        53–55, 59, 101, 124
Developmental
    disorders, 307, 308, 312
    trajectories, 266, 315
Deaf
    association, 291, 294

communities, 289, 290, 294, 296
children, 272, 277, 278
DNA and RNA, 59, 60
Domain-general capacities, 279
Down's syndrome, 315, 316

E

Ecology, 10, 11, 13
Ecological
    environment, 20
    inheritance, 17
Embryo development
    three modes of, 69, 72
Embryology, 2, 50, 51, 52, 54, 66, 67
Embryogenesis, 2, 3, 21, 35, 36, 49, 58,
        68, 70, 72, 119, 308
Emotions, 192–194, 196
Empathy, 190–194
Environmental levels, 22
Environments, 6, 20
    ecological, 20
    external, 20
    selective, 20
Epigenesis, 1, 13, 23, 24, 52, 88, 98, 100,
        103, 113, 118, 124, 136
Epigenetics, 70, 74 116, 120, 302, 317
Epigenetic
    inheritance, 42, 44, 72, 73, 75
    landscape, 55, 56
Equilibration, 36, 39, 93–95
Ethology, 11, 12
Evolutionary psychology, 124, 312, 318
Evolution
    modern synthesis of, 3, 11, 12, 40, 43,
        44, 48, 52, 54, 72, 80
    phenotype-centered, 3, 13, 14, 35, 81

F–G

Face processing, 310–312, 314, 315
Gene
    definition of, 60
    duplication, 63, 76, 115
    expression, 62, 100, 103

Gene regulation, 60, 66, 69
Genetics, 12, 52, 57, 58, 63, 70
Genetic assimilation, 33, 38, 42, 53, 54,
    56, 74, 80, 81, 88, 98, 100, 105,
    109, 111, 112
Genetic code, 44, 59, 65, 74
Genetic disorders, 307, 308, 312, 316
Genetic epistemology, 43, 87, 279
Genetic variation, 14, 37, 42, 46, 48
Genome, 34, 36, 38, 39, 60, 68, 80, 100,
    102, 231
Genomic imprinting, 71
Grammaticalization, 271, 278
Goal-related behaviors, 186, 189

H

Hemispherectomies, 223–225
Hemisphere advantage, 239–241
Hemisphere damage, 206, 209, 211, 212,
    214, 219, 220, 222, 224, 239
Heredity, 4, 10, 13, 17, 48, 75
Heterochrony, 51, 76, 78, 264
Homeorhesis, 35, 36, 55, 94
Homesigns, 269, 271, 275, 278, 280, 290,
    301
Hominid evolution, 3, 18, 133–135, 155,
    160, 161
Hominoid evolution, 160–162

I–J

Identity, 181, 183
    self-other, 190, 191, 194, 197
Information-processing capacity, 131,
    132, 134
Imitation, 9, 10, 24, 39, 128, 162, 192,
    194
Inheritance
    genetic, 96, 99
    germline, 74
    nongenetic, 110
    somatic, 74
Inheritance of acquired characteristics, 9,
    18, 45–47, 72, 91, 99, 109

Instincts, 4–8, 11, 35, 36, 115, 227
Interchangeability of environment and ge-
    netics, 15, 79
Intraselection, 103–107, 109–116
Jumping genes, 63–65

K–L

Knowledge
    logical-mathematical, 3–5, 25, 33, 36,
        90, 118, 264
    physical, 4, 5, 25
Lamarckism, 33, 37, 40, 45–47, 49, 50,
    72, 74, 88, 91, 95, 97, 118
Language acquisition, 92, 218, 276
    first language, 289, 292
    second language, 214, 216, 217, 238
Language change, 267, 288, 296, 302
Language comprehension, 220, 221, 226,
    239, 240, 316
Language development, 207, 219, 223,
    226, 239, 242, 243, 256, 266
Language emergence, 288, 304
Language
    expressive, 222, 241
Language production, 152, 220, 240, 300,
    315
Learning, 15, 16, 18, 95, 110, 115, 125,
    237, 239
    procedural, 153, 161–163
Linguistic environment, 295, 296, 302
Linguistic innovation, 256, 266, 268, 301
Limnaea
    see pond snail

M

Mental constructional abilities, 130, 136
Mental constructional capacities,
    130–133, 135, 136
Mirror neurons, 188, 191, 194, 195, 197
Mirror-matching mechanism, 191,
    195–197
Modules, 15, 123, 227, 231, 307–312
Modularity, 15, 78

Molecular biology, 57, 59, 66
Mutations, 11, 12, 14, 16, 18, 39, 44, 48,
    49, 53, 56, 65, 75, 80, 105
    varieties of, 77

N

Natural selection, 7, 9, 10, 13, 17 18, 21,
    44, 46, 47, 50, 55, 56, 89, 94, 105,
    110, 127, 163, 288, 302
Neo-Darwinian theory (NDT), 12, 33, 36,
    47, 87–89, 117
Neo-cortex, 126, 131, 136, 161
Neocortical expansion, 126
Neo-Lamarckism, 40, 45, 46
Neurobiology, 207, 227, 233
Neuroimaging, 154, 236, 255
Neuroconstructivism, 307, 310–312, 316,
    317
Nicaraguan signers
    cohorts of, 298–304
    historical periods, 292–294, 301
Norm of reaction, 39, 53

O–P

Ontogenetic
    development, 303, 304, 308, 317
    niches, 21
Organic selection, 8–10, 39, 88, 109–112
Phenocopy, 33, 39, 43, 55, 56, 89,
    99–105, 107, 108, 110, 114, 115
    definition, 42
    model, 25, 34, 39, 41, 43, 72, 80, 81
Phenotype, 1, 16, 101, 113, 115, 124
    alternative, 15
    continuity, 78
    extended, 13, 17, 24, 110
    field, 24
    plasticity, 10, 14, 23
    self-transforming, 23
Pidgin, 269, 271, 272, 278
Plasticity, 15, 214, 218, 226, 236, 237
    developmental, 207, 218
    epigenetic, 104

neural, 127, 136, 227, 318
phenotypic, 10, 14, 106, 127
Pond snails, 39–42
Primate
    evolution, 113, 158
Primate locomotion, 160, 162
Primate species, 3, 124, 131, 156, 158,
    162, 164, 266, 280
Protein synthesis, 60, 62, 66
Proto-language, 258–261, 273, 280

R

Recapitulation, 51, 52, 255, 266
Recapitulationism, 44, 50, 265
Reflective abstraction, 5, 36, 93
Regulation, 94–97, 107, 117
Representations, 181, 184, 186–188, 191,
    229, 230

S

Sensorimotor intelligence stages, 5, 35,
    146, 155, 163
Shared manifold, 94, 195, 197
Sign languages, 287, 292
Sign language
    American Sign Language (ASL), 277,
    291
    dictionaries, 29, 295
    Nicaraguan (NSL), 274, 275, 277, 287,
    288, 289, 295, 301
Stabilizing selection, 53, 76, 110, 113
Symbol-trained apes, 259, 264

T

Tool use, 123, 129, 133, 135
Transposable elements
    see Jumping genes

U–V–W

Upper Paleolithic technology, 134, 135
Verb
  regularization of, 265, 267–269

Visual processing, 184–185
Visuospatial cognition, 151, 161, 219
Visuospatial problem-solving, 147,
  161
Williams syndrome (WS), 307, 312–316

For Product Safety Concerns and Information please contact our EU
representative GPSR@taylorandfrancis.com
Taylor & Francis Verlag GmbH, Kaufingerstraße 24, 80331 München, Germany